Rare Event Simulation using Monte Carlo Methods

Rare Event Simulation using Monte Carlo Methods

Edited by

Gerardo Rubino

And

Bruno Tuffin

INRIA, Rennes, France

John Wiley & Sons, Ltd

This edition first published 2009

Registered office
John Wiley & Sons Ltd, The Atrium, Southern Gate, Chichester, West Sussex, PO19 8SQ, United Kingdom

For details of our global editorial offices, for customer services and for information about how to apply for permission to reuse the copyright material in this book please see our website at www.wiley.com.

Library of Congress Cataloging-in-Publication Data

Rare event simulation using Monte Carlo methods/edited by Gerardo Rubino, Bruno Tuffin.
p. cm.
Includes bibliographical references and index.
ISBN 978-0-470-77269-0 (cloth)
1. Limit theorems (Probability theory) 2. Monte Carlo method. 3. System analysis – Data processing. 4. Digital computer simulation. I. Rubino, Gerardo, 1955- II. Tuffin, Bruno.
QA273.67.R37 2009
519.2 – dc22

2009004186

A catalogue record for this book is available from the British Library

ISBN: 978-0-470-77269-0

Typeset in 10/12pt Times by Laserwords Private Limited, Chennai, India

Contents

Contributors

We list the contributors to this volume in alphabetical order, with their respective electronic addresses.

JOSÉ BLANCHET, University of Columbia, New York, USA
jose.blanchet@columbia.edu

HENK BLOM, NLR, The Netherlands
blom@nlr.nl

THOMAS BOOTH, Los Alamos National Laboratory, USA
teb@lanl.gov

HÉCTOR CANCELA, University of the Republic, Montevideo, Uruguay
cancela@fing.edu.uy

MOHAMED EL KHADIRI, University of Nantes/IUT of Saint-Nazaire, France
mohamed.elkhadiri@iutsn.univ-nantes.fr

PETER GLYNN, University of Stanford, USA
glynn@stanford.edu

PIERRE L'ECUYER, University of Montréal, Canada
lecuyer@iro.umontreal.ca

FRANÇOIS LEGLAND, INRIA, France
legland@irisa.fr

PASCAL LEZAUD, DGAC, France
lezaud@cena.fr

MICHEL MANDJES, CWI, The Netherlands
mmandjes@science.uva.nl

GERARDO RUBINO, INRIA, France
gerardo.rubino@inria.fr

DANIEL RUDOY, Harvard University, Cambridge, USA
rudoy@seas.harvard.edu

WERNER SANDMANN, Bamberg University
werner.sandmann@wiai.uni-bamberg.de

BRUNO TUFFIN, INRIA, France
bruno.tuffin@inria.fr

Preface

Rare event simulation has attracted a great deal of attention since the first development of Monte Carlo techniques on computers, at Los Alamos during the production of the first nuclear bomb. It has found numerous applications in fields such as physics, biology, telecommunications, transporting systems, and insurance risk analysis. Despite the amount of work on the topic in the last sixty years, there are still domains needing to be explored because of new applications. A typical illustration is the area of telecommunications, where, with the advent of the Internet, light-tailed processes traditionally used in queuing networks now have to be replaced by heavy-tailed ones, and new developments of rare event simulation theory are required.

Surprisingly, we found that not much was written on the subject, in fact only one book was devoted to it, with a special focus on large-deviations theory. The idea of writing this book therefore started from a collaborative project managed in France by Institut National de Recherche en Informatique et Automatique (INRIA) in 2005–2006 (see http://www.irisa.fr/armor/Rare/), with groups of researchers from INRIA, the University of Nice, the CWI in the Netherlands, Bamberg University in Germany, and the University of Montréal in Canada. In order to cover the broad range of applications in greater depth, we decided to request contributions from authors who were not members of the project. This book is the result of that effort.

As editors, we would like to thank the contributors for their effort in writing the chapters of this book. We are also grateful to John Wiley & Sons staff members, in particular Susan Barclay and Heather Kay, for their assistance and patience.

Gerardo Rubino and Bruno Tuffin

1

Introduction to rare event simulation

Gerardo Rubino and Bruno Tuffin

This monograph deals with the analysis by simulation of 'rare situations' in systems of quite different types, that is, situations that happen very infrequently, but important enough to justify their study. A rare event is an event occurring with a very small probability, the definition of 'small' depending on the application domain. These events are of interest in many areas. Typical examples come, for instance, from transportation systems, where catastrophic failures *must* be rare enough. For instance, a representative specification for civil aircraft is that the probability of failure must be less than, say, 10^{-9} during an 'average-length' flight (a flight of about 8 hours). Transportation systems are called *critical* in the dependability area because of the existence of these types of failures, that is, failures that can lead to loss of human life if they occur. Aircraft, trains, subways, all these systems belong to this class. The case of cars is less clear, mainly because the probability of a catastrophic failure is, in many contexts, much higher. Security systems in nuclear plants are also examples of critical systems. Nowadays we also call critical other systems where catastrophic failures may lead to significant loss of money rather than human lives (banking information systems, for example). In telecommunications, modern networks often offer very high speed links. Since information travels in small units or messages (packets in the Internet world, cells in asynchronous transfer mode infrastructures, etc.), the saturation of the memory of a node in the network, even during a small amount of time, may induce a huge amount of losses (in most cases, any unit arriving at

Rare Event Simulation using Monte Carlo Methods Edited by G. Rubino and B. Tuffin

a saturated node is lost). For this reason, the designer wants the overflow of such a buffer to be a rare event, with probabilities of the order of 10^{-9}. Equivalently, the probability of ruin is a central issue for the overall wealth of an insurance company: the (time) evolution of the reserves of the company is represented by a stochastic process, with initial value R_0; the reserves may decrease due to incoming claims, but also have a linear positive drift thanks to the premiums paid by customers. A critical issue is to estimate the probability of ruin, that is, the probability of the reserve process reaching zero. In biological systems, molecular reactions may occur on different time scales, and reactions with extremely small occurrence rates are therefore rare events. As a consequence, this stiffness requires the introduction of specific techniques to solve the embedded differential equations.

This chapter introduces the general area of rare event simulation, recalls the basic background elements necessary to understand the technical content of the following chapters, and gives an overview of the contents of those chapters.

1.1 Basics in Monte Carlo

Solving scientific problems often requires the computation of sums or integrals, or the solution of equations. Direct computations, also called analytic techniques, become quickly useless due to their stringent requirements in terms of complexity and/or assumptions on the model. In that case, approximation techniques can sometimes be used. On the other hand, standard numerical analysis procedures also require assumptions (even if less stringent) on the model, and suffer from inefficiency as soon as the mathematical dimension of the problem increases. A typical illustration is when using quadrature rules for numerical integration. Considering, for instance, the trapezoidal rule with n points in dimension s, the speed of convergence to the exact value is usually $O(n^{-2/s})$, therefore slow when s is large. The number of points necessary to reach a given precision increases exponentially with the dimension. To cope with those problems, we can use Monte Carlo simulation techniques, which are statistical approximation techniques, instead of the above mentioned deterministic ones.

Let us start with the basic concepts behind Monte Carlo techniques. Suppose that the probability γ of some event A is to be estimated. A model of the system is simulated n times (we say that we build an n-sample of the model) and at each realization we record whether A happens or not. In the simplest (and most usual) case, the n samples are independent (stochastically speaking) of each other. If X_i is the (Bernoulli) random variable $X_i = \mathbb{1}$ (A occurs in the nth sample) (i.e., $X_i = 1$ if A occurs in sample i, 0 if not), we estimate γ by $\widehat{\gamma} = (X_1 + \cdots + X_n)/n$. Observe that $\mathbb{E}(X_i) = \gamma$ and $\text{Var}(X_i) = \gamma(1 - \gamma)$ which we denote by σ^2 (for basic results on probability theory and to verify, for instance, typical results on Bernoulli random variables, the reader can consult textbooks such as [6]).

How far will the given estimator $\widehat{\gamma}$ be from the actual value γ? To answer this question, we can apply the central limit theorem, which says that $X_1 + \cdots + X_n$

is approximately normal (if n is 'large enough') [6]. Let us scale things first: $\mathbb{E}(X_1 + \cdots + X_n) = n\gamma$ and $\mathrm{Var}(X_1 + \cdots + X_n) = n\sigma^2$, so the random variable $Z = [n\sigma^2]^{-1/2}(X_1 + \cdots + X_n - n\gamma)$ has mean 0 and variance 1. The central limit theorem says that as $n \to \infty$, the distribution of Z tends to the standard normal distribution $\mathcal{N}(0, 1)$, whose cdf is $\Phi(x) = (2\pi)^{-1/2} \int_{-\infty}^{x} \exp(-u^2/2)du$. We then assume that n is large enough so that $Z \approx \mathcal{N}(0, 1)$ in distribution. This means, for instance, that $\mathbb{P}(-z \le Z \le z) = 2\Phi(z) - 1$ (using $\Phi(-z) = 1 - \Phi(z)$), which is equivalent to writing

$$\mathbb{P}\left(\left(\widehat{\gamma} - \frac{z\sigma}{\sqrt{n}}, \widehat{\gamma} + \frac{z\sigma}{\sqrt{n}}\right) \ni \gamma\right) \approx 2\Phi(z) - 1.$$

The random interval $I = (\widehat{\gamma} \mp z\sigma n^{-1/2})$ is called a *confidence interval* for γ, with level $2\Phi(z) - 1$. We typically consider, for instance, $z = 1.96$ because $2\Phi(1.96) - 1 = 0.95$ (or $z = 2.56$, for which $2\Phi(2.56) - 1 = 0.99$). The preceding observations lead to $\mathbb{P}(\gamma \in (\widehat{\gamma} \mp 1.96\sigma n^{-1/2})) \approx 0.95$. In general, for a confidence interval with level α, $0 < \alpha < 1$, we take $(\widehat{\gamma} \mp \Phi^{-1}((1 + \alpha)/2)\sigma n^{-1/2})$.

From the practical point of view, we build our n-sample (i.e., we perform our n system simulations), we estimate γ by $\widehat{\gamma}$ and, since σ^2 is unknown, we estimate it using $\widehat{\sigma}^2 = n\widehat{\gamma}(1 - \widehat{\gamma})/(n - 1)$. The reason for dividing by $n - 1$ and not by n is to have an unbiased estimator (which means $\mathbb{E}(\widehat{\sigma}^2) = \sigma^2$, as $\mathbb{E}(\widehat{\gamma}) = \gamma$), although from the practical point of view this is not relevant, since n will be usually large enough. Finally, the result of our estimation work will take the form $I = (\widehat{\gamma} \mp 1.96\widehat{\sigma}n^{-1/2})$, which says that '$\gamma$ is, with high probability (our *confidence level*), inside this interval'–and by 'high' we mean 0.95.

The speed of convergence is measured by the size of the confidence interval, that is, $2z\sigma n^{-1/2}$. This decreases as the inverse square root of the sample size, independently of the mathematical dimension s of the problem, and therefore faster than standard numerical techniques, even for small values of s.

Now, suppose that A is a rare event, that is to say, that $\gamma \ll 1$. For very small numbers, the absolute error (given by the size of the confidence interval, or by half this size) is not of sufficient interest: the accuracy of the simulation process is captured by the *relative error* instead, that is, the absolute error divided by the actual value: $\mathrm{RE} = zn^{-1/2}\sigma/\gamma$. This leads immediately to the main problem with rare events, because if $\gamma \ll 1$, then

$$\mathrm{RE} = z\frac{\sqrt{\gamma(1 - \gamma)}}{\sqrt{n}\gamma} \approx \frac{z}{\sqrt{n}\sqrt{\gamma}} \gg 1$$

(unless n is 'huge' enough). To illustrate this, let us assume that we want a relative error less than 10%, and that $\gamma = 10^{-9}$. The constraint $\mathrm{RE} \le 0.1$ translates into $n \ge 3.84 \times 10^{11}$. In words, this means that we need a few hundred billion experiments to get a modest 10% relative error in the answer. If the system being simulated is complex enough, this will be impossible, and something different must be done in order to provide the required estimation. More formally, if we want to assess a fixed RE but the event probability goes to zero, we need to increase the sample size as

$$n = \frac{z}{\mathrm{RE}^2 \gamma},$$

that is, in inverse proportion to γ. Such issues are related to robustness properties of the estimators with respect to rarity. The questions are: is it possible to define sampling strategies such that the sample size for getting a fixed RE does not increase when γ decreases? Can we define stronger or weaker definitions of robustness? These problems are addressed in Chapter 4. This monograph is devoted to the description of different techniques, some of them general, some specific to particular domains, that enable us to face this problem.

Another issue is the *reliability* of the confidence interval produced. One needs to pay attention to whether or not the coverage of the interval actually matches the theoretical one as $\gamma \to 0$. This has to be studied in general, but is especially true when the variance is estimated since this estimation is often even more sensitive to the rarity than the estimation of the mean itself. This topic is also considered in Chapter 4. Note that in the simple case discussed in this section of estimating the probability of a rare event with n independent samples, the random variable $n\widehat{\gamma}$ is binomial with parameters n and γ. In this case, there exist specific confidence interval constructions that can replace the one obtained by using the central limit theorem (see for instance [7], typical examples being the Wilson score interval or the Clopper–Pearson interval) with the advantage of yielding a more reliable confidence interval, but not solving the robustness issue: the relative error still grows to infinity as γ tends to zero. This illustrates also the difference between the two notions of robustness and reliability.

On the other hand, in some situations the estimators are built from a sequence of correlated samples. For our purposes here, these variations do not change the basic ideas, summarized in the fact that estimating γ needs specific efforts, in general depending on the problem at hand, and that this is due to the same essential fact, $\gamma \ll 1$.

1.2 Importance sampling

Importance sampling (IS) is probably the most popular approach in rare event analysis. The general setting is as follows. For the sake of greater generality than in the previous section, assume that the system is represented by some random variable X, and that the target is the expectation γ of some function ψ of X: $\gamma = \mathbb{E}(\psi(X))$, where $\gamma \ll 1$. In the previous case, $X = \mathbb{1}(A)$, the indicator function of some (rare) event A, and ψ is the identity function. Assume that X is a real random variable having a density, denoted by f, and let us denote by σ^2 the variance of $\psi(X)$, assumed to be finite. The standard estimator of γ is $\widehat{\gamma} = (\psi(X_1) + \cdots + \psi(X_n))/n$, where $X_1, \ldots, X_n$ are n independent copies of X sampled from f.

IS consists in sampling X from a different density $\widetilde{f}$ (we say that we perform a *change of measure*), with the only condition that $\widetilde{f}(x) > 0$ if $\psi(x) f(x) > 0$ (to keep the estimator unbiased). Write $\gamma = \mathbb{E}_f(\psi(X))$ to underline the density we are considering for X. Obviously, in general, $\gamma \neq \mathbb{E}_{\widetilde{f}}(\psi(X))$. Then, from

$$\gamma = \int \psi(x) f(x) dx = \int \psi(x) \frac{f(x)}{\widetilde{f}(x)} \widetilde{f}(x) dx,$$

we can write $\gamma = \mathbb{E}_{\widetilde{f}}(\psi(X) L(X))$ where function L is defined by $L(x) = f(x)/\widetilde{f}(x)$ on the set $\{x : \psi(x) f(x) > 0\}$ (where $\widetilde{f}$ is also strictly positive), and by $L(x) = 0$ otherwise. L is called the *likelihood ratio*. If we sample n copies of X using density $\widetilde{f}$, and we average the values obtained for function ψL, then we obtain a new unbiased estimator of γ, which we can call here $\widetilde{\gamma}$. That is, $\widetilde{\gamma} = (\psi(X_1)L(X_1) + \cdots + \psi(X_n)L(X_n))/n$, where $X_1, \ldots, X_n$ are independent copies of X having density $\widetilde{f}$. The 95% confidence interval associated with this estimator has the same form as before, $(\widetilde{\gamma} \mp 1.96 \widetilde{\sigma} n^{-1/2})$, where $\widetilde{\sigma}^2$ is the standard estimator of $\mathrm{Var}_{\widetilde{f}}(\psi(X)L(X))/n$,

$$\widetilde{\sigma}^2 = \frac{1}{n-1} \sum_{i=1}^{n} \psi^2(X_i) L^2(X_i) - \frac{n}{n-1} \widetilde{\gamma}^2 \quad \text{(here, } X_i\text{s sampled from } \widetilde{f}).$$

We will now see that a good IS scheme corresponds to a density $\widetilde{f}$ such that $\widetilde{f} \gg f$ in the appropriate part of the real line. For instance, if $\psi(X) = \mathbb{1}(X \in A)$, then a good new density $\widetilde{f}$ is such that $\widetilde{f} \gg f$ on A. Another typical situation is when γ small because when $|\psi|$ is large, f is very small. Then, we must look for some new density $\widetilde{f}$ which is not small when $|\psi| \gg 1$.

Let us, for instance, compare the width of the confidence intervals constructed with the crude and the IS estimators, in the case of $\psi(X) = \mathbb{1}(X \in A)$. For this purpose, we will compare the corresponding exact variances (assuming the associated estimators will be close enough to them). We have

$$\begin{aligned}
\mathrm{Var}(\widetilde{\gamma}) &= \frac{1}{n} \mathrm{Var}_{\widetilde{f}}(\mathbb{1}(X \in A) L(X)) \\
&= \frac{1}{n} \left[\mathbb{E}_{\widetilde{f}} \left(\mathbb{1}(X \in A) L^2(X) \right) - \gamma^2 \right] \\
&\ll \frac{1}{n} \left[\mathbb{E}_{\widetilde{f}} (\mathbb{1}(X \in A) L(X)) - \gamma^2 \right] \\
&= \frac{1}{n} \left[\int \mathbb{1}(X \in A) \frac{f(x)}{\widetilde{f}(x)} \widetilde{f}(x) dx - \gamma^2 \right] \\
&= \frac{1}{n} \left[\int \mathbb{1}(X \in A) f(x) dx - \gamma^2 \right]
\end{aligned}$$

$$= \frac{1}{n}\left[\mathbb{E}_f(\mathbb{1}(X \in A)) - \gamma^2\right]$$

$$= \frac{1}{n}\mathrm{Var}_f(\psi(X))$$

$$= \mathrm{Var}(\widehat{\gamma}).$$

This is the basic idea behind the IS method. Selecting a good change of measure can be difficult, but such a density can be found in many cases, leading to significant improvements in the efficiency of the methods. Chapter 2 analyzes this in detail. See also Chapters 4 and 5.

A relevant comment here is the following. Consider the case of $\psi > 0$. If the change of measure is precisely $\widetilde{f}(x) = f(x)\psi(x)/\gamma$, we have $L(x) = \gamma/\psi(x)$ when $f(x) > 0$, and thus

$$\mathrm{Var}(\widetilde{\gamma}) = \frac{1}{n}\mathrm{Var}_{\widetilde{f}}(L(X)\psi(X)) = \frac{1}{n}\mathrm{Var}_{\widetilde{f}}[\gamma] = 0.$$

This means that there is an optimal change of measure leading to a zero-variance estimator. The simulation becomes a kind of 'pseudo-simulation' leading to the exact value in only one sample (unbiased estimator with variance equal to zero). The bad news is that to make this change of measure, we need the value of γ, which was the target. But this observation leads to two remarks: first, there is a perfect change of measure, which suggests that there are other good and even very good densities out there waiting to be found; second, exploring in greater depth the optimal change of measure leading to the best possible estimator, in the case of specific families of problems, new IS schemes appear, having nice properties. This is also explored in Chapter 2.

Consider again the problem of estimating some rare event probability $\gamma = 10^{-9}$ with relative error less than $\delta = 0.1$. As we saw above, with the crude estimator, we have RE $\approx 1.96/\sqrt{n\gamma} < \delta$ leading to $n > 1.96^2/(\gamma\delta^2) \approx 384$ billion samples. In the case of the IS estimator using density $\widetilde{f}$, RE is $1.96\sqrt{\mathrm{Var}_{\widetilde{f}}(L(X)\psi(X))}/(\gamma\sqrt{n})$. This means that we need

$$n > \frac{1.96^2\mathrm{Var}_{\widetilde{f}}(L(X)\psi(X))}{\gamma^2\delta^2}$$

samples. Assuming that the computation cost in both techniques is similar (often true, though not always), the reduction factor in the effort necessary to reach the desired accuracy with the given confidence (the number of samples necessary) is thus approximately equal to $\gamma/\mathrm{Var}_{\widetilde{f}}(L(X)\psi(X))$ and it will be important if $\mathrm{Var}_{\widetilde{f}}(L(X)\psi(X)) \ll \gamma$, that is, if $\mathbb{E}_{\widetilde{f}}(L^2(X)\psi^2(X)) \ll \gamma$. A good change of measure will be then one for which this second moment is much less than γ.

To fix the ideas and better understand the difficulties, assume that the system is a discrete-time Markov chain Y such as the one depicted in Figure 1.1, with

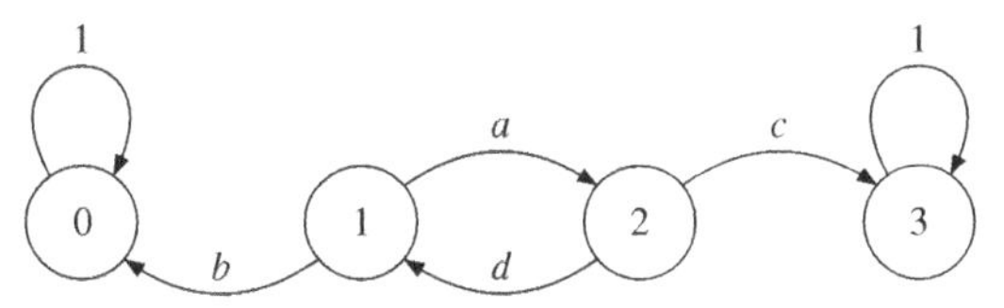

Figure 1.1 A small discrete-time Markov chain Y. We assume that $0 < a, b, c, d < 1$. The aim is to compute $\gamma = \mathbb{P}(Y(\infty) = 3 \mid Y(0) = 1)$. In this simple example, $\gamma = ac/(1 - ad)$.

state space S (in the example, $S = \{0, 1, 2, 3\}$). The chain starts in state 1, and we wish to evaluate the probability that it gets absorbed by state 3 (see relation (6.2) in Chapter 6 for an example of where this is useful). Needless to say, this computation is elementary here, and we can easily obtain the answer analytically: $\gamma = ac/(1 - ad)$. For instance, when a and c are small, the event '$Y(\infty) = 3$' is rare.

To put this problem in the previous setting, observe that it can be stated in the following manner. Let us call $\mathcal{P}$ the set of all possible paths in Y starting at state 1:

$$\mathcal{P} = \{\pi = (y_0, y_1, \ldots, y_K) \in S^{K+1}, K \geq 1,\\ \text{with } y_0 = 1, y_K = 0 \text{ or } 3, \text{ and } y_i \notin \{0, 3\} \text{ if } 1 \leq i < K\}.$$

We can now construct a new random variable Π defined on $\mathcal{P}$, to formalize the idea of a random path of Y, having distribution

$$\mathbb{P}(\Pi = \pi = (y_0, y_1, \ldots, y_K)) = p(\pi) = P_{y_0,y_1} P_{y_1,y_2} \cdots P_{y_{K-1}y_K},$$

where $P_{u,v}$ is the transition probability from state u to state v in chain Y. Finally, consider the real function ψ on $\mathcal{P}$ defined by $\psi(1, y_1, \ldots, y_K) = \mathbb{1}(y_K = 3)$. We are now in the context of the previous presentation about IS: $\gamma = \mathbb{P}(Y(\infty) = 3) = \mathbb{E}_p(\psi(\Pi))$. The crude estimator of γ is then $\widehat{\gamma} = (\psi(\Pi_1) + \cdots + \psi(\Pi_n))/n$ where the independently and identically distributed (i.i.d.) Π_i are sampled using p (i.e., using the transition probability matrix P).

Assume now that we change the dynamics of the chain, that is, the transition probabilities, from P to $\tilde{P}$ in the following way. For any two states u and v, if $P_{u,v} = 0$ then $\tilde{P}_{u,v} = 0$ as well, but when $P_{u,v} > 0$, $\tilde{P}_{u,v}$ takes some other strictly positive value, for the moment arbitrary (but note that this condition is not necessary in general to get an unbiased estimator, it suffices that $\tilde{P}_{u,v} > 0$ if transition from u to v is on a path for which ψ has a non-null value; see Chapter 2). If we denote by $\tilde{p}$ the new distribution of a random path ($\tilde{p}(y_0, y_1, \ldots, y_K) = \tilde{P}_{y_0,y_1} \tilde{P}_{y_1,y_2} \cdots \tilde{P}_{y_{K-1}y_K}$), we can write

$$\gamma = \sum_{\pi \in \mathcal{P}} \psi(\pi) p(\pi) = \sum_{\pi \in \mathcal{P}} \psi(\pi) L(\pi) \tilde{p}(\pi),$$

where $L(\pi) = p(\pi)/\tilde{p}(\pi)$ for all paths π such that $\psi(\pi) = 1$ (observe that in this example, p and $\tilde{p}$ are positive on the whole space $\mathcal{P}$).

This is the general setting of IS. What we have seen is that in order to obtain a new estimator with much smaller variance than the crude one, we need to choose some new dynamics $\tilde{P}$ such that the corresponding distribution $\tilde{p}$ satisfies $\tilde{p}(\pi) \gg p(\pi)$ *on all paths* π *such that* $\psi(\pi) = 1$, that is, on all paths that end at state 3.

Let us call $\mathcal{P}_s$ the set of *successful* paths, that is, the set of those paths in $\mathcal{P}$ ending with state 3 (in other words, $\mathcal{P}_s = \psi^{-1}(1)$). The analyst directly acts on the dynamics of the models, that is, on $\tilde{P}$ (imagine we are dealing with a huge Markov chain and not just with this simple four-state one), so the changes on $\tilde{p}$ are indirect. The first idea that probably comes in mind in order to increase the probability of the paths in $\mathcal{P}_s$ is to change a into some $a' > a$ and/or c into some $c' > c$. The problem is that this intuitively correct decision will not necessarily make all successful paths more probable under $\tilde{p}$ than under p, and finer analysis is needed. So, even in such a trivial example, IS design is not immediate. For instance, consider $a = c = \frac{1}{4}$ and suppose that we decide to make the event of interest $(Y(\infty) = 3)$ more frequent by changing a to $\tilde{a} = \frac{1}{2}$ and c to $\tilde{c} = \frac{3}{4}$. Observe that $\mathcal{P}_s = \{\pi_k, k \geq 1\}$ where $\pi_k = (1, (2, 1)^k, 2, 3)$ (the notation $(2, 1)^k$ meaning that the sequence $(2, 1)$ is repeated k times, $k \geq 0$). We have

$$p(\pi_k) = (ad)^k ac = \left(\frac{1}{4} \times \frac{3}{4}\right)^k \times \frac{1}{4} \times \frac{1}{4}, \quad \tilde{p}(\pi_k) = \left(\frac{1}{2} \times \frac{1}{4}\right)^k \times \frac{1}{2} \times \frac{3}{4}.$$

It can then be verified that $\tilde{p}(\pi_k) > p(\pi_k)$ for $k = 0, 1, 2, 3, 4$ but that, for $k \geq 5$, $\tilde{p}(\pi_k) < p(\pi_k)$. We see that even in such a simple model, finding an appropriate change of measure can be non-trivial. This is of course developed in Chapter 2, devoted to the IS method, and the reader can also look at Chapter 6, where the particular case of Markov models for dependability analysis is studied, and where the problem of finding appropriate changes of measures is discussed in contexts similar to that of the previous example.

Before leaving this example, consider the following IS scheme. Change a to $\tilde{a} = 1$ and c to $\tilde{c} = 1 - ad$. We can verify that $L(\pi_k) = ac/(1 - ad) = \gamma$ for all k, which means that this is the optimal change of measure, the one leading to a zero-variance estimator. This topic is explored in Chapter 2.

1.3 Splitting techniques

Splitting is based on a completely different idea than importance sampling. Suppose that the system is represented by some stochastic process X living in the state space S. Important examples of rare event analysis are when the target is $\gamma = \mathbb{P}(X \in A)$, X assumed to be in equilibrium, or $\gamma = \mathbb{P}(\tau_A < \tau_0)$, where

$\tau_0 = \inf\{t > 0 : X(t) = 0, X(t^-) \neq 0\}$ and $\tau_A = \inf\{t > 0 : X(t) \in A\}$, where A is some subset of S rarely visited by X.

A representative version of the splitting technique (the *fixed-splitting* version) involves considering a sequence of K embedded subsets of states $A_1 \supset A_2 \supset \cdots \supset A_K = A$, with the initial state $0 \notin A_1$. The idea is that reaching A_1 from 0 or A_k from A_{k-1} is not rare. Then, starting the simulation from state 0, when (if) the trajectory reaches A_1 at some state s_1, then n_1 copies of X are constructed (X is *split*), all starting at s_1 but evolving independently. The same behavior is repeated when some of these versions of X reach A_2, splitting any path reaching A_2 into n_2 copies, the same for A_3, etc., until (hopefully) reaching $A = A_K$ from A_{K-1}. For this reason, the A_k are sometimes called *levels* in these methods. When a path reaches A, either it stops, if we are analyzing $\mathbb{P}(\tau_A < \tau_0)$, or it continues its evolution, if the target is π_A. In words, the stochastic dynamics of X is kept untouched (no change of measure), but by making several copies of the 'good' trajectories (those reaching A_{k+1} from A_k), we increase the chances of visiting the rare set A. Of course, we must decide what to do with the trajectories that remain far from A, for instance those coming back to A_{k-1} having been born after a split when a trajectory entered A_k from A_{k-1}. The different specific splitting methods differ depending on whether we evaluate a probability of the form π_A or of the form $\mathbb{P}(\tau_A < \tau_0)$ (or other possible targets such as transient metrics, for example), in the way they deal with 'bad' trajectories (whether they discard them and how), and in the specific form of the estimators used. They all share the splitting idea just described.

To be more specific, suppose that the aim is to estimate $\gamma = \mathbb{P}(\tau_A < \tau_0) \ll 1$, and that we construct our sequence of levels $A_1, A_2, \ldots, A_K$. If $\tau_{A_i} = \inf\{t > 0 : X(t) \in A_i\}$, denote $p_i = \mathbb{P}(\tau_{A_i} < \tau_0 | \tau_{A_{i-1}} < \tau_0)$, for $i = 2, 3, \ldots, K$, and $p_1 = \mathbb{P}(\tau_{A_1} < \tau_0)$. We see that $\gamma = p_1 p_2 \cdots p_K$. In order to be able to observe many hits on the rare set A, we need these conditional probabilities to be not too small. Assume that a trajectory entering A_i from 'above', that is, from A_{i-1}, is split into n_i copies, where the sequence $n_0, n_1, \ldots, n_{K-1}$ is fixed (at the beginning, n_0 independent copies starting at state 0 begin their evolution). All trajectories are killed either when reaching A or when arriving back at 0. A trajectory that makes a move from some level A_j to A_{j-1}, calling A_0 the complement to the set A_1, is not split anymore.

Let H be the total number of hits on the set A. Then, the random variable

$$\widetilde{\gamma} = \frac{H}{n_0 n_1 \cdots n_{K-1}}$$

is an unbiased estimator of γ (this can be checked by taking conditional expectations of this last expression). The computation of the variance of $\widetilde{\gamma}$ is more involved. To simplify things in this short introduction, let us assume that the evolutions of the paths at each level are i.i.d., that is, have the same distribution. This happens, for instance, if X is a Markov process and the transitions from A_{j-1} to A_j enter the latter by the same state. Then, after some algebra, direct calculations give the following expression for the variance of the estimator:

$$\mathrm{Var}(\widetilde{\gamma}) = \gamma^2 \sum_{k=1}^{K} \frac{1 - p_k}{p_1 n_0 \cdots p_k n_{k-1}}.$$

The relative error obtained would be, as usual, proportional to the square root of this variance. An important point must be emphasized here. In order to increase the probability of reaching A many times, a first idea that may come to mind is to use many levels (thus taking K large) and to make many copies at each splitting step (thus taking n_i large). But then, the global cost would increase significantly, in particular due to all the 'bad' paths that do not lead to a hit. This means that the analysis of the efficiency of a splitting procedure, or a comparison study, needs to make specific assumptions on the underlying cost model adopted. In other words, just looking at the variance of the estimator will not be enough (these issues are discussed in depth in Chapter 3).

Returning to our variance, the analysis of the asymptotic behavior of $\widetilde{\gamma}$ when the number of sets K increases, and using the concepts of efficiency (recalled in Chapter 4), we conclude that the best situation is to have $p_i n_{i-1} = 1$ (ignoring the fact that n_i is an integer). To get an intuition for this, observe that when rarity increases, we will need more levels (otherwise, going from one level to the next can also become a rare event). In that case, if $p_i n_{i-1} > 1$ the number of paths will increase (stochastically) without bound, while if $p_i n_{i-1} < 1$ we will have an extinction phenomenon (as in branching processes, from which many of the results for the splitting methods are derived). The variance becomes $\mathrm{Var}(\widetilde{\gamma}) = \gamma^2 \sum_{k=1}^{K}(1 - p_k)$, and we can look for the minimum of this function of the p_i under the constraint $p_1 \cdots p_K = \gamma$, using Lagrange multipliers for instance, obtaining that it is reached when $p_i = p = \gamma^{1/K}$. The variance now becomes $\mathrm{Var}(\widetilde{\gamma}) = \gamma^2 K(1 - \gamma^{1/K}) \to -\ln(\gamma)$ as $K \to \infty$. Since $p = \gamma^{1/K} \to 1$ as $K \to \infty$, we have a relative freedom to choose a moderate value of K depending on our effective implementation.

Some remarks are necessary at this point. First, the fact that n_i is an integer, and that the asymptotic analysis suggests $n_i = 1/p_i$, needs to take some decision: rounding, randomizing things, etc. Second, as stated before, the cost model must be carefully taken into account here. Third, there are other variants of the same idea, leading to different expressions for the variance. For instance, in the fixed-effort model, we decide beforehand the number b_j of paths that will leave in each A_j. This leads to a different expression for the variance of the estimator, always in the i.i.d. context (namely, $\mathrm{Var}(\widetilde{\gamma}) = \gamma^2 \sum_{k=1}^{K}(1 - p_k)/(b_{k-1} p_k)$). Last, a supplementary remark here. If the recommendation is to try to obtain equal conditional probabilities p_k, this is not necessarily easy to implement. Process X may be a complex Markov chain and the subset A_j may itself be complex enough, such that there are many possible ways of reaching it from A_{j-1}. Then, tuning the parameters of the splitting method in order to follow the general guidelines emerging from a theoretical analysis is not an easy task. All these issues are discussed in Chapter 3.

1.4 About the book

This book is therefore specifically devoted to techniques for rare event simulation. A reader wishing to learn more about Monte Carlo simulation concepts in general is advised to consult [1, 3]. For other references on our main topic, we note that there exists another book on rare event simulation, by J.A. Bucklew [2], but that it focuses on importance sampling and its relation to large-deviations theory, in greater depth than in the present broader study. States of the art, published as papers or book chapters can otherwise be found in [4, 5].

This book consists of two parts, the first devoted to theory, the second to applications. Part I is composed of three chapters. In Chapter 2, an important principle for rare event simulation, the importance sampling family of methods, is described. These techniques were briefly introduced in Section 1.2 above. The idea is to simulate another model, different from the initial one, where the event of interest is not a rare one anymore. An unbiased estimator can be recovered by changing the random variable considered. Related material can be found in Chapters 4–7.

Chapter 3 is devoted to the other large class of methods, splitting techniques, where the idea is to keep the initial probabilistic structure in the model (no change of measure as in importance sampling), but to make clones/copies of the object being simulated when we are getting closer to the rare event. This procedure increases the chances of reaching the event, but a careful design of the method has to be performed when defining the number of splits/clones as well as when to split, to avoid an explosion in the number of objects considered and to properly reduce the variance. This family was introduced in Section 1.3 above. Related material will be found in Chapters 9 and 10.

The last chapter in Part I is Chapter 4, where the quality of the estimators of rare objects is discussed. The quality of an estimator of a very small probability or expectation has different aspects. One is the robustness of the method, which basically refers to its accuracy as the event of interest becomes rarer and rarer. Another important aspect is coverage, which refers to the validity, or reliability, of the confidence intervals, again as rarity increases. In both cases, these properties must construct a family of versions of the initial model where rarity can somehow be controlled, in general by a single parameter. The cost of the estimation technique is another factor that can play a crucial role in the global quality. The chapter reviews the notions of robustness and then focuses on one specific aspect of quality, the fact that in a rare event situation there may be an important difference between the theoretical properties of the estimators used and the effective behavior of their implementations, in relation to the notion of reliability of the confidence intervals produced. The chapter basically discusses some ideas about the problem of the diagnostics of the fact that the simulation is not working properly.

Part II is devoted to applications and is composed of seven chapters. Chapter 5 concerns models described in terms of queuing systems. Onc of the main areas of application of these models is telecommunications, where queues are a natural representation of the nodes in a communication network. There are specific domains where queues and networks of queues are very important modeling tools, such as teletraffic engineering, Internet-based applications, and call center design. Rarity can take different forms in this context. The typical example is the saturation of a buffer, that is, the fact that the number of units in a queue or a set of queues reaches some high value B. This is also an example of a parameter (the B threshold) that allows us to control rarity in the model, as mentioned in the previous paragraph: the higher the threshold, the rarer the event. Chapter 5 presents a nice description of the application of the theory of large deviations in probability to the design of a good importance sampling scheme. It additionally highlights the fact that large deviations are related to queues with so-called light-tailed (i.e., exponentially decreasing distribution tails) inputs, but also presents how to deal to deal with heavy-tailed processes that have become frequent in the Internet for instance. It then focuses on the techniques specific to Jackson queuing networks, that is, open networks of queues connected by Bernoulli probabilistic switches.

Chapter 6 is devoted to Monte Carlo methods for evaluating the probability of rare events and related metrics, when they are defined on the large family of Markov models mainly used for dependability analysis. Typical dependability metrics are the reliability at some point in time t of the system, its availability at t, the statistical properties of the random variable "interval availability on $[0, t]$" (such as its distribution), the mean time to failure, etc. For instance, in a highly dependable system, the mean time to failure may be a very large number, and its estimation has the same problem as the estimation of a very small probability. In this chapter, methods able to estimate this number by analyzing the probability of an associated rare event are discussed. We stress that both steady-state and transient dependability metrics are discussed.

Chapter 7 is the only one where the models are static, which means that we do not have a stochastic process evolving with time, on top of which the rare events are defined. Here, the system is considered at a fixed point in time (possibly at ∞). The main area representative of this type of models is that of dependability, where the system may have a huge number of states, but where the state space is decomposed into two classes, those states where the whole system works, and its complement where the system is not operational. The rare event is the event 'the system is down'. The static property plus the usual type of models in this area lead to some specificities, but the general approaches are still valid. The chapter reviews the main applications of these models and the most important simulation techniques available to deal with the rare event case in this context.

Chapter 8 has a slightly different focus. It deals with the relationships between rare event simulation and randomized approximation algorithms for counting. Indeed, there exist theoretical tools and efficiency measurement properties developed in counting problems which have a natural counterpart in rare event simulation. The aim of the chapter is therefore to review the methods and properties of algorithms for counting, in order to highlight this correspondence and new tools which can be useful for rare event simulation.

Chapter 9 is driven by an application of rare event simulation to the problem of safety verification in air traffic operations, in order to avoid catastrophic events such as collisions. The chapter makes use of splitting techniques to solve the problem. More precisely, it uses the ideas developed within the framework of the interacting particle system (IPS) algorithm to simulate a large-scale controlled stochastic hybrid system and illustrates the validity of the approach in specific scenarios.

Chapter 10 is about Monte Carlo (nuclear) particle transport simulation, probably the first application of Monte Carlo methods since the advent of computers, but it has a special focus on a critical and special application, shielding. The chapter explains the theoretical developments implemented in MCNP, the code developed at Los Alamos National Laboratory since the Second World War to simulate the operation of nuclear weapons. This code is now used worldwide for particle transport and interactions in the main research laboratories, radiotherapy centers and hospitals.

Chapter 11 describes efficient Monte Carlo techniques for simulating biological systems. The models presented are systems of differential equations representing biochemical reactions but where different reactions can occur at different time scales. In this situation, the reactions occurring at lower rates become rare events, and rare event techniques can be applied to obtain efficient simulations.

References

[1] S. Asmussen and P. W. Glynn. *Stochastic Simulation*. Springer, New York, 2007.

[2] J. A. Bucklew. *Introduction to Rare Event Simulation*. Springer, New York, 2004.

[3] G. S. Fishman. *Monte Carlo: Concepts, Algorithms and Applications*. Springer, New York, 1996.

[4] P. Heidelberger. Fast simulation of rare events in queueing and reliability models. *ACM Transactions on Modeling and Computer Simulation*, **5**(1): 43–85, 1995.

[5] S. Juneja and P. Shahabuddin. Rare event simulation techniques: An introduction and recent advances. In S. G. Henderson and B. L. Nelson, eds, *Simulation*, Handbooks in Operations Research and Management Science, pp. 291–350. Elsevier, Amsterdam, 2006.

[6] S. Ross. *A First Course in Probability*, 7th edition. Pearson Prentice Hall, Upper Saddle River, NJ, 2006.

[7] T. D. Ross. Accurate confidence intervals for binomial proportion and Poisson rate estimation. *Computers in Biology and Medicine*, **33**(3): 509–531, 2003.

Part I

THEORY

2

Importance sampling in rare event simulation

Pierre L'Ecuyer, Michel Mandjes and Bruno Tuffin

2.1 Introduction

As described in Chapter 1, crude (also called standard, or naive) Monte Carlo simulation is inefficient for simulating rare events. Recall that crude Monte Carlo involves considering a sample of n independent copies of the random variable or process at hand, and estimating the probability of a rare event by the proportion of times the rare event occurred over that sample. The resulting estimator can be considered useless when the probability of occurrence, γ, is very small, unless n is much larger than $1/\gamma$. Indeed, if for instance $\gamma = 10^{-9}$, a frequent target in rare event applications, this would require on average a sample of size $n = 10^9$ to observe just a single occurrence of the event, and much more if we expect a reliable estimation of the mean and variance to obtain a sufficiently narrow confidence interval.

Importance sampling (IS) has emerged in the literature as a powerful tool to reduce the variance of an estimator, which, in the case of rare event estimation, also means increasing the occurrence of the rare event. The generic idea of IS is to change the probability laws of the system under study to sample more frequently the events that are more 'important' for the simulation. Of course, using a new distribution results in a biased estimator if no correction is applied. Therefore the simulation output needs to be translated in terms of the original

Rare Event Simulation using Monte Carlo Methods Edited by G. Rubino and B. Tuffin

measure; this is done by multiplication with a so-called likelihood ratio. IS has received substantial theoretical attention; see *inter alia* [13, 12, 26] and, in the rare event context, [15] or the more up-to-date tutorial [16].

IS is one of the most widely used variance reduction technique in general, and for rare event estimation in particular. Typical and specific applications will be more extensively described in Part II of this book. The goal of this chapter is to give an overview of the technical framework and the main underlying ideas. It is organized as follows.

Section 2.2 reviews the very basic notions of IS. It describes what the ideal (zero-variance) estimator looks like, and why it is, except in situations where simulation is not needed, infeasible to implement it exactly. That section also provides illustrative examples and outlines some properties leading to a good IS estimator, the main message being that the zero-variance estimator has to be approximated as closely as possible.

In Section 2.3 the focus is on application of IS in the context of a Markov chain model. Since every discrete-event simulation model can be seen as a Markov chain (albeit over a high-dimensional state space), this setting is very general. We show how to define a zero-variance change of probabilities in that context. It is noted that, in general, the zero-variance change of probabilities must depend on the state of the chain. We compare this type of change of probabilities with a more restricted class of IS called *state-independent*, in which the probabilities are changed independently of the current state of the chain. This type of state-independent IS originates mainly from asymptotic approximations based on large-deviations theory [3, 15, 25], and has been developed in applications areas such as queuing and finance [10, 15, 16, 19]. However, in many situations, any good IS scheme *must* be state-dependent [3] (as state-independent IS leads to estimators with large, or even infinite, variance). Note that in computational physics (the application area from which it originates) and in reliability, IS has traditionally been state-dependent [5, 14, 16].

Finally, Section 2.4 describes various methods used to approximate the zero-variance (i.e., optimal) change of measure. Some just use intuitive approximations, whereas others are based on the asymptotic behavior of the system when the events of interest become rarer and rarer (this includes methods based on large-deviations theory, and other techniques as well). Another option is to use adaptive techniques that learn (and use) approximations of the zero-variance change of measure, or optimal parameter values within a class of parameterized IS strategies: the results of completed runs can be used as inputs of strategies for the next runs, but those IS strategies can also be updated at each step of a given run [16, 22].

The accuracy assessment of the resulting confidence interval, and the robustness properties of the estimator with respect to rarity, are the focus of the next chapter. To avoid overlap, we keep our discussion of those aspects to a minimum here.

2.2 Static problems

We wish to compute the expected value of a random variable $X = h(Y)$, $\mathbb{E}[h(Y)]$, where Y is assumed to be a random variable with density f (with respect to the Lebesgue measure) in the d-dimensional real space $\mathbb{R}^d$. (In our examples, we will have $d = 1$.) Then the crude Monte Carlo method estimates

$$\mathbb{E}[h(Y)] = \int h(y) f(y) dy \quad \text{by} \quad \frac{1}{n}\sum_{i=1}^{n} h(Y_i),$$

where $Y_1, \ldots, Y_n$ are independently and identically distributed copies of Y, and the integral is over $\mathbb{R}^d$.

IS, on the other hand, samples Y from another density $\tilde{f}$ rather than f. Of course, the same estimator $\frac{1}{n}\sum_{i=1}^{n} h(Y_i)$ then becomes biased in general, but we can recover an unbiased estimator by weighting the simulation output as follows. Assuming that $\tilde{f}(y) > 0$ whenever $h(y)f(y) \neq 0$,

$$\begin{aligned}\mathbb{E}[h(X)] &= \int h(y) f(y) dy = \int h(y) \frac{f(y)}{\tilde{f}(y)} \tilde{f}(y) dy \\ &= \int h(y) L(y) \tilde{f}(y) dy = \tilde{\mathbb{E}}[h(Y)L(Y)],\end{aligned}$$

where $L(y) = f(y)/\tilde{f}(y)$ is the *likelihood* ratio of the density $f(\cdot)$ with respect to the density $\tilde{f}(\cdot)$, and $\tilde{\mathbb{E}}[\cdot]$ is the expectation under density $\tilde{f}$. An unbiased estimator of $\mathbb{E}[h(Y)]$ is then

$$\frac{1}{n}\sum_{i=1}^{n} h(Y_i)L(Y_i), \tag{2.1}$$

where $Y_1, \ldots, Y_n$ are independently and identically distributed random variables sampled from $\tilde{f}$.

The case where Y has a discrete distribution can be handled analogously; it suffices to replace the densities by probability functions and the integrals by sums. That is, if $\mathbb{P}[Y = y_k] = p_k$ for $k \in \mathbb{N}$, then IS would sample n copies of Y, say $Y_1, \ldots, Y_n$, using probabilities $\tilde{p}_k$ instead of p_k, for $k \in \mathbb{N}$, where $\tilde{p}_k > 0$ whenever $p_k h(y_k) \neq 0$. An unbiased IS estimator of $\mathbb{E}[h(Y)]$ is again (2.1), but with $L(y_k) = p_k/\tilde{p}_k$. Indeed,

$$\tilde{\mathbb{E}}[h(Y)L(Y)] = \sum_{k\in\mathbb{N}} h(y_k)\frac{p_k}{\tilde{p}_k}\tilde{p}_k = \sum_{k\in\mathbb{N}} h(y_k)p_k = \mathbb{E}[h(Y)].$$

In full generality, if Y obeys some probability law (or measure) $\mathbb{P}$, and IS replaces $\mathbb{P}$ by another probability measure $\tilde{\mathbb{P}}$, we must multiply the original estimator by the likelihood ratio (or Radon--Nikodým derivative) $L = d\mathbb{P}/d\tilde{\mathbb{P}}$.

Clearly, the above procedure leaves us a huge amount of freedom: any alternative $\tilde{\mathbb{P}}$ yields an unbiased estimator (as long as the above-mentioned regularity conditions are fulfilled). Therefore, the next question is: based on what principle should we choose the IS measure $\tilde{\mathbb{P}}$? The aim is to find a change of measure for which the IS estimator has small variance, preferably much smaller than for the original estimator, and is also easy (and not much more costly) to compute (in that it should be easy to generate variates from the new probability law). We denote these two variances by

$$\tilde{\sigma}^2(h(Y)L(Y)) = \tilde{\mathbb{E}}[(h(Y)L(Y))^2] - (\mathbb{E}[h(Y)])^2$$

and

$$\sigma^2(h(Y)) = \mathbb{E}[(h(Y))^2] - (\mathbb{E}[h(Y)])^2,$$

respectively. Under the assumptions that the IS estimator has a normal distribution (which is often a good approximation–but not always), a confidence interval at level $1 - \alpha$ for $\mathbb{E}[h(Y)]$ is given by

$$\left[\frac{1}{n}\sum_{i=1}^{n} h(Y_i)L(Y_i) - z_{\alpha/2}\frac{\tilde{\sigma}(h(Y)L(Y))}{\sqrt{n}}, \frac{1}{n}\sum_{i=1}^{n} h(Y_i)L(Y_i) + z_{\alpha/2}\frac{\tilde{\sigma}(h(Y)L(Y))}{\sqrt{n}}\right]$$

where $z_{\alpha/2} = \Phi^{-1}(1 - \alpha/2)$ and Φ is the standard normal distribution function. For fixed α and n, the width of the confidence interval is proportional to the standard deviation (the square root of the variance). So reducing the variance by a factor K improves the accuracy by reducing the width of the confidence interval by a factor $\sqrt{K}$. The same effect is achieved if we multiply n by a factor K, but this requires (roughly) K times more work.

In the rare event context, one usually simulates until the relative accuracy of the estimator, defined as the ratio of the confidence-interval half-width and the quantity γ to be estimated, is below a certain threshold. For this, we need $\tilde{\sigma}^2(h(Y)L(Y))/n$ approximately proportional to γ^2. Thus, the number of samples needed is proportional to the variance of the estimator. In the case where γ is a small probability and $h(Y)$ is an indicator function, without IS, $\tilde{\sigma}^2(h(Y)L(Y)) = \sigma^2(h(Y)) = \gamma(1 - \gamma) \approx \gamma$, so the required n is roughly inversely proportional to γ and often becomes excessively large when γ is very small.

The optimal change of measure is to select the new probability law $\tilde{\mathbb{P}}$ so that

$$L(Y) = \frac{d\mathbb{P}}{d\tilde{\mathbb{P}}} = \frac{\mathbb{E}[|h(Y)|]}{|h(Y)|},$$

which means $\tilde{f}(y) = f(y)|h(y)|/\mathbb{E}[|h(Y)|]$ in the continuous case, and $\tilde{p}_k = p_k|h(y_k)|/\mathbb{E}[|h(Y)|]$ in the discrete case. Indeed, for any alternative IS measure

$\mathbb{P}'$ leading to the likelihood ratio L' and expectation $\mathbb{E}'$, we have

$$\tilde{\mathbb{E}}[(h(Y)L(Y))^2] = (\mathbb{E}[|h(Y)|])^2 = (\mathbb{E}'[|h(Y)|L'(Y)])^2 \leq \mathbb{E}'[(h(Y)L'(Y))^2].$$

In the special case where $h \geq 0$, the optimal change of measure gives $\tilde{\mathbb{E}}[(h(Y)L(Y))^2] = (\mathbb{E}[h(Y)])^2$, that is, $\tilde{\sigma}^2(h(Y)L(Y)) = 0$. Thus, IS provides a *zero-variance estimator*. We call the corresponding change from $\mathbb{P}$ to $\tilde{\mathbb{P}}$ the *zero-variance change of measure*. In many typical rare event settings, one indeed has $h \geq 0$; for example, this is obviously the case when the focus is on estimating the probability of a rare event (h is then an indicator function).

All of this is nice in theory, but in practice there is an obvious crucial drawback: implementing the optimal change of measure requires knowledge of $\mathbb{E}[|h(Y)|]$, the quantity that we wanted to compute; if we knew it, no simulation would be needed! But the expression for the zero-variance measure provides a hint on the general form of a 'good' IS measure, that is, a change of measure that leads to substantial variance reduction. As a rough general guideline, it says that $L(y)$ should be small when $|h(y)|$ is large.

In particular, if there is a constant $\kappa \leq 1$ such that $L(y) \leq \kappa$ for all y such that $h(y) \neq 0$, then

$$\tilde{\mathbb{E}}[(h(Y)L(Y))^2] \leq \kappa\tilde{\mathbb{E}}[(h(Y))^2 L(Y)] = \kappa\mathbb{E}[(h(Y)^2], \qquad (2.2)$$

so the second moment is guaranteed to be reduced at least by the factor κ. If h is also an indicator function, say $h(y) = 1_A(y)$ for some set A, and $\mathbb{E}[h(Y)] = \mathbb{P}[A] = \gamma$, then we have

$$\tilde{\sigma}^2(1_A(Y)L(Y)) = \tilde{\mathbb{E}}[(1_A(Y)L(Y))^2] - \gamma^2 \leq \kappa^2 - \gamma^2.$$

This implies that we always have $\kappa \geq \gamma$, but, evidently, we want to have κ as close as possible to γ.

In theoretical analysis of rare event simulation, it is customary to parameterize the model by a rarity parameter $\varepsilon > 0$ so that the important events occur (in the original model) with a probability that converges to 0 when $\varepsilon \to 0$. In that context, an IS estimator based on a change of measure that may depend on ε is said to have *bounded relative variance* (or *bounded relative error*) if $\tilde{\sigma}^2(h(Y)L(Y))/\mathbb{E}^2[h(Y)]$ is bounded uniformly in ε. This important property means that estimating $\mathbb{E}[h(Y)]$ with a given relative accuracy can be achieved with a bounded number of replications even if $\varepsilon \to 0$.

In the special case where $h(y) = 1_A(y)$ and $\gamma = \mathbb{P}[A]$, if we can find a constant κ' such that $L(y) \leq \kappa'\gamma$ when $y \in A$, then

$$\tilde{\sigma}^2(1_A(Y)L(Y)) \leq (\kappa'\gamma)^2 - \gamma^2 = \gamma^2((\kappa')^2 - 1),$$

which means that we have bounded relative variance: the relative variance remains bounded by $(\kappa')^2 - 1$ no matter how rare the event A is. This type of property will be studied in more detail in Chapter 4.

Example 1. To illustrate the ideas and the difficulty of finding a good IS distribution, we first consider a very simple example for which a closed-form expression is known. Suppose that the failure time of a system follows an exponential distribution with rate λ and that we wish to compute the probability γ that the system fails before T. We can write $h(y) = 1_A(y)$ where $A = [0, T]$, and we know that $\gamma = \mathbb{E}[1_A(Y)] = 1 - e^{-\lambda T}$. This quantity is small (i.e., A is a rare event) when λT is close to 0. The zero-variance IS here involves sampling Y from the same exponential density, but truncated to the interval $[0, T]$: $\tilde{f}(y) = \lambda e^{-\lambda y}/(1 - e^{-\lambda T})$ for $0 \leq y \leq T$.

But suppose that we insist on sampling from an exponential density with a different rate $\tilde{\lambda}$ instead of truncating the distribution. The second moment of that IS estimator will be

$$\tilde{\mathbb{E}}[(1_A(Y)L(Y))^2] = \int_0^T \left(\frac{\lambda e^{-\lambda y}}{\tilde{\lambda} e^{-\tilde{\lambda} y}}\right)^2 \tilde{\lambda} e^{-\tilde{\lambda} y} dy = \frac{\lambda^2}{\tilde{\lambda}(2\lambda - \tilde{\lambda})}(1 - e^{-(2\lambda - \tilde{\lambda})T}).$$

Figure 2.1 displays the variance ratio $\tilde{\sigma}^2(1_A(Y)L(Y))/\sigma^2(1_A(Y))$ as a function of $\tilde{\lambda}$, for $T = 1$ and $\lambda = 0.1$. The variance is minimized with $\tilde{\lambda} \approx 1.63$, that is, with a 16-fold increase in the failure rate, and its minimal value is about 5.3% of the value with $\tilde{\lambda} = \lambda$. If we increase $\tilde{\lambda}$ too much, then the variance increases again. With $\tilde{\lambda} > 6.01$ (approximately), it becomes larger than with $\tilde{\lambda} = \lambda$. This is due to the fact that for very large values of $\tilde{\lambda}$, the likelihood ratio takes huge values when Y is smaller than T but close to T.

Suppose now that $A = [T, \infty)$ instead, that is, $\gamma = \mathbb{P}[Y \geq T]$. The zero-variance density is exponential with rate λ, truncated to $[T, \infty)$. If we use an exponential with rate $\tilde{\lambda}$ instead, the second moment of the IS estimator is

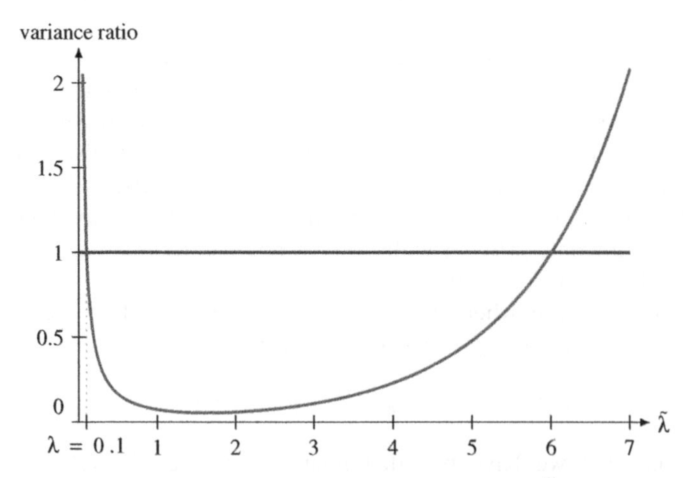

Figure 2.1 Variance ratio (IS vs non-IS) as a function of $\tilde{\lambda}$ for Example 1 with $\lambda = 0.1$ and $A = [0, 1]$.

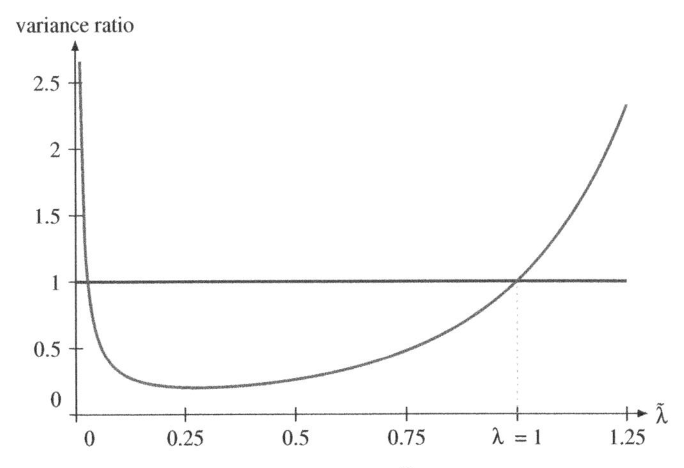

Figure 2.2 Variance ratio as a function of $\tilde{\lambda}$ for Example 1 with $\lambda = 1$ and $A = [3, \infty)$.

finite if and only if $0 < \tilde{\lambda} < 2\lambda$, and is

$$\tilde{\mathbb{E}}[(1_A(Y)L(Y))^2] = \int_T^\infty \left(\frac{\lambda e^{-\lambda y}}{\tilde{\lambda} e^{-\tilde{\lambda} y}}\right)^2 \tilde{\lambda} e^{-\tilde{\lambda} y} dy = \frac{\lambda^2}{\tilde{\lambda}(2\lambda - \tilde{\lambda})} e^{-(2\lambda - \tilde{\lambda})T}.$$

In this case, the variance is minimized for $\tilde{\lambda} = \lambda + 1/T - (\lambda^2 + 1/T^2)^{1/2} < \lambda$. When $\tilde{\lambda} > 2\lambda$, the variance is infinite because the squared likelihood ratio grows exponentially with y at a faster rate than the exponential rate of decrease of the density. Figure 2.2 shows the variance ratio (IS vs non-IS) as a function of $\tilde{\lambda}$, for $T = 3$ and $\lambda = 1$. We see that the minimal variance is attained with $\tilde{\lambda} \approx \lambda/4$.

Another interesting situation is if $A = [0, T_1] \cup [T_2, \infty)$ where $0 < T_1 < T_2 < \infty$. The zero-variance density is again exponential truncated to A, which is now split in two pieces. If we just change λ to $\tilde{\lambda}$, then the variance associated with the first (second) piece increases if $\tilde{\lambda} < \lambda$ ($\tilde{\lambda} > \lambda$). So one of the two variance components increases, regardless of how we choose $\tilde{\lambda}$. One way of handling this difficulty is to use a mixture of exponentials: take $\tilde{\lambda}_1 < \lambda$ with probability p_1 and $\tilde{\lambda}_2 > \lambda$ with probability $p_2 = 1 - p_1$. We now have three parameters to optimize: $\tilde{\lambda}_1$, $\tilde{\lambda}_2$, and p_1.

Example 2. Now let X be binomially distributed with parameters (n, p), and suppose we wish to estimate $\gamma = \mathbb{P}[X \geq na]$ for some constant $a > 0$, where na is assumed to be an integer. Again, the zero-variance IS samples from this binomial distribution truncated to $[na, \infty)$. But if we restrict ourselves to the class of non-truncated binomial changes of measure, say with parameters $(n, \tilde{p})$, following the same line of reasoning as in Example 1, we want to find the $\tilde{p}$ that

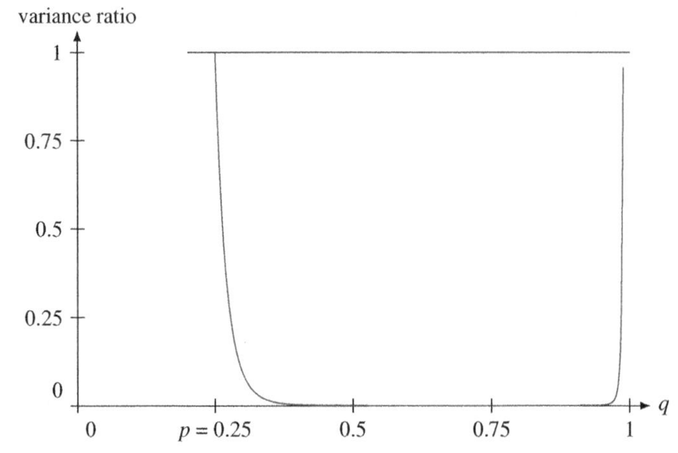

Figure 2.3 Variance ratio of IS vs non-IS estimators, as a function of $\tilde{p}$, for Example 2 with $n = 20$, $p = 1/4$, and $A = [15, \infty)$.

minimizes the second moment

$$\sum_{i=na}^{n} \binom{n}{i} \left(\frac{p^2}{\tilde{p}}\right)^i \left(\frac{(1-p)^2}{1-\tilde{p}}\right)^{n-i}.$$

Figure 2.3 shows the variance ratio as a function of $\tilde{p}$, for $a = 3/4$, $p = 1/4$, and $n = 20$. It shows that the best choice of $\tilde{p}$ lies around a. If $a > p$ is fixed and $n \to \infty$, then large-deviations theory tells us that γ decreases exponentially with n and that the optimal $\tilde{p}$ (asymptotically) is $\tilde{p} = a$. The intuitive interpretation is that when $a > p$, conditional on the event $\{X/n \geq a\}$, most of the density of X/n is concentrated very close to a when n is large. By selecting $\tilde{p} = a$, IS mimics this conditional density. We also see from the plot that the variance ratio is a very flat function of $\tilde{p}$ in a large neighborhood of a; the ratio is approximately 1.2×10^{-5} from 0.72 to 0.78, and is still 5.0×10^{-5} at 0.58 and 0.88. On the other hand, the IS variance blows up quickly when $\tilde{p}$ approaches 1 or goes below p.

2.3 Markov chains

Having dealt in the previous section with the case of a general random variable, we focus here on the specific case where this random variable is a function of the sample path of a Markov chain. We introduce IS in this context, both for discrete-time and continuous-time chains, as well as the form of the corresponding zero-variance change of measure. Approximation algorithms for this change of measure are discussed in the next section.

2.3.1 Discrete-time Markov chains

Consider now a discrete-time Markov chain (DTMC), say $\{Y_j, j \geq 0\}$, with discrete state space $\mathcal{Y}$ (possibly infinite and high-dimensional). The chain evolves up to a stopping time τ defined as the first time the chain hits a given set of states, $\Delta \subset \mathcal{Y}$; that is, $\tau = \inf\{j \geq 0 : Y_j \in \Delta\}$. We assume that $\mathbb{E}[\tau] < \infty$. The chain has a transition probability matrix whose elements are $P(y, z) = \mathbb{P}[Y_j = z \mid Y_{j-1} = y]$ for all $y, z \in \mathcal{Y}$, and the initial probabilities are $\pi_0(y) = \mathbb{P}[Y_0 = y]$ for all $y \in \mathcal{Y}$. We consider the random variable $X = h(Y_0, \ldots, Y_\tau)$, where h is a given function of the trajectory of the chain, with values in $[0, \infty)$. Let $\gamma(y) = \mathbb{E}_y[X]$ denote the expected value of X when $Y_0 = y$, and define $\gamma = \mathbb{E}[X] = \sum_{y \in \mathcal{Y}} \pi_0(y)\gamma(y)$, the expected value of X for the initial distribution π_0.

Our discussion could be generalized to broader classes of state spaces. For a continuous state space $\mathcal{Y}$, the transition probabilities would have to be replaced by a probability transition kernel and the sums by integrals, and we would need some technical measurability assumptions. Any discrete-event simulation model for which we wish to estimate the expectation of some random variable $X = h(Y_0, \ldots, Y_\tau)$ as above can fit into this framework. For simplicity, we stick to a discrete state space.

The basic idea of IS here is to replace the probabilities of sample paths $(y_0, \ldots, y_n)$,

$$\mathbb{P}[(Y_0, \ldots, Y_\tau) = (y_0, \ldots, y_n)] = \pi_0(y_0) \prod_{j=1}^{n} P(y_{j-1}, y_j),$$

where $n = \min\{j \geq 0 : y_j \in \Delta\}$, by new probabilities $\tilde{\mathbb{P}}[(Y_0, \ldots, Y_\tau) = (y_0, \ldots, y_n)]$ such that $\tilde{\mathbb{E}}[\tau] < \infty$ and $\tilde{\mathbb{P}}[\cdot] > 0$ whenever $\mathbb{P}[\cdot]h(\cdot) > 0$. This is extremely general.

To be more practical, we might wish to restrict ourselves to changes of measure under which $\{Y_j, j \geq 0\}$ remains a DTMC with the same state space $\mathcal{Y}$. That is, we replace the transition probabilities $P(y, z)$ by new transition probabilities $\tilde{P}(y, z)$ and the initial probabilities $\pi_0(y)$ by $\tilde{\pi}_0(y)$. The new probabilities must be chosen so that any sample path having a positive contribution to γ must still have a positive probability, and $\tilde{\mathbb{E}}[\tau] < \infty$. The likelihood ratio becomes

$$L(Y_0, \ldots, Y_\tau) = \frac{\pi_0(Y_0)}{\tilde{\pi}_0(Y_0)} \prod_{j=1}^{\tau} \frac{P(Y_{j-1}, Y_j)}{\tilde{P}(Y_{j-1}, Y_j)}$$

and we have

$$\gamma = \tilde{\mathbb{E}}[X L(Y_0, \ldots, Y_\tau)].$$

A question that comes to mind is whether there is a zero-variance change of measure in this setting. What is it?

To answer this question, following [5, 17, 18, 20], we restrict ourselves to the case where the cost X is additive:

$$X = \sum_{j=1}^{\tau} c(Y_{j-1}, Y_j) \tag{2.3}$$

for some function $c : \mathcal{Y} \times \mathcal{Y} \to [0, \infty)$. Note that in this case, we can multiply the term $c(Y_{j-1}, Y_j)$ by the likelihood ratio only up to step j. This gives the estimator

$$\tilde{X} = \sum_{i=1}^{\tau} c(Y_{i-1}, Y_i) \prod_{j=1}^{i} \frac{P(Y_{j-1}, Y_j)}{\tilde{P}(Y_{j-1}, Y_j)}.$$

We now show that in this setting, if we take $\tilde{P}(y, z)$ proportional to

$$P(y, z)[c(y, z) + \gamma(z)]$$

for each $y \in \mathcal{Y}$, then we have zero variance. (Without the additivity assumption (2.3), to get zero variance, the probabilities for the next state must depend in general on the entire history of the chain.) Suppose that

$$\tilde{P}(y, z) = \frac{P(y, z)(c(y, z) + \gamma(z))}{\sum_{w \in \mathcal{Y}} P(y, w)(c(y, w) + \gamma(w))} = \frac{P(y, z)(c(y, z) + \gamma(z))}{\gamma(y)}, \tag{2.4}$$

where the denominator acts as a normalization constant (the probabilities add up to 1 from the first equality; the second equality results from conditioning with respect to a one-step transition). Then

$$\begin{aligned}
\tilde{X} &= \sum_{i=1}^{\tau} c(Y_{i-1}, Y_i) \prod_{j=1}^{i} \frac{P(Y_{j-1}, Y_j)}{\tilde{P}(Y_{j-1}, Y_j)} \\
&= \sum_{i=1}^{\tau} c(Y_{i-1}, Y_i) \prod_{j=1}^{i} \frac{P(Y_{j-1}, Y_j)\gamma(Y_{j-1})}{P(Y_{j-1}, Y_j)(c(Y_{j-1}, Y_j) + \gamma(Y_j))} \\
&= \sum_{i=1}^{\tau} c(Y_{i-1}, Y_i) \prod_{j=1}^{i} \frac{\gamma(Y_{j-1})}{c(Y_{j-1}, Y_j) + \gamma(Y_j)} \\
&= \gamma(Y_0)
\end{aligned}$$

by induction on the value taken by τ, using the fact that $\gamma(Y_\tau) = 0$. In other words, the estimator is a constant, so it has zero variance.

Another way to show this property is by looking at the variance and using the classical decomposition

$$\tilde{\sigma}^2[X|Y_0] = \tilde{\sigma}^2[\tilde{\mathbb{E}}[X|Y_0, Y_1]|Y_0] + \tilde{\mathbb{E}}[\tilde{\sigma}^2[X|Y_0, Y_1]|Y_0]. \tag{2.5}$$

Define $v(y) = \tilde{\sigma}^2[\tilde{X}|Y_0 = y]$. Then

$$\begin{aligned}
v(Y_0) &= \tilde{\sigma}^2[\tilde{\mathbb{E}}[\tilde{X} \mid Y_1]|Y_0] + \tilde{\mathbb{E}}[\tilde{\sigma}^2[\tilde{X} \mid Y_1]|Y_0] \\
&= \tilde{\sigma}^2[(c(Y_0, Y_1) + \gamma(Y_1))L(Y_0, Y_1)|Y_0] + \tilde{\mathbb{E}}[L^2(Y_0, Y_1)v(Y_1)|Y_0] \\
&= \tilde{\mathbb{E}}[(c(Y_0, Y_1)+\gamma(Y_1))^2L^2(Y_0, Y_1)|Y_0]-\gamma^2(Y_0) + \tilde{\mathbb{E}}[L^2(Y_0, Y_1)v(Y_1)|Y_0] \\
&= \tilde{\mathbb{E}}[((c(Y_0, Y_1) + \gamma(Y_1))^2 + v(Y_1))L^2(Y_0, Y_1)|Y_0] - \gamma^2(Y_0).
\end{aligned}$$

From the change of measure, we have

$$\tilde{\mathbb{E}}[(c(Y_0, Y_1) + \gamma(Y_1))^2L^2(Y_0, Y_1)|Y_0] = \gamma^2(Y_0),$$

leading to

$$\begin{aligned}
v(Y_0) &= \tilde{\mathbb{E}}[((c(Y_0, Y_1) + \gamma(Y_1))^2 + v(Y_1))L^2(Y_0, Y_1)|Y_0] - \gamma^2(Y_0) \\
&= \tilde{\mathbb{E}}[v(Y_1)L^2(Y_0, Y_1)|Y_0].
\end{aligned}$$

Applying induction, we again obtain

$$v(Y_0) = \tilde{\mathbb{E}}\left[v(Y_\tau)\prod_{j=1}^{\tau} L(Y_{i-1}, Y_i)\right] = 0$$

because $v(Y_\tau) = 0$.

The change of measure (2.4) is actually the *unique* Markov chain implementation of the zero-variance change of measure. To see that, suppose we are in state $Y_j = y \notin \Delta$. Since

$$\begin{aligned}
v(y) &\geq \tilde{\sigma}^2[\tilde{\mathbb{E}}[\tilde{X} \mid Y_1]|Y_0 = y] \\
&= \tilde{\sigma}^2[(c(y, Y_1) + \gamma(Y_1))P(y, Y_1)/\tilde{P}(y, Y_1) \mid Y_0 = y],
\end{aligned}$$

zero-variance implies that $(c(y, Y_1) + \gamma(Y_1))P(y, Y_1)/\tilde{P}(y, Y_1) = K_y$ for some constant K_y that does not depend on Y_1. But since the probabilities $\tilde{P}(y, Y_1)$ must sum to 1 for any fixed y, the constant K_y must take the value $\gamma(y)$ as in (2.4). The same argument can be repeated at each step of the Markov chain.

It is important to emphasize that in (2.4) the probabilities are changed in a way that depends in general on the current state of the chain.

Again, knowing the zero-variance IS measure requires knowledge of $\gamma(y)$ for all y–that is, of the values we are trying to estimate. In practice, we can try to approximate the zero-variance IS by replacing γ by an accurate proxy, and using this approximation in (2.4) [1, 4, 5, 17]. Some methods restrict themselves to a parametric class of IS distributions and try to optimize the parameters, instead of trying to approximate the zero-variance IS. We will return to this in Section 2.4.

Example 3. Consider a Markov chain with state space $\{0, 1, \ldots, B\}$, for which $P(y, y+1) = p_y$ and $P(y, y-1) = 1 - p_y$, for $y = 1, \ldots, B-1$, and $P(0, 1) = P(B, B-1) = 1$. Note that a birth-and-death process with bounded state space has an embedded DTMC of this form. We take $\Delta = \{0, B\}$ and define $\gamma(y) = \mathbb{P}[Y_\tau = B \mid Y_0 = y]$. This function γ satisfies the recurrence equations

$$\gamma(y) = p_y \gamma(y+1) + (1 - p_y)\gamma(y-1)$$

for $y = 1, \ldots, B-1$, with the boundary conditions $\gamma(0) = 0$ and $\gamma(B) = 1$. This gives rise to a linear system of equations that is easy to solve. In the case where $p_y = p < 1$ for $y = 1, \ldots, B-1$, this is known as the gambler's ruin problem, and $\gamma(y)$ is given by the explicit formula $\gamma(y) = (1 - \rho^{-y})/(1 - \rho^{-B})$ if $\rho = p/(1-p) \neq 1/2$, and $\gamma(y) = y/B$ if $\rho = 1/2$.

Suppose, however, for the sake of illustration, that we wish to estimate $\gamma(1)$ by simulation with IS. The zero-variance change of measure in this case replaces each p_y, for $1 \leq y < B$, by

$$\tilde{p}_y = \frac{p_y \gamma(y+1)}{\gamma(y)} = \frac{p_y \gamma(y+1)}{p_y \gamma(y+1) + (1 - p_y)\gamma(y-1)}.$$

Since $\gamma(0) = 0$, this gives $\tilde{p}_1 = 1$, which means that this change of measure cuts the link that returns to 0, so it brings us to B with probability 1. For the special case where $p_y = p$ for $y = 1, \ldots, B-1$, by plugging the formula for $\gamma(y)$ into the expression for $\tilde{p}_y$, we find that the zero-variance probabilities are

$$\tilde{p}_y = \frac{1 - \rho^{-y-1}}{1 - \rho^{-y}} \, p.$$

Note that all the terms $1 - \rho^{-B}$ have canceled out, so the new probabilities $\tilde{p}_y$ do not depend on B. On the other hand, they depend on y even though the original probabilities p did not depend on y.

One application that fits this framework is an $M/M/1$ queue with arrival rate λ and service rate $\mu > \lambda$. Let $\rho = \lambda/\mu$ and $p = \lambda/(\lambda + \mu)$. Then $\gamma(y)$ represents the probability that the number of customers in the system reaches level B before the system empties, given that there are currently y customers.

2.3.2 Continuous-time Markov chains

We now examine how the previous framework applies to continuous-time Markov chains (CTMC). Following [13], let $Y = \{Y(t), t \geq 0\}$ be a CTMC evolving in $\mathcal{Y}$ up to some stopping time $T = \inf\{t \geq 0 : Y(t) \in \Delta\}$, where $\Delta \subset \mathcal{Y}$. The initial distribution is π_0 and the jump rate from y to z, for $z \neq y$, is $a_{y,z}$. Let $a_y = \sum_{z \neq y} a_{y,z}$ be the departure rate from y. The aim is to estimate $\mathbb{E}[X]$, where $X = h(Y)$ is a function of the entire sample path of the CTMC up to its stopping time T. A sample path for this chain is determined uniquely by the sequence $(Y_0, V_0, Y_1, V_1, \ldots, Y_\tau, V_\tau)$ where Y_j is the jth visited state of the chain, V_j the time spent in that state, and τ is the index of the jump that corresponds to the stopping time (the first jump that hits Δ). Therefore $h(Y)$ can be re-expressed as $h^*(Y_0, V_0, Y_1, V_1, \ldots, Y_n, V_n)$, and a sample path $(y_0, v_0, y_1, v_1, \ldots, y_n, v_n)$ has density (or likelihood)

$$p(y_0, v_0, \ldots, y_n, v_n) = \prod_{j=0}^{n-1} \frac{a_{y_j, y_{j+1}}}{a_{y_j}} \prod_{j=0}^{n} a_{y_j} \exp[-a_{y_j} v_j]$$

$$= \prod_{j=0}^{n-1} a_{y_j, y_{j+1}} \exp\left[-\sum_{j=0}^{n} a_{y_j} v_j\right],$$

each term $a_{y_j, y_{j+1}}/a_{y_j}$ being the probability of moving from y_j to y_{j+1} and $a_{y_j} \exp[-a_{y_j} v_j]$ the density for leaving y_j after a sojourn time v_j. Then we have

$$\mathbb{E}[X] = \sum_{y_0, \ldots, y_n} \int_0^\infty \cdots \int_0^\infty h^*(y_0, v_0, \ldots, y_n, v_n) p(y_0, v_0, \ldots, y_n, v_n) dv_0 \cdots dv_n.$$

Suppose that the cost function has the form

$$X = h(Y) = \sum_{j=1}^{\tau} c'(Y_{j-1}, V_{j-1}, Y_j)$$

where $c' : \mathcal{Y} \times [0, \infty) \times \mathcal{Y} \to [0, \infty)$. In this case, a standard technique that always reduces the variance, and often reduces the computations as well, is to replace the estimator X by

$$X_{\text{cmc}} = \mathbb{E}[X \mid Y_0, \ldots, Y_\tau] = \sum_{j=1}^{\tau} c(Y_{j-1}, Y_j),$$

where $c(Y_{j-1}, Y_j) = \mathbb{E}[c'(Y_{j-1}, V_{j-1}, Y_j) \mid Y_{j-1}, Y_j]$ [9]. In other words, we would never generate the sojourn times. We are now back in our previous DTMC setting and the zero-variance transition probabilities are given again by (2.4).

Consider now the case of a *fixed* time horizon T, which therefore no longer has the form $T = \inf\{t \geq 0 : Y(t) \in \Delta\}$. We then have two options: either we again reformulate the process as a DTMC, or retain a CTMC formulation. In the first case, we can redefine the state as (Y_j, R_j) at step j, where R_j is the remaining clock time (until we reach time T), as in [6]. Then the zero-variance scheme is the same as for the DTMC setting if we replace the state Y_j there by (Y_j, R_j), and if we redefine Δ. We then have a non-denumerable state space, so the sums must be replaced by combinations of sums and integrals. In this context of a finite time horizon, effective IS schemes will typically use *non-exponential* (often far from exponential) sojourn time distributions. This means that we will no longer have a CTMC under IS. Assume now that we want to stick with a

CTMC formulation, and that we restrict ourselves to choosing a Markovian IS measure with new initial distribution $\tilde{\pi}_0$ and new generator $\tilde{A}$ such that $\tilde{\pi}_0(y) > 0$ (or $\tilde{a}_{y,z} > 0$) whenever $\pi_0(y) > 0$ (or $a_{y,z} > 0$). Let τ be the index j of the first jump to a state $Y_j = Y(t_j)$ at time t_j such that $t_j \geq T$. Then, similarly to the discrete-time case, it can be shown that, provided τ is a stopping time with finite expectation under $(\tilde{\pi}_0, \tilde{A})$,

$$\mathbb{E}[X] = \tilde{\mathbb{E}}[XL_\tau],$$

with L_τ the likelihood ratio given by

$$L_\tau = \frac{\pi_0(Y_0)}{\tilde{\pi}_0(Y_0)} \prod_{j=0}^{\tau-1} \frac{a_{Y_j,Y_{j+1}}}{\tilde{a}_{Y_j,Y_{j+1}}} \exp\left[\sum_{j=0}^{\tau}(\tilde{a}_{Y_j} - a_{Y_j})V_j\right].$$

In the above formula, we can also replace the likelihood $a_{Y_{\tau-1},Y_\tau} \exp[-a_{Y_{\tau-1}} V_{\tau-1}]$ of the last occurrence time V_τ by the probability that this event occurs after the remaining time $T - \sum_{j=0}^{\tau-1} V_j$, which is $\exp[-a_{Y_{\tau-1}}(T - \sum_{j=0}^{\tau-1} V_j)]$, the same being done for the IS measure: instead of considering the exact occurrence time after time T, we consider its expected value given that it happens after T. This reduces the variance of the estimator, because it replaces it by its conditional expectation.

2.3.3 State-independent vs state-dependent changes of measure

In the context of simulating a Markov chain, we often distinguish two types of IS strategies:

- state-independent IS, where the change of measure does not depend on the current state of the Markov chain;
- state-dependent IS where, at each step of the Markov chain, a new IS change of measure is used that takes into account the current state of the Markov chain. In the case where the state of the chain must contain the current simulation time (e.g., if the simulation stops at a fixed clock time in the model), then the change of measure will generally depend on the current time.

Example 4. In Example 3, even though the birth-and-death process had original transition probabilities p and $1 - p$ that did not depend on the current state y, the zero-variance probabilities $\tilde{p}_y$ did depend on y (although not on B). These probabilities satisfy the equations

$$\tilde{p}_y(1 - \tilde{p}_{y-1}) = p(1 - p)$$

for $y \geq 2$, with boundary condition $\tilde{p}_1 = 1$. For $p < 1/2$, we have $1 - p < \tilde{p}_y < \tilde{p}_{y-1} < 1$ for all $y > 2$, and $\tilde{p}_y \to 1 - p$ when $y \to \infty$. That is, the optimal change of measure is very close to simply permuting p and $1 - p$, that is, taking $\tilde{p} = 1 - p > 1/2$. For the $M/M/1$ queue, this means exchanging the arrival rate and the service rate, which gives an unstable queue (i.e., the event under consideration is not rare anymore). This simple permutation is an example of a state-independent change of measure; it does not depend on the current state y.

With $\tilde{p} = 1 - p$, the likelihood ratio associated with any sample path that reaches level B before returning to 0 is ρ^{B-1}, so, when estimating $\gamma(1)$, the second moment is reduced at least by that factor, as shown by Inequality (2.2) This reduction can be quite substantial. Moreover, the probability $\tilde{\gamma}(1)$ of reaching B under the new measure must satisfy $\tilde{\gamma}(1)\rho^{B-1} = \gamma(1)$, which implies that

$$\tilde{\gamma}(1) = \gamma(1)\rho^{1-B} = \frac{1-\rho}{1-\rho^B}.$$

Then the relative variance is

$$\frac{\tilde{\gamma}(1)\rho^{2B-2}}{\gamma^2(1)} - 1 = \frac{1-\rho^B}{1-\rho} - 1 \approx \frac{\rho}{1-\rho}$$

when B is large. We have the remarkable result that the number of runs needed to achieve a predefined precision remains bounded in B, that is, we have bounded relative error as $B \to \infty$, even with a state-independent change of measure.

Example 5. Suppose now that our birth-and-death process evolves over the set of non-negative integers and let $\gamma(y)$ be the probability that the process ever reaches 0 if it starts at $y > 0$. This $\gamma(y)$ can be seen as the probability of ruin if we start with y euros in hand and win (lose) one euro with probability p $(1-p)$ at each step. For $p \leq 1/2$, $\gamma(y) = 1$, so we assume that $p > 1/2$. In this case, we have that $\gamma(1) = (1-p) + p\gamma(2) = (1-p) + \gamma^2(1)$. For $j \geq 2$, $\gamma(j+1) = \gamma(1)\gamma(j)$ because the probability of reaching 0 from $j+1$ is the probability of eventually reaching j from $j+1$, which equals $\gamma(1)$, multiplied by the probability of reaching 0 from j. From this, we see that $\gamma(1) = (1-p)/p$. Still from $\gamma(j+1) = \gamma(1)\gamma(j)$, we find easily that the zero-variance probabilities are $\tilde{p}_j = 1 - p$ for all $j \geq 1$. In this case, the zero-variance change of measure is state-independent.

Example 6. We return to Example 2, where we wish to estimate $\gamma = \mathbb{P}[X \geq na]$, for X binomially distributed with parameters (n, p), and for some constant $a > p$. If we view X as a sum of n independent Bernoulli random variables and define Y_j and the partial sum of the first j variables, then $X = Y_n$ and we have a Markov chain $\{Y_j, j \geq 0\}$. We observed in Example 2 that when we restricted ourselves to a state-independent change of measure that replaced p by $\tilde{p}$ for this Markov chain, the variance was approximately minimized by taking $\tilde{p} = a$. In fact, this choice turns out to be optimal asymptotically when $n \to \infty$ [24]. But even this optimal choice fails to provide a bounded relative error. That is, a state-independent change of p cannot provide a bounded relative error in this case. The only way of getting a bounded relative error is via state-dependent IS. However, when p is replaced by $\tilde{p} = a$, the relative error increases only very slowly when $n \to \infty$: the second moment decreases exponentially at the same exponential rate as the square of the first moment. When this property holds, the estimator is said to have *logarithmic efficiency*. In this example, it holds for no other value of $\tilde{p}$. All these results have been proved in a more general setting by Sadowsky [24].

2.4 Algorithms

A general conclusion from the previous section is that to accurately approximate the zero variance IS estimator, a key ingredient is a good approximation of the function $\gamma(\cdot)$. In fact, there are several ways of finding a good IS strategy. Most of the good methods can be classified into two large families: those that try to directly approximate the zero-variance change of measure via an approximation of the function $\gamma(\cdot)$, and those that restrict a priori the change of measure to a parametric class, and then try to optimize the parameters. In both cases, the choice can be made either via simple heuristics, or via a known asymptotic approximation for $\gamma(y)$, or by adaptive methods that learn (statistically) either the function $\gamma(\cdot)$ or the vector or parameters that minimizes the variance. In the remainder of this section, we briefly discuss these various approaches.

In the scientific literature, IS has often been applied in a very heuristic way, without making any explicit attempt to approximate the zero-variance change of measure. One heuristic idea is simply to change the probabilities so that the system is pushed in the direction of the rare event, by looking at what could increase its occurrence. However, Example 1 shows very well how pushing too much can have the opposite effect; in fact, it can easily lead to an infinite variance. Changes of measure that may appear promising a priori can eventually lead to a variance increase. In situations where the rare event can be reached in more than one direction, pushing in one of those directions may easily inflate the variance by reducing the probability or density of paths that lead to the rare event via other directions. The last part of Example 1 illustrates a simplified case of this. Other illustrations can be found in [2, 3, 11], for example. Generally speaking, good heuristics should be based on a reasonable understanding of the shape of $\gamma(\cdot)$ and/or the way the likelihood ratio will behave under IS. We give examples of these types of heuristics in the next subsection.

2.4.1 Heuristic approaches

Here, the idea is to use a heuristic approximation of $\gamma(\cdot)$ in the change of measure (2.4).

Example 7. We return to Example 3, with $p_y = p$. Our aim is to estimate $\gamma(1)$. Instead of looking at the case where B is large, we focus on the case where p

is small, $p \to 0$ for fixed B. This could be seen as a (simplified) dependability model where each transition from y to $y+1$ represents a component failure, each transition from y to $y-1$ corresponds to a repair, and B is the minimal number of failed components for the whole system to be in a state of failure. If $p \ll 1$, each failure transition (except the first) is rare and we have $\gamma(1) \ll 1$ as well. Instead of just blindly increasing the failure probabilities, we can try to mimic the zero-variance probabilities (2.4) by replacing $\gamma(\cdot)$ in this expression by an approximation, with $c(y,z)=0$, $\gamma(0)=0$ and $\gamma(B)=1$. Which approximation $\hat{\gamma}(y)$ could we use instead of $\gamma(y)$? Based on the asymptotic estimate $\gamma(y) = p^{B-y} + o(p^{B-y})$, taking $\hat{\gamma}(y) = p^{B-y}$ for all $y \in \{1, \ldots, B-1\}$, with $\hat{\gamma}(0)=0$ and $\hat{\gamma}(B)=1$, looks like a good option. This gives

$$\tilde{P}(y, y+1) = \frac{p^{B-y}}{p^{B-y} + (1-p)p^{B-y+1}} = \frac{1}{1+(1-p)p}$$

for $y = 2, \ldots, B-2$. Repairs then become rare while failures are no longer rare.

We can extend the previous example to a multidimensional state space, which may correspond to the situation where there are different types of components, and a certain subset of the combinations on the numbers of failed components of each type corresponds to the failure state of the system. Several IS heuristics have been proposed for this type of setting [16] and some of them are examined in Chapter 6. One heuristic suggested in [20] approximates $\gamma(y)$ by considering the probability of the most likely path to failure. In numerical examples, it provides a drastic variance reduction with respect to previously known IS heuristics.

2.4.2 Learning the function $\gamma(\cdot)$

Various techniques that try to approximate the function $\gamma(\cdot)$, often by adaptive learning, and plug the approximation (2.4), have been developed in the literature [16]. Old proposals of this type can be found in the computational physics literature, for example; see the references in [5]. We outline examples of such techniques taken from recent publications.

One simple type of approach, called *adaptive Monte Carlo* in [8, 17], proceeds iteratively as follows. At step i, it replaces the exact (unknown) value $\gamma(x)$ in (2.4) by a guess $\gamma^{(i)}(x)$, and it uses the probabilities

$$\tilde{P}^{(i)}(y,z) = \frac{P(y,z)(c(y,z)+\gamma^{(i)}(z))}{\sum_{w \in \mathcal{Y}} P(y,w)(c(y,w)+\gamma^{(i)}(w))} \tag{2.6}$$

in n_i independent simulation replications, to obtain a new estimation $\gamma^{(i+1)}(y)$ of $\gamma(y)$, from which a new transition matrix $\tilde{P}^{(i+1)}$ is defined. These iterations could go on until we feel that the probabilities have converged to reasonably good estimates.

A second type of approach is to try to approximate the function $\gamma(\cdot)$ stochastically. The *adaptive stochastic approximation* method proposed in [1] for the simulation of discrete-time finite-state Markov chains falls in that category. One starts with a given distribution for the initial state y_0 of the chain, an initial transition matrix $\tilde{P}^{(0)}$ (which may be the original transition matrix of the chain), and an initial guess $\gamma^{(0)}(\cdot)$ of the value function $\gamma(\cdot)$. The method simulates a single sample path as follows. At each step n, given the current state y_n of the chain, if $y_n \notin \Delta$, we use the current transition matrix $\tilde{P}^{(n)}$ to generate the next state y_{n+1}, we update the estimate of $\gamma(y_n)$ by

$$\gamma^{(n+1)}(y_n) = (1 - a_n(y_n))\gamma^{(n)}(y_n) \\ + a_n(y_n)\left[c(y_n, y_{n+1}) + \gamma^{(n)}(y_{n+1})\frac{P(y_n, y_{n+1})}{\tilde{P}^{(n)}(y_n, y_{n+1})}\right],$$

where $\{a_n(y), n \geq 0\}$ is a sequence of *step sizes* such that $\sum_{n=1}^{\infty} a_n(y) = \infty$ and $\sum_{n=1}^{\infty} a_n^2(y) < \infty$ for each state y, and we update the probability of the current transition by

$$\tilde{p}^{(n+1)}(y_n, y_{n+1}) = \max\left(P(y_n, y_{n+1})\frac{c(y_n, y_{n+1}) + \gamma^{(n+1)}(y_{n+1})}{\gamma^{(n+1)}(y_n)}, \delta\right)$$

where $\delta > 0$ is a constant whose role is to ensure that the likelihood ratio remains bounded (to rule out the possibility that it takes huge values). For the other states, we take $\gamma^{(n+1)}(y) = \gamma^{(n)}(y)$ and $\tilde{P}^{(n+1)}(y, z) = P^{(n)}(y, z)$. We then normalize via

$$P^{(n+1)}(y_n, y) = \frac{\tilde{P}^{(n+1)}(y_n, y)}{\sum_{z \in \mathcal{Y}} \tilde{P}^{(n+1)}(y_n, z)}$$

for all $y \in \mathcal{Y}$. When $y_n \in \Delta$, that is, if the stopping time is reached at step n, y_{n+1} is generated again from the initial distribution, the transition matrix and the estimate of $\gamma(\cdot)$ are kept unchanged, and the simulation is resumed. In [1], batching techniques are used to obtain a confidence interval.

Experiments reported in [1] show that these methods can be quite effective when the state space has small cardinality. However, since they require the approximation $\gamma^{(n)}(y)$ to be stored for each state y, their direct implementation quickly becomes impractical as the number of states increases (e.g., for continuous state spaces or for multidimensional state spaces such as those of Example 7).

In the case of large state spaces, one must rely on interpolation or approximation instead of trying to estimate $\gamma(y)$ directly at each state y. One way of doing this is by selecting a set of k predefined basis functions $\gamma_1(y), \ldots, \gamma_k(y)$, and searching for a good approximation of $\gamma(\cdot)$ within the class of linear combinations of the form $\hat{\gamma}(y) = \sum_{j=1}^{k} \alpha_j \gamma_j(y)$, where the weights $(\alpha_1, \ldots, \alpha_k)$ can be learned or estimated in various ways, for instance by stochastic approximation. It therefore involves a parametric approach, where the parameter is the vector of weights.

2.4.3 Optimizing within a parametric class

Most practical and effective IS strategies in the case of large state spaces restrict themselves to a parametric class of IS measures, either explicitly or implicitly, and try to estimate the parameter vector that minimizes the variance. More specifically, we consider a family of measures $\{\tilde{\mathbb{P}}_\theta, \theta \in \Theta\}$, which may represent a family of densities $\tilde{f}_\theta$, or a family of probability vectors $\tilde{p}_\theta$ for a discrete distribution, or the probability measure associated with the transition matrix $\tilde{P}_\theta$ or the transition kernel of a Markov chain. Then, we look for a θ that minimizes the variance of the IS estimator under $\tilde{\mathbb{P}}_\theta$, or some other measure of distance to the zero-variance measure, over the set Θ. Of course, a key issue is a clever selection of this parametric class, so that it includes good IS strategies within the class. The value of θ can be selected either via a separate prior analysis, for example based on asymptotically valid approximations, or can be learned adaptively. We briefly discuss these two possibilities in what follows.

Non-adaptive parameter selection

Examples 2 and 6 illustrate the popular idea of fixing θ based on an asymptotic analysis. The parametric family there is the class of binomial distributions with parameters $(n, \tilde{p})$. We have $\theta = \tilde{p}$. Large-deviations theory shows that twisting the binomial parameter p to $\tilde{p} = a$ is asymptotically optimal [24]. This choice works quite well in practice for this type of example. On the other hand, we also saw that it cannot provide a bounded relative variance. Several additional examples illustrating the use of large-deviations theory to select a good change of measure can be found in [3, 15, 16], for example.

Adaptive learning of the best parameters

The value of θ that minimizes the variance can be learned adaptively in various ways. For example, the adaptive stochastic approximation method described earlier can be adapted to optimize θ stochastically. Another type of approach is based on *sample average approximation*: write the variance or the second moment as a mathematical expectation that depends on θ, replace the expectation by a sample average function of θ obtained by simulation, and optimize this sample function with respect to θ. These simulations are performed under an IS measure $\tilde{P}$ that may differ from P and does not have to belong to the selected family. The optimizer $\hat{\theta}^*$ is used in a second stage to estimate the quantity of interest using IS.

A more general way of formulating this optimization problem is to replace the variance by some other measure of distance between $\tilde{\mathbb{P}}_\theta$ and the optimal (zero-variance) change of measure $\tilde{\mathbb{P}}^*$, which is known to satisfy $d\tilde{\mathbb{P}}^* = (|X|/\mathbb{E}[|X|])d\mathbb{P}$ when we wish to estimate $\gamma = \mathbb{E}[X]$. Again, there are many ways of measuring this distance and some are more convenient than others.

Rubinstein [23] proposed and motivated the use of the Kullback--Leibler (or cross-entropy) 'distance', defined by

$$\mathcal{D}(\tilde{\mathbb{P}}^*, \tilde{\mathbb{P}}_\theta) = \tilde{\mathbb{E}}^* \left[\log \frac{d\tilde{\mathbb{P}}^*}{d\tilde{\mathbb{P}}_\theta} \right]$$

(this is not a true distance, because it is not symmetric and does not satisfy the triangle inequality, but this causes no problem), and called the resulting technique the *cross-entropy* (CE) method [7, 22, 23]. Easy manipulations lead to

$$\mathcal{D}(\tilde{\mathbb{P}}^*, \tilde{\mathbb{P}}_\theta) = \mathbb{E}\left[\frac{|X|}{\mathbb{E}[|X|]} \log \left(\frac{|X|}{\mathbb{E}[|X|]} d\mathbb{P} \right) \right] - \frac{1}{\mathbb{E}[|X|]} \mathbb{E}\left[|X| \log d\tilde{\mathbb{P}}_\theta \right].$$

Since only the last expectation depends on θ, minimizing the above expression is equivalent to solving

$$\max_{\theta\in\Theta} \mathbb{E}\left[|X| \log d\tilde{\mathbb{P}}_\theta \right] = \max_{\theta\in\Theta} \tilde{\mathbb{E}}\left[\frac{d\mathbb{P}}{d\tilde{\mathbb{P}}} |X| \log d\tilde{\mathbb{P}}_\theta \right]. \tag{2.7}$$

The CE method basically solves the optimization problem on the right-hand side of (2.7) by sample average approximation, replacing the expectation $\tilde{\mathbb{E}}$ in (2.7) by a sample average over simulations performed under $\tilde{\mathbb{P}}$.

How should we select $\tilde{\mathbb{P}}$? In the case of rare events, it is often difficult to find a priori a distribution $\tilde{\mathbb{P}}$ under which the optimizer of the sample average approximation does not have too much variance and is sufficiently reliable. For this reason the CE method is usually applied in an iterative manner, starting with a model under which the rare events are not so rare, and increasing the rarity at each step. We start with some $\theta_0 \in \Theta$ and a random variable X_0 whose expectation is easier to estimate than X, and having the same shape. At step $i \geq 0$, n_i independent simulations are performed using IS with parameter θ_i, to approximate the solution of (2.7) with $\tilde{\mathbb{P}}$ replaced by $\tilde{\mathbb{P}}_{\theta_i}$ and X replaced by X_i, where X_i becomes closer to X as i increases, and eventually becomes identical when $i = i_0$, for some finite i_0. In Example 3, for instance, we could have $X_i = 1_{Y_\tau = B_i}$ with $B_i = a + ib$ for some fixed positive integers a and b such that $B = a + i_0 b$ for some i_0. The solution of the corresponding sample average problem is

$$\theta_{i+1} = \arg\max_{\theta\in\Theta} \frac{1}{n_i} \sum_{j=1}^{n_i} |X_i(\omega_{i,j})| \log(d\tilde{\mathbb{P}}_\theta(\omega_{i,j})) \frac{d\mathbb{P}}{d\tilde{\mathbb{P}}_{\theta_i}}(\omega_{i,j}), \tag{2.8}$$

where $\omega_{i,j}$ represents the jth sample at step i. This θ_{i+1} is used for IS at the next step.

A quick glance at (2.8) shows that the specific choice of the Kullback–Leibler distance is convenient for the case where $\tilde{\mathbb{P}}_\theta$ is from an exponential family, because the log and the exponential cancel, simplifying the solution to (2.8) considerably.

In some specific contexts, the parametric family can be a very rich set of IS measures. For example, in the case of a DTMC over a finite state space, one can define the parametric family as the set of all transition probability matrices over that state space [21]. In this case, CE serves as a technique to approximate the zero-variance change of measure, but at the higher cost of storing an entire transition matrix instead of just the vector $\gamma(\cdot)$.

References

[1] I. Ahamed, V. S. Borkar, and S. Juneja. Adaptive importance sampling for Markov chains using stochastic approximation. *Operations Research*, **54**(3): 489–504, 2006.

[2] S. Andradóttir, D. P. Heyman, and T. J. Ott. On the choice of alternative measures in importance sampling with Markov chains. *Operations Research*, **43**(3): 509–519, 1995.

[3] S. Asmussen. Large deviations in rare events simulation: Examples, counterexamples, and alternatives. In K.-T. Fang, F. J. Hickernell, and H. Niederreiter, eds, *Monte Carlo and Quasi-Monte Carlo Methods* 2000, pp. 1–9. Springer, Berlin, 2002.

[4] N. Bolia, S. Juneja, and P. Glasserman. Function-approximation-based importance sampling for pricing American options. In *Proceedings of the 2004 Winter Simulation Conference*, pp. 604–611. IEEE Press, Piscataway, NJ, 2004.

[5] T. E. Booth. Generalized zero-variance solutions and intelligent random numbers. In *Proceedings of the 1987 Winter Simulation Conference*, pp. 445–451. IEEE Press, 1987.

[6] P. T. De Boer, P. L'Ecuyer, G. Rubino, and B. Tuffin. Estimating the probability of a rare event over a finite horizon. In *Proceedings of the 2007 Winter Simulation Conference*, pp. 403–411. IEEE Press, 2007.

[7] P. T. De Boer, V. F. Nicola, and R. Y. Rubinstein. Adaptive importance sampling simulation of queueing networks. In *Proceedings of the 2000 Winter Simulation Conference*, pp. 646–655. IEEE Press, 2000.

[8] P. Y. Desai and P. W. Glynn. A Markov chain perspective on adaptive Monte Carlo algorithms. In *Proceedings of the 2001 Winter Simulation Conference*, pp. 379–384. IEEE Press, 2001.

[9] B. L. Fox and P. W. Glynn. Discrete-time conversion for simulating semi-Markov processes. *Operations Research Letters*, **5**: 191–196, 1986.

[10] P. Glasserman. *Monte Carlo Methods in Financial Engineering*. Springer, New York, 2004.

[11] P. Glasserman and Y. Wang. Counterexamples in importance sampling for large deviations probabilities. *Annals of Applied Probability*, **7**(3): 731–746, 1997.

[12] P. W. Glynn. Efficiency improvement techniques. *Annals of Operations Research*, **53**: 175–197, 1994.

[13] P. W. Glynn and D. L. Iglehart. Importance sampling for stochastic simulations. *Management Science*, **35**: 1367–1392, 1989.

[14] G. Goertzel and M. H. Kalos. Monte Carlo methods in transport problems. *Progress in Nuclear Energy, Series I*, **2**: 315–369, 1958.

[15] P. Heidelberger. Fast simulation of rare events in queueing and reliability models. *ACM Transactions on Modeling and Computer Simulation*, **5**(1): 43–85, 1995.

[16] S. Juneja and P. Shahabuddin. Rare event simulation techniques: An introduction and recent advances. In S. G. Henderson and B. L. Nelson, eds, *Simulation*, Handbooks in Operations Research and Management Science, pp. 291–350. Elsevier, Amsterdam, 2006.

[17] C. Kollman, K. Baggerly, D. Cox, and R. Picard. Adaptive importance sampling on discrete Markov chains. *Annals of Applied Probability*, **9**(2): 391–412, 1999.

[18] I. Kuruganti and S. Strickland. Importance sampling for Markov chains: computing variance and determining optimal measures. In *Proceedings of the 1996 Winter Simulation Conference*, pp. 273–280. IEEE Press, 1996.

[19] P. L'Ecuyer and Y. Champoux. Estimating small cell-loss ratios in ATM switches via importance sampling. *ACM Transactions on Modeling and Computer Simulation*, **11**(1): 76–105, 2001.

[20] P. L'Ecuyer and B. Tuffin. Effective approximation of zero-variance simulation in a reliability setting. In *Proceedings of the 2007 European Simulation and Modeling Conference*, pp. 48–54, Ghent, Belgium, 2007. EUROSIS.

[21] A. Ridder. Importance sampling simulations of Markovian reliability systems using cross-entropy. *Annals of Operations Research*, **134**: 119–136, 2005.

[22] R. Rubinstein and D. P. Kroese. *A Unified Approach to Combinatorial Optimization, Monte Carlo Simulation, and Machine Learning*. Springer, Berlin, 2004.

[23] R. Y. Rubinstein. Optimization of computer simulation models with rare events. *European Journal of Operations Research*, **99**: 89–112, 1997.

[24] J. S. Sadowsky. On the optimality and stability of exponential twisting in Monte Carlo estimation. *IEEE Transactions on Information Theory*, **IT-39**: 119–128, 1993.

[25] D. Siegmund. Importance sampling in the Monte Carlo study of sequential tests. *Annals of Statistics*, **4**: 673–684, 1976.

[26] R. Srinivasan. *Importance Sampling – Applications in Communications and Detection*. Springer, Berlin, 2002.

3

Splitting techniques

Pierre L'Ecuyer, François Le Gland, Pascal Lezaud and Bruno Tuffin

3.1 Introduction

As already explained in previous chapters, rare event simulation requires acceleration techniques to speed up the occurrence of the rare events under consideration, otherwise it may take unacceptably large sample sizes to get enough positive realizations, or even a single one, on average. On the other hand, accelerating too much can be counterproductive and even lead to a variance explosion and/or an increase in the computation time. Therefore, an appropriate balance must be achieved, and this is not always easy. This difficulty was highlighted in the previous chapter when discussing the importance sampling (IS) technique, the idea of which is to change the probability laws driving the model in order to make the events of interest more likely, and to correct the bias by multiplying the estimator by the appropriate likelihood ratio.

In this chapter, we review an alternative technique called *splitting*, which accelerates the rate of occurrence of the rare events of interest. Here, we do not change the probability laws driving the model. Instead, we use a selection mechanism to favor the trajectories deemed likely to lead to those rare events. The main idea is to decompose the paths to the rare events of interest into shorter subpaths whose probability is not so small, encourage the realizations that take these subpaths (leading to the events of interest) by giving them a chance to reproduce (a bit like in selective evolution), and discourage the realizations that

Rare Event Simulation using Monte Carlo Methods Edited by G. Rubino and B. Tuffin

go in the wrong direction by killing them with some positive probability. The subpaths are usually delimited by levels, much like the level curves on a map. Starting from a given level, the realizations of the process (which we also call *trajectories* or *chains* or *particles*) that do not reach the next level will not reach the rare event, but those that do are split (cloned) into multiple copies when they reach the next level, and each copy pursues its evolution from then on. This creates an artificial drift toward the rare event by favoring the trajectories that go in the right direction. In the end, an unbiased estimator can be recovered by multiplying the contribution of each trajectory by the appropriate weight. The procedure just described is known as *multilevel splitting*.

If we assume, for instance, that we are simulating a stochastic process (usually a Markov chain) and that the rare event of interest occurs when we reach a given subset of states before coming back to the initial state, then the levels can be defined by a decreasing (embedded) sequence of state sets that all contain the rare set of interest. In general, these levels are defined via an *importance function* whose aim is to represent how close a state is from this rare set. Several strategies have been designed to determine the levels, to decide the number of splits at each level, and to handle the trajectories that tend to go in the wrong direction (away from the rare event of interest). The amount of splitting when reaching a new level is an important issue; with too much splitting, the population of chains will explode, while with too little splitting, too few trajectories are likely to reach the rare event.

There is also the possibility of doing away with the levels, by following a strategy that can either split the trajectory or kill it at any given step. One applies splitting (sometimes with some probability) if the weighted importance function is significantly larger at the current (new) state than at the previous state, and we apply Russian roulette (we kill the chain with some probability), when the weighted importance function becomes smaller. Russian roulette can also be viewed as splitting the chain into zero copies. The expected number of clones after the split (which is less than 1 in the case of Russian roulette) is usually taken as the ratio of the importance function value at the new state to that at the old state [13, 22].

The most important difficulty in general is to find an appropriate importance function. This function defines the levels (or the amount of splitting if we get rid of levels), and a poor choice can easily lead to bad results. In this sense, its role is analogous to the importance measure whose choice is critical in IS (see the previous chapter).

One important advantage of splitting compared with IS is that there is no need to modify the probability laws that drive the system. This means (among other things) that the computer program that implements the simulation model can just be a black box, as long as it is possible to make copies (clones) of the model, and to maintain weights and obtain the current value of the importance function for each of those copies. It is also interesting to observe that for splitting implementations where all chains always have the same weight at any given level, the empirical distribution of the states of the chains when they hit a given level provides an unbiased estimate of the theoretical entrance distribution of the chain at that level (the distribution of the state when it hits that level for the first time) under the original probabilities. With splitting implementations where chains may have different weights, and with IS, this is true only for the weighted (and rescaled) empirical distributions, where each observation keeps its weight when we define the distribution. There are also situations where it is simpler and easier to construct a good importance function for splitting than for IS, because IS can be more sensitive to the behavior of the importance function near the boundaries of the state space, as explained in [9, 12] (see also Section 3.2.3).

One limitation of splitting with respect to IS is the requirement to decompose the state space into subsets (or layers) determined by the levels of some importance function, such that the probability of reaching the next level starting from the current one is not so small. When such a decomposition can be found, splitting can be efficiently applied. However, there are situations where the most probable paths that lead to the rare event have very few steps (or transitions), and where rarity comes from the fact that each of these steps has a very low probability. For example, in a reliability setting, suppose that the rare event is a system failure and that the most likely way that this failure occurs is by a failure of two components of the same type, which happens from two transitions of the Markov chain, where each transition has a very small probability. In such a situation, splitting cannot be effectively applied, at least not directly. It would require a trick to separate the rare transitions into several phases. IS, on the other hand, can handle this easily by increasing the probability of occurrence of these rare transitions. It is also important to recognize that in the case of large models (such as a large queuing system with many state variables), the state-cloning operations can easily induce a significant overhead in CPU time.

This chapter is organized as follows. Section 3.2 describes the general principles of splitting techniques and the main versions (or implementations) found in the literature. Section 3.3 provides an asymptotic analysis of the method in a simplified setting that involves assuming that reaching the next level from the current one can be modeled by a Bernoulli random variable independent of the current state (given that we have just reached the current level). This is equivalent to assuming that there is a single entrance state at each level. We then discuss how much we should split and how many levels we should define to minimize the variance, or its work-normalized version (the variance multiplied by the expected computing time), in an asymptotic setting. In Section 3.4 we provide an analysis based on interacting particle systems, following the general framework of [10]. This permits us to obtain a central limit theorem in a general setting, in an asymptotic regime where the number of initial trajectories (or particles) increases to infinity. While previous results focused on a specific case of splitting where the number of trajectories at each level is fixed, we additionally provide versions of the central limit theorem for other splitting implementations. Section 3.5 applies different versions of the splitting technique to a simple example of a tandem queue, used earlier by several authors. It illustrates the effectiveness of the method, and also the difficulties and the critical issue of finding an appropriate importance function.

Note that both IS and splitting techniques were introduced and investigated with the Monte Carlo method as early as in the mid 1940s in Los Alamos [21, 22, 29]. The main relevant issues, such as an analysis of the optimal splitting strategies and the definition of the importance function, were already identified at that time.

3.2 Principles and implementations

3.2.1 Mathematical setting

Assume that the dynamics of the system under consideration is described by a strong Markov process $X = \{X(t), t \geq 0\}$ with state space E, where the time index t can be either continuous (on the real line) or discrete (on the non-negative integers $t = 0, 1, 2, \ldots$). In the continuous-time case, we assume that all the trajectories are right-continuous with left-hand limits (càdlàg). Let $B \subset E$ be some closed critical region which the system could enter with a positive but very small probability, for example 10^{-10} or less. Our objective is to compute the probability of the critical event,

$$\gamma = \mathbb{P}[T_B \leq T], \quad \text{where} \quad T_B = \inf\{t \geq 0 : X(t) \in B\}$$

denotes the entrance time into the critical region B, and where T is an almost surely finite stopping time.

Note that this can always be transformed into a model where the stopping time T is defined as the first hitting time of some set Δ by the process X, that is,

$$T = \inf\{t \geq 0 : X(t) \in \Delta\}.$$

For this, it suffices to put enough information in the state of the Markov process X so that T and every statistic that we want to compute are measurable with respect to the filtration generated by X up to time T. From now on, we assume that T is a stopping time of that form. As an important special case, this covers the situation where T is a deterministic finite time horizon: it suffices to include either the current clock time, or the time that remains on the clock before the time horizon is reached, in the definition of the state $X(t)$. For example, if we are interested in the probability that some Markov process $\{Y(t), t \geq 0\}$ hits some set C before some deterministic time t_1, then we can define $X(t) = (t, Y(t))$ for all t, $B = (0, t_1) \times C$, and $\Delta = B \cup ([t_1, \infty) \times E)$. Here, T_B is the first time Y hits C if this happens before time t_1, $T_B = \infty$ otherwise, and $T = \min(t_1, T_B)$. Alternatively, it may be more convenient to define $X(t) = (t_1 - t, Y(t))$, where $t_1 - t$ is the time that remains on the clock before reaching the horizon t_1, B is the same as before, and $\Delta = B \cup ((-\infty, 0] \times E)$. For situations of this type, we will assume (when necessary) that the state $X(t)$ always contains the clock time

t, and that the sets B and Δ depend on the time horizon t_1. More generally, we could also have one or more clocks with random time-varying speeds.

Our results could be generalized to situations where the objective is to compute the entrance distribution in the critical region, or the probability distribution of critical trajectories, that is,

$$\mathbb{E}[\phi(X(T_B)) \mid T_B \leq T] \quad \text{or} \quad \mathbb{E}[f(X(t), 0 \leq t \leq T_B) \mid T_B \leq T],$$

respectively, for some measurable functions ϕ and f. For simplicity, we focus our development here on the problem of estimating γ, which suffices to illustrate the main issues and tools.

The fundamental idea of splitting is based on the assumption that there exist some identifiable intermediate subsets of states that are visited much more often than the rare set B, and that must be crossed by sample paths on their way to B. In splitting, the step-by-step evolution of the system follows the original probability measure. Entering the intermediate states, usually characterized by crossing a threshold determined by a control parameter, triggers the splitting of the trajectory. This control is generally defined via a so-called *importance function* h [16] which should satisfy $B = \{x \in E : h(x) \geq L\}$ for some level L.

Multilevel splitting uses an increasing sequence of values $L_0 \leq \ldots \leq L_k \leq \ldots \leq L_n$ with $L_n = L$, and defines the decreasing sequence of sets

$$E \supset B_0 \supset \ldots \supset B_k \supset \ldots \supset B_n = B,$$

with

$$B_k = \{x \in E : h(x) \geq L_k\},$$

for any $k = 0, 1, \ldots, n$. Note that in the case of a deterministic time horizon, $h(x)$ will usually depend on the current time, which is contained in the state x. Similarly, we can define the entrance time

$$T_k = \inf\{t \geq 0 : X(t) \in B_k\}$$

into the intermediate region B_k, and the event $A_k = \{T_k \leq T\}$, for $k = 0, 1, \ldots, n$. Again, these events form a decreasing sequence

$$A_0 \supset \ldots \supset A_k \supset \ldots \supset A_n = \{T_B \leq T\},$$

and the product formula

$$\begin{aligned}\mathbb{P}[T_B \leq T] &= \mathbb{P}(A_n) = \mathbb{P}(A_n \cap \ldots \cap A_k \cap \ldots \cap A_0) \\ &= \mathbb{P}(A_n \mid A_{n-1}) \cdots \mathbb{P}(A_k \mid A_{k-1}) \cdots \mathbb{P}(A_1 \mid A_0)\, \mathbb{P}(A_0) \quad (3.1)\end{aligned}$$

clearly holds, where ideally each conditional probability on the right-hand side of (3.1) is 'not small'. The idea is to estimate each of these conditional probabilities somehow separately, although not completely independently, according to a branching splitting technique.

Suppose for now that all the chains have the same weight at any given level. A population of N_0 independent trajectories of the Markov process is created (their initial states can be either deterministic or generated independently from some initial distribution), and each trajectory is simulated until it enters the first intermediate region B_0 or until time T is reached, whichever occurs first. Let R_0 be the number of trajectories that have managed to enter the first intermediate region B_0 before time T. The fraction $\widehat{p}_0 = R_0/N_0$ is an unbiased estimate of $\mathbb{P}(A_0) = \mathbb{P}[T_0 \leq T]$. At the next stage, N_1 replicas (or offspring) of these R_0 successful trajectories are created, so as to maintain a sufficiently large population; this is done by cloning some states if $N_1 > R_0$ or choosing them randomly otherwise. Each new trajectory is simulated until it enters the second intermediate region B_1 or until time T is reached, whichever occurs first. Again, the fraction $\widehat{p}_1 = R_1/N_1$ of the R_1 successful trajectories that have managed to enter the second intermediate region B_1 before time T is a natural estimate of $\mathbb{P}(A_1 \mid A_0) = \mathbb{P}[T_1 \leq T \mid T_0 \leq T]$. The procedure is repeated again until the last step, in which each trajectory is simulated until it enters the last (and critical) region $B_n = B$ or until time T is reached, whichever occurs first. The fraction of the successful trajectories that have managed to enter the last (and critical) region $B_n = B$ before time T is a natural estimate of $\mathbb{P}(A_n \mid A_{n-1}) = \mathbb{P}[T_n \leq T \mid T_{n-1} \leq T]$. In other words, the probability of the rare event is estimated as the product of estimates of the transition probabilities from one intermediate region to the next intermediate region, where the transition probability at level k is estimated as the fraction $\widehat{p}_k = R_k/N_k$ of the number R_k of successful trajectories that have managed to enter the next intermediate region before time T over the number N_k of trials. If $R_k = 0$ at any given stage k, we define $\hat{p}_{k'} = 0$ for all $k' > k$.

It is worth noting that the resulting estimator is unbiased, although the successive estimates are dependent because the result at level $k + 1$ depends on the entrance states in region B_k [15, 26]. Indeed, by induction, assuming that $\mathbb{E}[\widehat{p}_0 \cdots \widehat{p}_{k-1}] = p_1 \cdots p_{k-1}$ with $p_k = \mathbb{P}(A_k \mid A_{k-1})$, we have

$$
\begin{aligned}
\mathbb{E}[\widehat{p}_0 \cdots \widehat{p}_k] &= \mathbb{E}[\widehat{p}_0 \cdots \widehat{p}_{k-1} \mathbb{E}[\widehat{p}_k \mid N_0, \ldots, N_{k-1}, R_0, \ldots, R_{k-1}]] \\
&= \mathbb{E}[\widehat{p}_0 \cdots \widehat{p}_{k-1} (N_{k-1} p_k)/N_{k-1}] \qquad (3.2) \\
&= p_0 \cdots p_k = \gamma.
\end{aligned}
$$

All the implementations described below are also unbiased [27].

The *entrance distribution* to B_k is the probability distribution μ_k of $X(T_k)$, the first entrance state into B_k, conditional on $T_k \leq T$. An important observation is that each of the R_k trajectories that hits B_k before T hits it for the first time at a state having distribution μ_k. Using the same conditioning argument as in (3.2), one can see that for any measurable set $C \subseteq B_k$, the proportion of these R_k trajectories that hit B_k for the first time in C is a random variable (actually a ratio of two random variables) with expectation $\mu_k(C)$. That is, the empirical distribution $\widehat{\mu}_k^N$ of these R_k entrance states into B_k is an unbiased estimator of

μ_k. Then the N_{k+1} states obtained after the splitting are essentially a bootstrap sample from this empirical distribution. However, the R_k entrance states into B_k are not independent (in fact, they can be strongly dependent in some cases, especially when k is large), and this complicates the convergence analysis of this empirical distribution. We will return to this in Section 3.4. We recognize that the empirical distribution is undefined when $R_k = 0$. This is rarely a problem in practice and we neglect this possibility here.

In a more general setting where the chains can have different weights, we also define the *weight* of a trajectory as follows. A starting trajectory has weight 1. Each time it is split, its weight is divided by the number of offspring (or its expected value when this number is random). As a consequence, the above estimator of γ is just the sum of weights of the successful trajectories, divided by the number of trajectories that were originally started. If Russian roulette is applied, the weights can also increase when a chain survives the roulette. In that case, an unbiased estimator of γ is the sum of (final) weights of the chains that reach B at time $T_B \leq T$, and an unbiased estimator of the entrance distribution to B_k is the weighted empirical distribution of the states of the chains that hit B_k, at the first step when they hit it.

3.2.2 Implementations

There are many different ways of implementing the splitting idea. First, various types of strategies can be used to determine the number of retrials (i.e., clones) of a chain at each level, including the following:

- In a *fixed-splitting* implementation, each trajectory that has managed to reach the intermediate region B_{k-1} before time T receives the same deterministic number O_{k-1} of offspring. Then, $N_k = R_{k-1}O_{k-1}$ is a random variable. One advantage is that this can be implemented in a depth-first fashion, recursively: at level k, each chain is simulated until $\min(T, T_k)$. If A_k occurs, each clone is completely simulated by looking at all its offspring, before going to the next clone. Thus, it suffices to store a single entrance state at each level.

- In a *fixed-effort* implementation, a fixed and predetermined number N_k of offspring are allocated to the collection of successful trajectories that have managed to reach the intermediate region B_{k-1} before time T. To determine the starting point of the offspring, all the entrance states must be known, which means that the algorithm must be applied sequentially, level by level. Several strategies are then possible to assign the offspring to a successful trajectory. In the *random assignment*, the N_k starting states are selected at random, with replacement, from the R_{k-1} available states. In the *fixed assignment*, each successful trajectory is split approximately the same number of times, resulting in a smaller variance [1]. This is applied by first assigning $\lfloor N_k/R_{k-1} \rfloor$ offspring (or splits) to each state, and then assigning the remaining $N_k \bmod R_{k-1}$ offspring to distinct trajectories chosen at random (without replacement), so these chosen trajectories would have $\lfloor N_k/R_{k-1} \rfloor + 1$ offspring assigned to them [25, 26].

- In a *fixed success* implementation [24], a different perspective is considered. The idea is to create and simulate sufficiently many offspring, from time T_{k-1} onward, so that a fixed and predetermined number H_k of trajectories actually manage to reach the intermediate region B_k before time T. The issue here is to control the computational effort, because the number N_k of replicas needed to achieve exactly H_k successes is random. On the positive side, this implementation sorts out the extinction problem automatically, by construction, that is, the simulation will always run until it reaches the rare event a sufficient number of times. On the other hand, the computing effort may have a large variance.

- In a *fixed probability of success* implementation proposed in [8], the level sets are constructed recursively such that the probability of reaching one level from the previous level is approximately q, where q a fixed constant such that $0 < q < 1$. In this variant, assuming $\mathbb{E}[T] < \infty$, each of the $N = N_0$ chains is simulated until it reaches the recurrent set Δ. Let us denote by $X^i(\cdot)$ the trajectory of the ith chain, T^i its stopping time, and $S_{N,i} = \sup_{0 \le t \le T^i} h(X^i(t))$ the maximum value of the importance function over its entire trajectory. Sort in increasing order the values $(S_{N,1}, \ldots, S_{N,N})$, to obtain $S_{N,(1)} \le \ldots \le S_{N,(N)}$. The $K = \lfloor Nq \rfloor$ chains yielding the largest values $S_{N,(N-K+1)}, \ldots, S_{N,(N)}$ are kept, and in order to maintain a population of N chains, $N - K$ new trajectories are simulated with initial state the state at which the value $S_{N,(N-K)}$ was recorded, and until they reach Δ. Combining the maximum value of the importance function of these $N - K$ new trajectories with the K values recorded previously, we obtain a new sample of N values that we sort again in increasing order. We repeat the procedure while $S_{N,(N-K)} \le L$, that is, while at least $N - K$ chains have not reached B. The number n of iterations of the algorithm, and the number R of chains reaching B when the algorithm stops, are random variables, and the estimator of the probability of the rare event is $(K/N)^n(R/N)$. This estimator is biased but consistent. It also achieves the same asymptotic variance as $N \to \infty$ as the fixed-effort algorithm, with a probability q of going from any given level to the next.

Another important issue, from a practical viewpoint, is the computational effort required at each level. If T is the return time to a given set A of 'initial' states, for example, then the average time before either reaching the next level or going back to A is likely to increase significantly when k increases. Several techniques can be designed to alleviate this problem. A simple heuristic is to pick a positive integer β and just kill (truncate) the chains that go down by β levels or more below the current level L_{k-1}, based on the idea that they are very unlikely to come up again and reach level k. This reduces the computation time, but on the other hand introduces a bias. One way to deal with this bias is to apply the Russian roulette principle [22] and modify the weight of the chain accordingly. Several versions of this are proposed in [25, 26], including the following (where we also select a positive integer β and assume that the chain tries to reach level k):

- *Probabilistic truncation* applies Russian roulette each time a trajectory crosses a level $k - 1 - j$ downward, for any $j \geq \beta$. We select real numbers $r_{k-1,j} \in [1, \infty)$ for $j = \beta, \ldots, k - 1$. Whenever a chain crosses level $k - 1 - j$ downward from level $k - 1$, for $j \geq \beta$, it is killed with probability $1 - 1/r_{k-1,j}$. If it survives, its weight is multiplied by $r_{k-1,j}$. When a chain of weight $w > 1$ reaches level k, it is cloned into $w - 1$ additional copies and each copy is given weight 1 (if w is not an integer, we make $\lfloor w \rfloor$ additional copies with probability $\delta = w - \lfloor w \rfloor$ and $\lfloor w - 1 \rfloor$ additional copies with probability $1 - \delta$). The latter is done to reduce the variance introduced by the weights.

- *Periodic truncation* [26] reduces the variability due to the Russian roulette in probabilistic truncation by adopting a more systematic selection of the chains that we retain. Otherwise it works similarly to probabilistic truncation and also uses positive integers $r_{k-1,j}$. It also uses a random integer $D_{k-1,j}$ generated uniformly in $\{1, \ldots, r_{k-1,j}\}$, for each k and $j \geq \beta$. When a chain crosses a level $k - 1 - j$ downward, if it is the $(ir_{k-1,j} + D_{k-1,j})$th chain that does that for some integer i, it is retained and its weight is multiplied by $r_{k-1,j}$, otherwise it is killed.

- *Tag-based truncation* [26] fixes beforehand the level at which a chain would be killed. Each chain is *tagged* to level $k - 1 - j$ with probability $q_{k-1,j} = (r_{k-1,j} - 1)/(r_{k-1,\beta} \cdots r_{k-1,j})$ for $j = \beta, \ldots, k - 1$, and it is killed if it reaches that level. With the remaining probability, it is never killed. By properly choosing integers $r_{k-1,j}$, the proportion of chains tagged to level L_{k-1-j} can be *exactly* $q_{k-1,j}$, while the probability of receiving a given tag is the same for all chains.

To get rid of the weights, which carry additional variance, we can let the chain resplit when it crosses some levels upward after having gone down and having its weight increased. The idea is to keep the weights close to 1. The above truncation schemes can be adapted to fit that framework.

One of the best-known versions of splitting is the RESTART method [31–33]. Here, when a chain hits a level upward, fixed splitting is used (i.e., the chain is split by a fixed factor), but one of the copies is tagged as the *original* for that level. Truncation is used to reduce the work: when a non-original copy hits its creation level downward, it is killed. Only the original chain continues its path (to avoid starvation). The weight of the original chain accounts (in some sense) for those that are killed, to keep the estimator unbiased. This rule applies recursively, and the method is implemented in a depth-first fashion as follows: whenever there is a split, all the copies are simulated completely, one after the other, then simulation continues for the original chain. The gain in work reduction is counterbalanced by the loss in terms of a higher variance in the number of chains, and a stronger positive correlation between the chains due to resplits [15].

In the discrete-time situation, another implementation does not make use of levels, but applies *splitting and Russian roulette* at each step of the simulation [3, 5, 13, 28]. The number of splits and the killing probabilities are determined in terms of the importance function h. Define $\alpha = \alpha(x, y) = h(y)/h(x)$ to be the ratio of importance values for a transition from x to y. If $\alpha \geq 1$, the chain is split into C copies where $\mathbb{E}[C] = \alpha$, whereas if $\alpha < 1$ it is killed with probability $1 - \alpha$ (this is Russian roulette). A weight is again associated with each chain to keep the estimator unbiased: whenever a chain of weight w is split into C copies, the weight of all the copies is set to $w/\mathbb{E}[C]$. When Russian roulette is applied, the weight of a surviving chain is multiplied by $1/(1 - \alpha)$.

Yet another version, again in the discrete-time situation, mixes splitting and Russian roulette with IS. The *weight* of a chain is redefined as the weight due to splitting and Russian roulette (as above) times the *likelihood ratio* accumulated so far (see the previous chapter on IS). To reduce the variance of the weights, the idea of *weight windows* was introduced in [2], and further studied in [4, 14, 27]. The goal is to keep the weights of chain inside a given predefined window, with the aim of reducing the variance. This is done by controlling the *weighted importance* of each chain, defined as the product of its weight w and the value of the importance function $h(x)$ at its current state, so that it remains close to $\gamma = \mathbb{P}[T_B \leq T]$ for the trajectories for which $T_B \leq T$. If these windows are selected correctly (this requires a good prior approximation of γ), the main source of variance will then be the random *number* of chains that reach B [4]. To proceed, we select three real numbers $0 < a_{\min} < a < a_{\max}$. Whenever the weighted importance $\omega = wh(x)$ of a chain falls below $a_{\min}$, Russian roulette is applied, killing the chain with probability $1 - \omega/a$. If the chain survives, its weight is set (increased) to $a/h(x)$. If the weighted importance ω rises above $a_{\max}$, we split the chain into $c = \lceil \omega/a_{\max} \rceil$ copies and give (decreased) weight w/c to each copy. If $a = (a_{\min} + a_{\max})/2 \approx \mathbb{P}[T_B \leq T]$, this number has expectation N_0 (approximately), the initial number of chains.

3.2.3 Major issues to address

Having described the general principles and some known versions of splitting, we now discuss several key issues that need to be addressed for an efficient implementation of splitting.

First, how should the importance function h be defined? This is definitely the most important and most difficult question to address. For multilevel splitting, in the simple case where the state space is one-dimensional and included in $\mathbb{R}$, the final time is an almost surely finite stopping time, and the critical region has the form $B = [b, \infty)$, then all strictly increasing functions h are equivalent if we assume that we have the freedom to select the levels (it suffices to move the levels to obtain the same subsets B_k). So we can just take $h(x) = x$, for instance. Otherwise, especially if the state space is multidimensional, the question is much more complicated. Indeed, the importance function is a one-dimensional projection of the state space. Under simplifying assumptions, it is shown in [16] and below that ideally, to minimize the residual variance of the estimator from the current stage onward, the probability of reaching the next level should be the same at each possible entrance state to the current level. This is equivalent to having $h(x)$ proportional to $\mathbb{P}[T_B \leq T \mid X(0) = x]$. But if we knew these probabilities, we would know the exact solution and there would be no need for simulation. In this sense, this is a similar issue to that of the optimal (zero-variance) change of measure in IS. The idea is then to use an approximation of $\mathbb{P}[T_B \leq T \mid X(0) = x]$ or an adaptive (learning) technique. One way to learn the importance function was proposed in [4]: the state space is partitioned in a finite number of regions and the importance function h is assumed to be constant in each region. The 'average' value of $\mathbb{P}[T_B \leq T \mid X(0) = x]$ in each region is estimated by the fraction of chains that reach B among those that have entered this region. These estimates are combined to define the importance function for further simulations, which are used in turn to improve the estimates, and so on. We will see in Section 3.5, on a simple tandem queue, that the choice of the importance function is really a critical issue; an intuitively appealing (but otherwise poor) selection can lead to high inefficiency.

It is important to emphasize that the above analysis considers only the variance and not the computing time (the work). If we take the work into account (which we should normally do) then taking $h(x)$ proportional to $\mathbb{P}[T_B \leq T \mid X(0) = x]$ is not necessarily optimal, because the expected work to reach B may depend substantially on the current state x.

In a rare event setting, it is important to understand how a proposed importance function would behave asymptotically as a function of the rare event probability γ when $\gamma \to 0$, that is, in a *rare event asymptotic* regime. This type of analysis is pursued in [9], in a framework where γ is assumed to be well approximated by a large-deviation limit, for which the rate of decay is described by the solution of the Hamilton–Jacobi–Bellman (HJB) nonlinear partial differential equations associated with some control problem. The authors show that a good importance function must be a viscosity subsolution of the HJB equations, multiplied by an appropriate scalar selected so that the probability of reaching a given level k from the previous level $k-1$ is $1/O_{k-1}$ when $L_k = k-1$. In the context of fixed splitting, this condition is necessary and sufficient for the expected total number of particles not to grow exponentially with $-\log\gamma$. Moreover, if the subsolution also has its maximal possible value at a certain point, then the splitting scheme is asymptotically optimal, in the sense that the relative variance grows slower than exponentially in $-\log\gamma$.

Second, how should the number of offspring be chosen? In fixed splitting, the question is how to select the number O_k of offspring at each level. If we do not split enough, reaching the next level (and the rare event) becomes unlikely. On the other hand, if we split too much, the number of trajectories will explode exponentially with the number of levels, which will result in computational problems. A compromise has to be found. In the next subsection, we investigate this issue in a simplified setting. In fixed-effort splitting, no explosion is possible, as a fixed total number N_k of offspring are allocated at level k to the collection of successful trajectories that have managed to reach B_k. Nonetheless, deciding how many offspring to create, as well as the number of successful trajectories in the case of a fixed-performance implementation, are important issues.

Finally, given the importance function h, how many intermediate regions should be introduced and how should the increasing sequence of thresholds be defined? The next subsection investigates this point. However, the precise optimal strategy depends on the implementation considered. There is also the option to learn the levels, as is done in the fixed-probability-of-success method of [8].

3.3 Analysis in a simplified setting: a coin-flipping model

Suppose we have already selected an importance function and one of the splitting implementations discussed in the previous section. For a given total computation budget, we would like to find the number and the locations of the thresholds, or equivalently the numbers n, $p_0, \ldots, p_n$, that minimize the variance of the estimator. We are also interested in convergence results for the variance and the work-normalized variance, under various asymptotic regimes, such as when $N \to \infty$ while n and $p_0, \ldots, p_n$ are fixed, or when $\gamma \to 0$ and $n \to \infty$. Here we study these questions and provide partial answers under a very simplified (but tractable) model, for the fixed-effort and fixed-splitting strategies. The main focus is on the asymptotic behavior when $N \to \infty$. Our simplified setting is a coin-flipping model uniquely characterized by the initial probability $p_0 = \mathbb{P}(A_0)$ (i.e., the occurrence of the event A_0 depends only on the outcome of a $\{0, 1\}$ Bernoulli trial with parameter p_0), and by the transition probabilities $p_k = \mathbb{P}(A_k \mid A_{k-1})$ (i.e., the occurrence of A_k, conditional on A_{k-1}, depends only on the outcome of a $\{0, 1\}$ Bernoulli trial with parameter p_k), for $k = 1, \ldots, n$. This model is equivalent to assuming that there is only a single entrance state at each level.

For the work-normalized analysis, we need to make some assumptions on how much work it takes, on average, to run a trajectory from a given level $k - 1$ until it reaches either the next level or the set $A = \Delta \setminus B$ (i.e., the stopping time T without reaching B). If there is a natural drift toward A, it appears reasonable to assume that the chains will reach A in $O(1)$ expected time, independently of n, if A and B (and therefore γ) are fixed. If we use truncation and/or Russian roulette, we still have $O(1)$ expected time. Then the total expected work for all stages is proportional to $\sum_{k=0}^{n} \mathbb{E}[N_k]$. This is the assumption we will make everywhere in this section, unless stated otherwise. If $\mathbb{E}[N_k] = N$ for all k, then this sum is $N(n + 1)$. For simplicity, we will further assume that the constant of proportionality (in the $O(1)$ expected time mentioned above) is 1.

In a different asymptotic regime, where $\gamma \to 0$ and $n \to \infty$ jointly, and if truncation and/or Russian roulette are not applied, the average time to reach A should increase when $\gamma \to 0$, typically as $O(-\ln \gamma)$, in which case the total work will be proportional to $(-\ln \gamma)(n+1)\sum_{k=0}^{n} N_k$. If we further assume that $p_0, \ldots, p_n$ are all equal to a fixed constant p, then $\gamma = p^{(n+1)}$, so $-\ln \gamma = -(n+1)\ln p$ and the total work is proportional to $(-\ln p)(n+1)^2 \sum_{k=0}^{n} N_k$. As it turns out, the extra linear factor $(n+1)$ has a negligible role in the asymptotic behavior [18, 20, 26].

3.3.1 Fixed effort

Several analytical studies have been performed for the fixed-effort model. In [25], an asymptotic analysis is performed for the case where $N_k = N$ and $p_k = p = \gamma^{1/(n+1)}$ for all k, in the simplified setting adopted here. In this setting, $R_0, R_1, \ldots, R_n$ are independent binomial random variables with parameters n and p. Then we have [15, 25, 26]:

$$\begin{aligned}
\mathrm{Var}(\widehat{p}_0 \cdots \widehat{p}_n) &= \prod_{k=0}^{n} \mathbb{E}(\widehat{p}_k^2) - \gamma^2 \\
&= (p^2 + p(1-p)/N)^{n+1} - p^{2(n+1)} \\
&= \frac{(n+1)p^{2n+1}(1-p)}{N} + \frac{n(n+1)p^{2n}(1-p)^2}{2N^2} \\
&\quad + \cdots + \frac{(p(1-p))^{n+1}}{N^{n+1}}.
\end{aligned}$$

If we assume that $N \gg n(1-p)/p$, the first term

$$(n+1)p^{2n+1}(1-p)/N \approx (n+1)\gamma^{2-1/(n+1)}/N$$

dominates this variance expression. Given that the expected work is $N(n+1)$, the *work-normalized variance* is proportional to $[(n+1)\gamma^{2-1/(n+1)}/N]N(n+1) = (n+1)^2\gamma^{2-1/(n+1)}$, asymptotically, when $N \to \infty$. Minimizing with respect to n yields a minimum value at $n+1 = -\frac{1}{2}\ln \gamma$, which corresponds to $p = e^{-2}$. If we assume that the constant of proportionality is 1, as we said earlier, then the resulting work-normalized relative variance is $(\ln \gamma)^2 e^2/4$.

For the asymptotic regime where $\gamma \to 0$ while p and N are fixed (so $n \to \infty$), the first term no longer dominates the variance expression, because the assumption $N \gg n(1-p)/p$ is no longer valid. In this case, the relative error and its work-normalized version both increase to infinity at a logarithmic rate [26].

3.3.2 Fixed splitting

In a fixed-splitting setting, the algorithm is equivalent to a simple Galton–Watson branching process, where each successful trial for which the event A_k occurs receives the same (deterministic) number O_k of offspring, for $k = 0, \ldots, n-1$. Each p_k is estimated by $\widehat{p}_k = R_k/N_k$. An unbiased estimator of $\gamma = \mathbb{P}(A_n) = p_0\, p_1 \cdots p_n$ is then given by

$$\widehat{p}_0 \cdots \widehat{p}_n = \frac{R_0}{N_0}\frac{R_1}{N_1} \cdots \frac{R_n}{N_n} = \frac{R_n}{N_0\, O_0 \cdots O_{n-1}} = \gamma \frac{R_n}{N m_n},$$

where $N = N_0$ and $m_k = m_{k-1} O_{k-1} p_k = p_0 O_0 p_1 \cdots p_{k-1} O_{k-1} p_k$ for $k = 1, \ldots, n$, with $m_0 = p_0$ by definition. The second equality in the display follows from the relation $N_k = R_{k-1} O_{k-1}$, which holds for $k = 1, \ldots, n$, and means that the probability of the rare event is equivalently estimated as the fraction of the number R_n of successful trials, for which the rare event A_n occurs, over the maximum possible number of trials, $N_0 O_0 \cdots O_{n-1}$.

The relative variance of this estimator is

$$\frac{\mathrm{Var}(\widehat{p}_0 \cdots \widehat{p}_n)}{\gamma^2} = \frac{1}{N} \sum_{k=0}^{n} \frac{1 - p_k}{m_k}.$$

Moreover, by the strong law of large numbers, $R_k/N \to m_k$ almost surely for any $k = 0, 1, \ldots, n$ when $N \to \infty$, and in particular $\widehat{p}_0 \cdots \widehat{p}_n \to \gamma$ almost surely when $N \to \infty$. We also have the central limit theorem

$$\sqrt{N} \left(\frac{\widehat{p}_0 \cdots \widehat{p}_n}{\gamma} - 1 \right) \Longrightarrow \mathcal{N}\left(0, \sum_{k=0}^{n} \frac{1 - p_k}{m_k}\right)$$

in distribution as $N \to \infty$, where $\mathcal{N}(\mu, \sigma^2)$ denotes a normal random variable with mean μ and variance σ^2.

The performance analysis below follows [23]. Under our assumptions, the total work is approximately $C = \sum_{k=0}^{n} N_k$, which satisfies

$$\frac{C}{N} = \sum_{k=0}^{n} \frac{N_k}{N} = \sum_{k=0}^{n} \frac{N_k}{R_k}\frac{R_k}{N} \to \sum_{k=0}^{n} \frac{m_k}{p_k},$$

almost surely when $N \to \infty$.

Suppose that we are allowed a fixed *expected* total computing budget c, that is, we have the constraint $\mathbb{E}[C] \le c$. What is the optimal way of selecting N, n, $p_0, \ldots, p_n$ and $O_0, \ldots, O_{n-1}$, to minimize the variance given this fixed budget? Assuming that we use all the budget and that N is large enough so that we can approximate C/N by its almost sure limit as $N \to \infty$, and neglecting the fact that n and the O_k must be integers, this optimization problem can be formulated as:

$$\min \frac{1}{N} \sum_{k=0}^{n} \frac{1 - p_k}{m_k} \quad \text{subject to} \quad N \sum_{k=0}^{n} \frac{m_k}{p_k} = c.$$

Solving this in terms of N and $O_0, \ldots, O_{n-1}$, with the other variables fixed, yields

$$N = c\frac{(1/p_0 - 1)^{1/2}}{\sum_{k=0}^{n}(1/p_k - 1)^{1/2}} \quad \text{and} \quad O_k = \left(\frac{p_{k+1}(1 - p_{k+1})}{p_k(1 - p_k)}\right)^{1/2} \frac{1}{p_{k+1}},$$

for $k = 0, \ldots, n-1$. This gives the relative variance

$$\frac{\text{Var}(\widehat{p}_0 \cdots \widehat{p}_n)}{\gamma^2} = \frac{1}{c}\left(\sum_{k=0}^{n}(1/p_k - 1)^{1/2}\right)^2.$$

Next, minimizing with respect to $p_0, p_1, \ldots, p_n$ for a given n gives that the transition probabilities should all be equal to the same value, $p_k = p = \gamma^{1/(n+1)}$ for all k. This implies that the branching rates should all be the same, $O_k = O = 1/p$, which corresponds to the critical regime of the Galton–Watson branching process, for which $Op = 1$. It also implies that the initial population size should be equal to $N = c/(n+1)$. In this optimal case, the work-normalized relative variance becomes

$$c\frac{\text{Var}(\widehat{p}_0 \cdots \widehat{p}_n)}{\gamma^2} = (n+1)^2(\gamma^{-1/(n+1)} - 1) = (n+1)^2\frac{(1-p)}{p}.$$

Finally, minimizing with respect to n gives

$$n = \frac{-\ln\gamma}{\ln(1+u^*)} - 1 \approx -0.6275\ln\gamma - 1,$$

where $u^* \approx 3.9214$ is the unique positive minimum of the mapping $u \mapsto u/(\ln(1+u))^2$. Thus, the transition probabilities should all be equal to $p = 1/(1+u^*)$. The resulting work-normalized variance is

$$c\frac{\text{Var}(\widehat{p}_0 \cdots \widehat{p}_n)}{\gamma^2} = \frac{u^*(\ln\gamma)^2}{(\ln(1+u^*))^2} \approx 1.5449(\ln\gamma)^2,$$

which is slightly smaller than the value $(\ln\gamma)^2 e^2/4 \approx 1.8473(\ln\gamma)^2$ obtained in the fixed-effort case.

Consider now an asymptotic regime where $p_k = p$ is fixed and $n \to \infty$, so that $\gamma = p^{n+1} \to 0$. Suppose that $O_k = 1/p$, that is, $N_{k+1} = R_k/p$, for $k = 0, \ldots, n-1$. Then the relative variance

$$\frac{\text{Var}(\widehat{p}_0 \cdots \widehat{p}_n)}{\gamma^2} = \frac{1}{N}\frac{(n+1)(1-p)}{p}$$

is unbounded when $n \to \infty$.

However, the *asymptotic logarithmic relative variance* (see Chapter 4) is

$$\lim_{n\to\infty}\frac{\ln[\text{Var}(\widehat{p}_0 \cdots \widehat{p}_n)/\gamma^2]}{\ln\gamma} = \lim_{n\to\infty}\frac{\ln[1 + (1/N)(n+1)(1-p)/p]}{(n+1)\ln p} = 0,$$

which means that the splitting estimator is asymptotically efficient under the assumptions made. Asymptotic results of this type were shown in [18, 20] in a more general setting where the probability transition matrix for the first-entrance state at level k, given the first-entrance state at level $k-1$, converges to a matrix with spectral radius $\rho < 1$, which implies that $p_k \to \rho$ when $k \to \infty$. In [20], the authors also show that in their setting, the multilevel splitting estimator is also work-normalized asymptotically efficient if and only if $O_k = 1/\rho$ for all k. This result holds if the expected computing time at level k is proportional to N_k, and it still holds if this expected time increases polynomially in k.

It is important to emphasize that for practical applications, γ and p are unknown, so the condition $O_k = 1/p$ (exactly) for all k cannot really be satisfied. Then, the population of chains is likely to either decrease too much and perhaps extinguish (so no chain will reach B) or explode (so the amount of work will also explode). This suggests that when γ is very small, fixed splitting is likely to lead to a large relative variance of the estimator and also a huge variance in the computing costs. For this reason, the more robust fixed-effort approach is usually preferable.

3.4 Analysis and central limit theorem in a more general setting

We will now relax the 'coin-flipping' assumption of the previous section, so that the probability of hitting the next level L_{k+1} may now depend on the entrance state into the current set B_k. This is certainly more realistic. For example, there are situations where we might enter B_k and B_{k+1} simultaneously, in which case this probability is 1.

We study the performance of some of the splitting implementations introduced in Section 3.2.2 in the framework of multilevel Feynman–Kac distributions and their approximation in terms of interacting particle systems [6, 7, 10, 11]. We state a central limit theorem and provide expressions for the asymptotic variance for the various implementations, in the large-sample asymptotic regime in which γ is fixed and $N = N_0 \to \infty$ (assuming that this implies $\mathbb{E}[N_k] \to \infty$ for all k). This provides some insight into the issues raised in Section 3.2.3. The results are stated here without proof; most of the proofs can be found in the references cited above. We emphasize that this analysis is not for a rare event asymptotic regime, for which $\gamma \to 0$; for this, we refer the reader to [9, 20]. For simplicity, throughout this section, we make the assumption that, at any level, all the chains have the same weight (so no Russian roulette is allowed, for example).

3.4.1 Empirical entrance distributions

As we pointed out earlier, splitting can be used to estimate expectations of more general functions of the sample paths than just the probability γ. In particular, we argued in Section 3.2.1 that when all the particles have the same weight, the entrance distribution at any level does not depend on the choice of importance function, is the same as for the original chain, and can be estimated without bias by the empirical entrance distribution at that level, which we shall denote by $\hat{\mu}_k^N$. This empirical distribution is already available at no extra cost when running the simulation.

More specifically, recall that N_k particles are simulated in stage k, and R_k of them hit B_k at the end of that stage. Let $\{\xi_k^i, i = 1, \ldots, N_k\}$ be the states of the N_k chains at the end of stage k, and let $I_k = \{i : \xi_k^i \in B_k\}$ be the subset of those states that have successfully hit B_k by their stopping time T. Note that I_k has cardinality R_k. We have

$$\hat{\mu}_k^N = \frac{1}{R_k} \sum_{i \in I_k} \delta_{\xi_k^i}, \tag{3.3}$$

where δ_x represents the Dirac mass at x.

Proposition 1. *For any measurable set $C \subseteq B_k$, $\mathbb{E}[\hat{\mu}_k^N(C)] = \mu_k(C)$. This implies that for any measurable function ϕ,*

$$\mathbb{E}[\mathbb{E}[\phi(X(T_k)) \mid T_k \leq T]] = \mathbb{E}\left[\frac{1}{R_k} \sum_{i \in I_k} \phi(\xi_k^i)\right].$$

3.4.2 Large-sample asymptotics

We saw that the empirical entrance distribution $\hat{\mu}_k^N$ provides an unbiased estimate of μ_k, but what about the convergence (and speed of convergence) of $\hat{\mu}_k^N$ to μ_k when $N \to \infty$? The next proposition answers this question by providing a central limit theorem, which can be proved using the technology developed in [10].

Proposition 2. *Let $\phi : E \to \mathbb{R}$ be a bounded and continuous function and $0 \leq k \leq n$. Then there is a constant $v_k(\phi)$, which depends on ϕ and on the splitting implementation, such that*

$$\sqrt{N}\left(\frac{1}{R_k} \sum_{i \in I_k} \phi(\xi_k^i) - \mathbb{E}[\phi(X(T_k)) \mid T_k \leq T]\right) \Longrightarrow \mathcal{N}(0, v_k(\phi)),$$

in distribution when $N \to \infty$, where $\mathcal{N}(0, \sigma^2)$ is a normal random variable with mean 0 and variance σ^2. The result also extends to unbounded functions ϕ under appropriate uniform integrability conditions.

We also have a central limit theorem for the probability of reaching level k before T. When $k = n$, this gives a central limit theorem for the estimator of γ, the probability of the rare event.

Proposition 3. *For $0 \leq k \leq n$, there is a constant V_k that depends on the splitting implementation, such that*

$$\sqrt{N}\left(\frac{\widehat{p}_0 \cdots \widehat{p}_k}{p_0 \cdots p_k} - 1\right) \Longrightarrow \mathcal{N}(0, V_k)$$

in distribution when $N \to \infty$.

By combining these two propositions, we also obtain a central limit result for the unconditional average cost, when a cost is incurred when we hit B_k:

$$\frac{\sqrt{N}}{p_0 \cdots p_k}\left(\frac{\widehat{p}_0 \cdots \widehat{p}_k}{R_k}\sum_{i \in I_k}\phi(\xi_k^i) - \mathbb{E}[\phi(X(\min(T, T_k)))]\right) \Longrightarrow \mathcal{N}(0, v_k(\phi)),$$

in distribution when $N \to \infty$, if we assume that $\phi(x) = 0$ when $x \notin B_k$. By taking $\phi(x)$ equal to the indicator that $x \in B_k$, and $v_k(\phi) = V_k$, we recover the result of the second proposition.

An intuitive argument to justify these central limit theorems is that although the particles have dependent trajectories to a certain extent, the amount of dependence remains bounded, in some sense, when $N \to \infty$. The idea (roughly) is that the trajectories that start from the same initial state in stage 0 have some dependence, but those that start from different initial states are essentially independent (in fixed splitting they are totally independent whereas in fixed effort they are almost independent when N is large). When $N \to \infty$ while everything else is fixed, the number of initial states giving rise to one or more successful trajectories eventually increases approximately linearly with N, while the average number of successful trajectories per successful initial state converges to a constant. So the amount of independence increases (asymptotically) linearly with N, and this explains why these central limit theorems hold.

In what follows, we derive expressions for the asymptotic variance V_n, for selected splitting implementations. Straightforward modifications can provide expressions for V_k, for $0 \leq k < n$. One may also rightfully argue that instead of normalizing by $\sqrt{N}$ in the central limit theorem, we should normalize by $\sqrt{C_N}$ where $C_N = \sum_{k=0}^{n} N_k$, the total number of particle levels simulated, which could be seen as the total amount of computation work if we assume that simulating one particle for one level represents one unit of work. This makes sense if we assume that the expected work is the same at each level. If $C_N/N \to C$ in probability as $N \to \infty$, which is typically the case (in particular, $C_N/N = C = n+1$ exactly in the fixed-effort implementations), then using Slutsty's lemma yields

$$\sqrt{C_N}\,(\widehat{p}_0 \cdots \widehat{p}_n/\gamma - 1) \Longrightarrow \mathcal{N}(0, CV_n)$$

in distribution as $N \to \infty$. So normalizing by C_N instead of N only changes the variance by the constant factor C.

Define the function h_B by

$$h_B(x) = \mathbb{P}[T_B \leq T \mid X(t) = x],$$

for $x \in E$. This function turns out to be an optimal choice of importance function when we wish to estimate γ. We also define

$$\nu_k = \frac{\mathrm{Var}[h_B(X(T_k)) \mid T_k \leq T]}{\mathbb{E}^2[h_B(X(T_k)) \mid T_k \leq T]} = \frac{\int_E h_B^2(x)d\mu_k(x)}{\left(\int_E h_B(x)d\mu_k(x)\right)^2} - 1,$$

the relative variance of the random variable $h_B(X(T_k))$ conditional on $T_k \leq T$ (i.e., when $X(T_k)$ is generated from μ_k). These ν_k depend only on the original model, and not on the splitting implementation.

In the *fixed-splitting* implementation, we have

$$V_n = \sum_{k=0}^{n} \frac{1-p_k}{m_k} + \sum_{k=0}^{n-1} \frac{\nu_k}{m_k}\left(1 - \frac{1}{O_k}\right),$$

where m_k is defined recursively by $m_0 = p_0$ and $m_k = m_{k-1}O_{k-1}p_k$ for $k = 1, \ldots, n$. This coincides for $n = 1$ with equation (2.21) in [15]. We also have

$$\frac{C_N}{N} = \sum_{k=0}^{n} \frac{N_k}{N} \longrightarrow C = \sum_{k=0}^{n} \frac{m_k}{p_k},$$

in probability as $N \to \infty$.

In the *fixed-effort* implementation with *random assignment* using *multinomial resampling*, it is shown in [7] that

$$V_n = \sum_{k=0}^{n} \left(\frac{1}{p_k} - 1\right) + \sum_{k=0}^{n-1} \frac{\nu_k}{p_k}.$$

In the *fixed-effort* implementation with *fixed assignment* using *residual resampling*, if $1/p_k$ is not an integer, for any $k = 0, 1, \ldots, n$, then

$$V_n = \sum_{k=0}^{n} \left(\frac{1}{p_k} - 1\right) + \sum_{k=0}^{n-1} \frac{\nu_k}{p_k}\left(1 - p_k(1 - r_k)\right).$$

If $\frac{1}{2} < p_k < 1$ then $r_k = 1 - p_k$ and it is shown in [7] that

$$V_n = \sum_{k=0}^{n} \left(\frac{1}{p_k} - 1\right) + \sum_{k=0}^{n-1} \frac{\nu_k}{p_k}\left(1 - p_k^2\right).$$

In each of the three cases considered above, the asymptotic variance splits as the sum of two terms, a first term that depends on the transition probabilities only, that is, indirectly on the thresholds only, and a second term that depends on the entrance distributions also, that is, indirectly on the importance function h that defines the shape of the intermediate regions. If for any given k the function h_B is constant on the support of the entrance distribution μ_k, then $\nu_k = 0$ and the second term vanishes, so only the first term remains and we obtain

$$CV_n = \left(\sum_{k=0}^{n} \frac{m_k}{p_k}\right) \sum_{k=0}^{n} \frac{1 - p_k}{m_k} \quad \text{and} \quad CV_n = (n+1) \sum_{k=0}^{n} \frac{1 - p_k}{p_k},$$

in the fixed-splitting case and in the (two different implementations of the) fixed-effort case, respectively. Note that if the continuous-time Markov chain has almost surely continuous trajectories, then the support of the entrance distribution μ_k is $\{x \in E : h(x) = L_k\}$, and a sufficient condition for $\nu_k = 0$ is to take $h = h_B$ as importance function. In this special case, the model reduces to the coin-flipping model already studied in Section 3.3.

3.5 A numerical illustration

The example described in this section is simple, but it has been widely used, because it provides a good illustration of the impact of the choice of importance function [20, 15]. We consider an open tandem Jackson queuing network with two queues. The arrival rate at the first queue is $\lambda = 1$ and the mean service time is $\rho_i = 1/\mu_i$ at queue i, for $i = 1, 2$. The corresponding discrete-time Markov chain is given by $X = \{X_j, j \geq 0\}$, where $X_j = (X_{1,j}, X_{2,j})$ is the number of customers in each of the two queues immediately after the jth event, where an event is an arrival or a service completion at a given queue. Our aim is to estimate the probability of reaching $B = \{(x_1, x_2) : x_2 \geq L\}$, the set of states for which the second queue has length at least L, before reaching $A = \{(0, 0)\}$. The final stopping time is $T = \min(T_A, T_B)$.

To illustrate the impact and difficulty of the choice of the importance function h, some choices are compared in [25, 26] for the case where $\rho_1 < \rho_2$, and in [19, 17, 16] for $\rho_1 > \rho_2$. Consider the three choices

$$\begin{aligned} h_1(x_1, x_2) &= x_2, \\ h_2(x_1, x_2) &= (x_2 + \min(0, x_2 + x_1 - L))/2, \\ h_3(x_1, x_2) &= x_2 + \min(x_1, L - x_2 - 1) \times (1 - x_2/L). \end{aligned}$$

The function h_1 is the simplest choice and is motivated (naively) by the fact that the set B is defined in terms of x_2 only. The second choice h_2 counts L minus half the minimal number of steps required to reach B from the current state, because we need at least $L - \min(0, x_2 + x_1 - L)$ arrivals at the first queue and $L - x_2$ transfers to the second queue. The function h_3 is inspired by [30], where $h(x_1, x_2) = x_2 + x_1$ is used when $\rho_1 < \rho_2$. This h was modified as follows. We have $h_3(x) = x_1 + x_2$ when $x_1 + x_2 \leq L - 1$ and $h_3(x) = L$ when $x_2 \geq L$. In between, when $L - x_1 - 1 \leq x_2 \leq L$, we interpolate linearly in x_2 for any fixed x_1.

In [25, 26], the authors compare these functions in the fixed-effort case, with several truncation implementations. For a numerical example with $\rho_1 = 1/4$, $\rho_2 = 1/2$, and $L = 30$, for instance, they estimate the constants V_n and CV_n defined in the previous section, and find that they are much higher for h_1 than for h_2 and h_3. Using h_3 yields just slightly better results than h_2. The truncation and resplit increases the variance slightly, but it also decreases the computation time, and overall it improves the work-normalized variance CV_n roughly by a factor of 3. Detailed results can be found in [26].

When $\rho_1 > \rho_2$, the first queue is the bottleneck of the system, and the most likely sample paths to B are those where the first queue builds up first, and then there is a transfer of customers from the first to the second queue. But h_1 does not favor these types of paths. Instead, it favors the paths where x_1 remains small, because the customers in the first queue are transfered quickly to the second queue. As a result, splitting with h_1 can give a variance that is even larger than with standard Monte Carlo in this case [19]. This problem can be solved by a better choice of h [33].

Variants of this example with $B = \{(x_1, x_2) : x_1 + x_2 \geq L\}$ and $B = \{(x_1, x_2) : \min(x_1, x_2) \geq L\}$ are examined in [9], where the authors design importance functions by finding subsolutions to the HJB equations associated with a control problem. These importance functions perform extremely well when γ is very small.

References

[1] J. E. Baker. Reducing bias and inefficiency in the selection algorithm. In J. J. Grefenstette, ed., *Proceedings of the 2nd International Conference on Genetic Algorithms, Cambridge MA 1987*, pp. 14–21. Lawrence Erlbaum Associates, Mahwah, NJ, 1987.

[2] T. E. Booth. Automatic importance estimation in forward Monte Carlo calculations. *Transactions of the American Nuclear Society*, **41**: 308–309, 1982.

[3] T. E. Booth. Monte Carlo variance comparison for expected-value versus sampled splitting. *Nuclear Science and Engineering*, **89**: 305–309, 1985.

[4] T. E. Booth and J. S. Hendricks. Importance estimation in forward Monte Carlo calculations. *Nuclear Technology/Fusion*, **5**: 90–100, 1984.

[5] T. E. Booth and S. P. Pederson. Unbiased combinations of nonanalog Monte Carlo techniques and fair games. *Nuclear Science and Engineering*, **110**: 254–261, 1992.

[6] F. Cérou, P. Del Moral, F. Le Gland, and P. Lezaud. Limit theorems for the multilevel splitting algorithm in the simulation of rare events. In *Proceedings of the 2005 Winter Simulation Conference, Orlando 2005*, pp. 682–691, December 2005.

[7] F. Cérou, P. Del Moral, F. Le Gland, and P. Lezaud. Genetic genealogical models in rare event analysis. *ALEA, Latin American Journal of Probability and Mathematical Statistics*, **1**: 181–203, 2006. Paper 01–08.

[8] F. Cérou and A. Guyader. Adaptive multilevel splitting for rare event analysis. *Stochastic Analysis and Applications*, **25**(2): 417–443, 2007.

[9] T. Dean and P. Dupuis. Splitting for rare event simulation: A large deviation approach to design and analysis. *Stochastic Processes and their Applications*, 2008. In press.

[10] P. Del Moral. *Feynman–Kac Formulae. Genealogical and Interacting Particle Systems with Applications*. Springer, New York, 2004.

[11] P. Del Moral and P. Lezaud. Branching and interacting particle interpretation of rare event probabilities. In H. Blom and J. Lygeros, eds, *Stochastic Hybrid Systems: Theory and Safety Critical Applications*, Lecture Notes in Control and Information Sciences 337, pp. 277–323. Springer, Berlin, 2006.

[12] P. Dupuis, A. D. Sezer, and H. Wang. Dynamic importance sampling for queueing networks. *Annals of Applied Probability*, **17**: 1306–1346, 2007.

[13] S. M. Ermakov and V. B. Melas. *Design and Analysis of Simulation Experiments*, volume 339 of *Mathematics and its Applications*. Kluwer Academic Publishers, Dordrecht, 1995.

[14] B. L. Fox. *Strategies for Quasi-Monte Carlo*, volume 22 of *International Series in Operations Research & Management Science*. Kluwer Academic Publishers, Norwell, MA, 1999.

[15] M. J. J. Garvels. The splitting method in rare event simulation. PhD thesis, Faculty of Mathematical Sciences, University of Twente, Enschede, October 2000.

[16] M. J. J. Garvels, J.-K. C. W. van Ommeren, and D. P. Kroese. On the importance function in splitting simulation. *European Transactions on Telecommunications*, **13**(4): 363–371, 2002.

[17] P. Glasserman, P. Heidelberger, and P. Shahabuddin. Asymptotically optimal importance sampling and stratification for pricing path dependent options. *Mathematical Finance*, **9**(2): 117–152, 1999.

[18] P. Glasserman, P. Heidelberger, P. Shahabuddin, and T. Zajic. Splitting for rare event simulation: Analysis of simple cases. In *Proceedings of the 1996 Winter Simulation Conference, San Diego* 1996, pp. 302–308, December 1996.

[19] P. Glasserman, P. Heidelberger, P. Shahabuddin, and T. Zajic. A large deviations perspective on the efficiency of multilevel splitting. *IEEE Transactions on Automatic Control*, **AC-43**(12): 1666–1679, 1998.

[20] P. Glasserman, P. Heidelberger, P. Shahabuddin, and T. Zajic. Multilevel splitting for estimating rare event probabilities. *Operations Research*, **47**(4): 585–600, 1999.

[21] H. Kahn. Modifications of the Monte Carlo method. Paper P–132, Rand Corporation, Santa Monica, CA, 1949.

[22] H. Kahn and T. E. Harris. Estimation of particle transmission by random sampling. *National Bureau of Standards Applied Mathematical Series*, **12**: 27–30, 1951.

[23] A. Lagnoux. Rare event simulation. *Probability in the Engineering and Informational Sciences*, **20**(1): 45–66, 2006.

[24] F. Le Gland and N. Oudjane. A sequential algorithm that keeps the particle system alive. In Henk Blom and John Lygeros, eds, *Stochastic Hybrid Systems: Theory and Safety Critical Applications*, Lecture Notes in Control and Information Sciences 337, pp. 351–389. Springer, Berlin, 2006.

[25] P. L'Ecuyer, V. Demers, and B. Tuffin. Splitting for rare-event simulation. In *Proceedings of the 2006 Winter Simulation Conference*, Monterey 2006, pp. 137–148, December 2006.

[26] P. L'Ecuyer, V. Demers, and B. Tuffin. Rare events, splitting, and quasi-Monte Carlo. *ACM Transactions on Modeling and Computer Simulation*, **17**(2), 2007. Article 9.

[27] P. L'Ecuyer and B. Tuffin. Splitting and weight windows to control the likelihood ratio in importance sampling. In *Proceedings of the 1st International Conference on Performance Evaluation Methodologies and Tools (ValueTools), Pisa 2006*, volume 180 of *ACM International Conference Proceeding Series*. ACM, October 2006. Article 21.

[28] V. B. Melas. On the efficiency of the splitting and roulette approach for sensitivity analysis. In *Proceedings of the 1997 Winter Simulation Conference, Atlanta 1997*, pp. 269–274, December 1997.

[29] H. Soodak. Pile kinetics. In C. Goodman, ed., *The Science and Engineering of Nuclear Power*, Volume 2, pp. 89–102. Addison-Wesley, Reading, MA, 1949.

[30] J. Villén-Altamirano. Rare event RESTART simulation of two-stage networks. *European Journal of Operations Research*, **179**(1): 148–159, 2007.

[31] M. Villén-Altamirano and J. Villén-Altamirano. RESTART: A straightforward method for fast simulation of rare events. In *Proceedings of the 1994 Winter Simulation Conference, Orlando 1994*, pp. 282–289, December 1994.

[32] M. Villén-Altamirano and J. Villén-Altamirano. Analysis of RESTART simulation: Theoretical basis and sensitivity study. *European Transactions on Telecommunications*, **13**(4): 373–385, 2002.

[33] M. Villén-Altamirano and J. Villén-Altamirano. On the efficiency of RESTART for multidimensional state systems. *ACM Transactions on Modeling and Computer Simulation*, **16**(3): 251–279, 2006.

4

Robustness properties and confidence interval reliability issues

Peter W. Glynn, Gerardo Rubino and Bruno Tuffin

4.1 Introduction

In this chapter, we discuss the robustness and reliability of the estimators of the probability of a rare event (or, more generally, of the expectation of some function of rare events) with respect to rarity: is the estimator accurate as rarity increases? (recall that accuracy, when estimating small probabilities, focuses on relative rather than absolute errors). And what about the reliability (i.e., the coverage) of the associated confidence interval?

If we parameterize the model with a (small) real ε such that the probability of the rare event considered decreases to zero as $\varepsilon \to 0$, we need to control the quality of the estimator as rarity increases, with respect to accuracy and coverage. An estimator will be said to be *robust* (in different senses defined hereafter) if its quality (i.e., the gap with respect to the true value) is not significantly affected when $\varepsilon \to 0$. Similarly, an estimator is always accompanied with a confidence interval. A *reliable* estimator is then an estimator for which the confidence interval coverage does not deteriorate as $\varepsilon \to 0$. Those two notions are different: one focuses on the error itself, the other on quality of the error estimation.

Rare Event Simulation using Monte Carlo Methods Edited by G. Rubino and B. Tuffin

To better illustrate this, let us start with the standard or crude estimator of the probability of a rare event. Let ε be this probability and $(X_i)_{1\leq i\leq n}$ be independently and identically distributed random variables such that $X_i = 1$ if the rare event occurs at the ith trial and 0 otherwise. The standard estimator of ε is $\widehat{\gamma}_n^{\mathrm{STD}} = n^{-1}\sum_{i=1}^n X_i$. The sum $\sum_{i=1}^n X_i$ is a binomial random variable with variance $n\varepsilon(1-\varepsilon)$, and the resulting confidence interval for ε, centered at $\widehat{\gamma}_n^{\mathrm{STD}}$, at confidence level $1-\alpha$, is

$$\left[\widehat{\gamma}_n^{\mathrm{STD}} - z_{1-\alpha/2}\frac{\sqrt{\varepsilon(1-\varepsilon)}}{\sqrt{n}}, \widehat{\gamma}_n^{\mathrm{STD}} + z_{1-\alpha/2}\frac{\sqrt{\varepsilon(1-\varepsilon)}}{\sqrt{n}}\right]$$

where $z_{1-\alpha/2} = \Phi^{-1}(1-\alpha/2)$ and Φ is the standard normal cumulative distribution function. The relative half-width RE of the confidence interval is therefore $z_{1-\alpha/2}\sqrt{1-\varepsilon}/\sqrt{n\varepsilon}$. For a fixed sample size n, this means that, as $\varepsilon \to 0$, the relative error of the estimation goes to infinity. Therefore, the accuracy of the estimator deteriorates as $\varepsilon \to 0$. The *absolute* error given by the confidence interval half-width $z_{\alpha/2}\sqrt{\varepsilon(1-\varepsilon)}/\sqrt{n}$ tends to 0 with ε, but at the much smaller rate $\sqrt{\varepsilon}$ than ε, so it does not give a good idea of the order of magnitude of the probability of interest. In other words, in order to get a fixed relative half-width RE $= \delta$ of the confidence interval as $\varepsilon \to 0$, one would have to increase the sample size (which usually means the simulation computating time) as

$$n = (z_{1-\alpha/2})^2\frac{1-\varepsilon}{\delta^2\varepsilon},$$

that is, in inverse proportion to ε. The aim of rare event simulation is to construct estimators for which the relative error is kept under control as the event probability decreases to zero. Such estimators are said to be *robust*, and families of *robusness properties* will be discussed in this chapter.

But looking only at the (theoretical) relative error, or some of its closely related notions introduced below, may be hazardous, or may only provide partial views of the possible problems. When evaluating γ using some unbiased estimator $\widehat{\gamma}_n = n^{-1}\sum_{i=1}^n X_i$, where the X_i are independently and identically (generally) distributed random variables with mean μ and variance σ^2, not only is $\mathbb{E}(\widehat{\gamma}_n) = \gamma$ unknown in practice, but so is its variance $\mathrm{Var}(\widehat{\gamma}_n) = \sigma_n^2 = \sigma^2/n$. Generally σ^2 is estimated by the unbiased $\widehat{\sigma}_n^2$:

$$\sigma^2 \approx \widehat{\sigma}_n^2 = \frac{1}{n-1}\sum_{i=1}^n (X_i - \widehat{\gamma}_n)^2.$$

This estimator is at least as sensitive to rarity as $\widehat{\gamma}_n$ itself.

Returning to the crude estimation of a probability ε by the average of Bernoulli random variables $\widehat{\gamma}_n^{\mathrm{STD}}$, if n is much smaller than $1/\varepsilon$, the rare event will most likely not be observed (on average, an occurrence appears after $1/\varepsilon$ replications), leading to a confidence interval $(0, 0)$ because $\widehat{\gamma}_n = \widehat{\sigma}_n^2 = 0$. With the (very unlikely) assumption that we end up with exactly one occurrence of the rare event, $\widehat{\gamma}_n = 1/n$ overestimates the event, and the variance is also overestimated by $\widehat{\sigma}_n^2 = 1/n$. We then get a large confidence interval, with a very high coverage in this (very unlikely) case. This highlights not only the problem of robustness of the estimator, but also the problem of the *reliability*, meaning the error in terms of coverage, of the confidence interval produced. As stated before, the two notions are different: robustness is about the actual error with respect to the true value, while reliability is about the coverage of the confidence interval, both as the probability of the rare event goes to zero.

Note that for binomial random variables, such as the one we were looking at, we know how to generate a more reliable confidence interval even for small probabilities ε. For instance, the Wilson score interval gives an interval

$$\left(\frac{\widehat{\gamma}_n^{\mathrm{STD}} + \frac{1}{2n} z_{1-\alpha/2}^2 \pm z_{1-\alpha/2}\sqrt{\frac{\widehat{\gamma}_n^{\mathrm{STD}}(1-\widehat{\gamma}_n^{\mathrm{STD}})}{n} + \frac{z_{1-\alpha/2}^2}{4n^2}}}{1 + \frac{1}{n} z_{1-\alpha/2}^2}\right)$$

(but note that there exist other interval constructions; see [11] for a description and some comparisons). This interval is known to yield a better reliability, but is very conservative for fixed n as ε decreases. The relative half-width of the confidence interval, on the other hand, is still growing to infinity as ε tends to zero.

This chapter investigates the robustness properties and reliability issues in rare event simulation. Section 4.2 quickly reviews the known robustness properties in the literature, including bounded relative error (also called bounded relative variance), and logarithmic efficiency (also called asymptotic optimality). Section 4.3 discusses the efficiency of an estimator when computation time is taken into account. Section 4.4 discusses the related notion of reliability of the corresponding confidence interval. We start by illustrating in Section 4.4.1 that bad rare event estimations are not always checked by looking at intervals of the form (0, 0), but can be much more difficult to detect. We then present two reliability measures. Section 4.5 summarizes the chapter by setting out some practical rules for detecting the presence of problems associated with the reliability of the observed confidence interval. Section 4.6 concludes the chapter.

4.2 Classical asymptotic robustness properties

This section describes the basic asymptotic robustness properties that can be found in the literature. For a recent survey, the reader is advised to look at [7], where more definitions are covered and discussed in detail.

As noted before, if we want to investigate the robustness properties of estimators with respect to rarity, it is very useful to parameterize the model. Let $\gamma = \gamma(\varepsilon)$ be the expectation (or probability if we restrict ourselves to integrating indicator functions) we are trying to estimate, parameterized by ε and such that $\gamma(\varepsilon) \to 0$ as $\varepsilon \to 0$. In this way the event can be arbitrarily small by playing with the value of ε, which allows the behavior of the estimator to be captured as rarity increases.

Consider an unbiased estimator $\widehat{\gamma}_n$ of γ, built from a sample having size n. The bounded relative error (BRE) is defined in [14]. It basically states that the relative half-width confidence interval already studied above is bounded uniformly in ε, for a fixed sample size n. This asserts that the relative error is not sensitive to the rarity of the event and is then the typical desirable property.

Definition 1. *Let σ_n^2 denote the variance of the estimator $\widehat{\gamma}_n$, $\sigma_n = \sqrt{\sigma_n^2}$ and let z_δ denote the $1 - \delta/2$ quantile of the standard normal distribution ($z_\delta = \Phi^{-1}(1 - \delta/2)$ where Φ is the standard normal cumulative distribution). Recall that the relative error* RE *associated with $\widehat{\gamma}_n$ is defined by the half-width confidence interval*

$$\mathrm{RE} = z_\delta \frac{\sigma_n}{\gamma}. \tag{4.1}$$

We say that we have a bounded relative error if RE *remains bounded as $\varepsilon \to 0$ (i.e., uniformly in ε).*

This property has been extensively studied and is often seen as the key property to verify [6, 8].

The aforementioned crude estimator is a typical illustration of one not verifying BRE. Additionally, increasing the occurrence of the rare event might not be sufficient. On the other hand, some estimators do possess the BRE property. Those two assertions are verified by the next two examples.

Consider the following example taken from [16], which can be seen as a simple case of the Markovian dependability models described in Chapter 6.

Example 1. A system consists of two types of components with two components of each type. Failure rates are $o(\varepsilon)$ for some parameter ε, and the transition probabilities of the embedded discrete-time Markov chain are as described in Figure 4.1, where (i, j) denotes the state with i (j) operational components of type 1 (2). The states where the system is down are shaded gray. We see that the system is functioning as soon as there is at least one component of each class that is operational.

Associated with each transition we put the first term of the development of the corresponding probability in powers of ε. We want to estimate the probability γ that, starting from $(2, 2)$, we reach a down state before returning to $(2, 2)$.

Given the target γ, we can simplify the model by collapsing or aggregating the failed states into a single one which we make absorbing. The resulting chain is shown in Figure 4.2.

Since $\gamma \ll 1$ because $\varepsilon \ll 1$ (we will see that $\gamma \approx 2\varepsilon^2$), we use the importance sampling (IS) method, and specifically the failure biasing scheme (see Section 6.3.2), with transition probabilities described in Figure 4.3. Basically, for each functioning state different from the initial $(2, 2)$, we increase the probability of failure to the constant q and use individual probabilities proportional to the original ones. The parameter q is chosen between $1/2$ and 1, for instance,

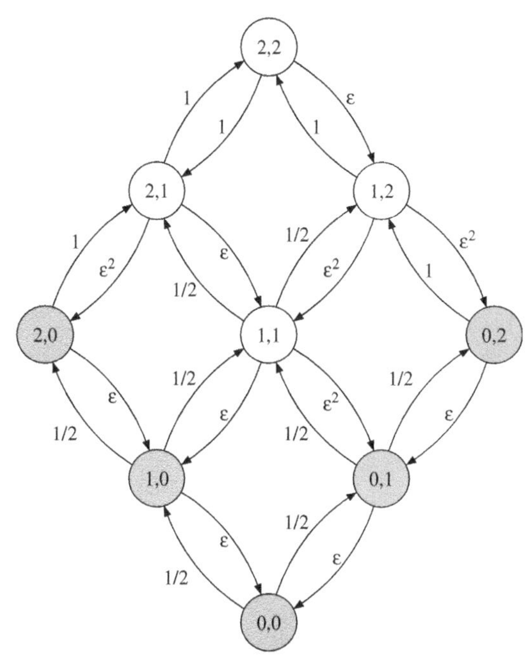

Figure 4.1 The evolution of a four-component system with two classes of components, subject to failures and repairs. The scheme shows the canonically embedded discrete-time Markov chain, where we give the simplest equivalents of the transition probabilities as $\varepsilon \to 0$.

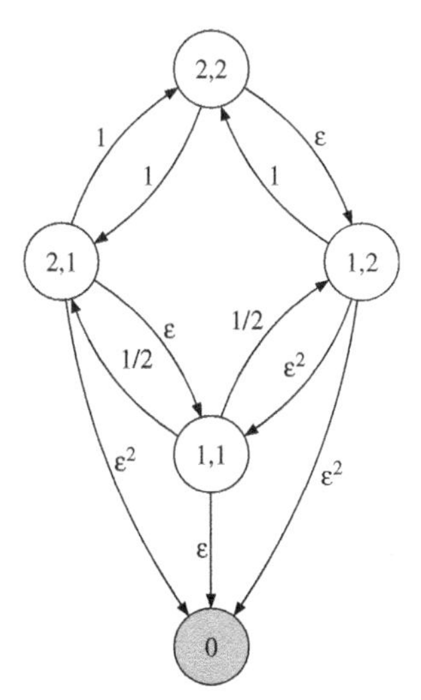

Figure 4.2 The result of aggregating the failed states in previous chain into a single absorbing one.

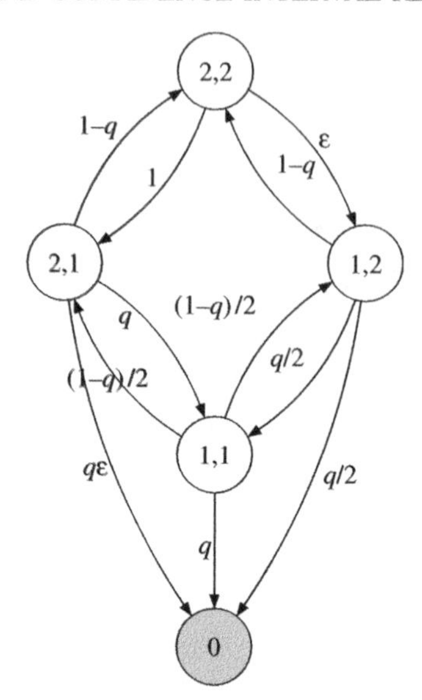

Figure 4.3 The result of changing the measure according to the failure biasing scheme with parameter q, again indicating the equivalents of the transition probabilities.

$q = 0.8$. The idea is, more generally, to enforce the transition probability associated with a failure to some $\Theta(1)$ value, instead of $o(1)$.

As seen in Chapter 1, the probability γ is given by

$$\gamma = \sum_{\pi \in \mathcal{P}_F} p(\pi),$$

where $\mathcal{P}_F$ is the set of all paths starting at (2,2), ending at a down state, and not visiting either (2,2) or a failed state in between, and $p(\pi)$ is the probability of path π under the original measure.

In this simple chain, there are six elementary paths in $\mathcal{P}_F$ (an elementary path is a path not visiting the same state more than once): $\pi_1 = ((2,2),(2,1),(0))$; $\pi_2 = ((2,2),(2,1),(1,1),(0))$; $\pi_3 = ((2,2),(2,1),(1,1),(1,2),(0))'$ $\pi_4 = ((2,2),(1,2),(0))$; $\pi_5 = ((2,2),(1,2),(1,1),(0))$; $\pi_6 = ((2,2),(1,2),(1,1),(2,1),(0))$. Their corresponding probabilities are $p(\pi_1) \approx \varepsilon^2$, $p(\pi_2) \approx \varepsilon^2$, $p(\pi_3) \approx \varepsilon^3/2$, $p(\pi_4) \approx \varepsilon^3$, $p(\pi_5) \approx \varepsilon^4$, $p(\pi_6) \approx \varepsilon^5/2$.

Observe that any other path include cycles that always strictly increase the order of the path probability in ε. This means that there are only a finite number of paths having the same order k in ε for any k, and thus, that $\gamma = 2\varepsilon^2 + o(\varepsilon^2)$ because of the two dominant paths π_1 and π_2 [14].

Let us now consider the IS scheme. To explore its performance, we must evaluate the variance of the IS estimator $\widehat{\gamma}_n^{\mathrm{IS}}$. For this purpose, denoting by Π a generic random path and by $\widetilde{p}(\pi)$ the probability of path π under the new measure, we write

$$\mathrm{Var}(\widehat{\gamma}_n^{\mathrm{IS}}) = \frac{1}{n}\left\{\widetilde{\mathbb{E}}[L^2(\Pi)1(\Pi \in \mathcal{P}_F)] - \gamma^2\right\} = \frac{1}{n}\left[\sum_{\pi \in \mathcal{P}_F} \frac{p^2(\pi)}{\widetilde{p}(\pi)} - \gamma^2\right],$$

where $\widetilde{\mathbb{E}}$ denotes the expectation with respect to the IS measure. Looking at the probability of the six paths under the IS measure, the dominant term in this sum comes from π_1; it is in ε^3, and we get

$$\mathrm{Var}(\widehat{\gamma}_n^{\mathrm{IS}}) = \frac{\varepsilon^3}{nq} + o(\varepsilon^3).$$

The relative error of the IS estimator is $\mathrm{RE} = 1.96\sqrt{\mathrm{Var}(\widehat{\gamma}_n^{\mathrm{IS}})}/(\widehat{\gamma}_n^{\mathrm{IS}}\sqrt{n})$. We see that RE is proportional to $1/\sqrt{\varepsilon}$ and thus goes to infinity as $\varepsilon \to 0$.

Example 2. Consider a system failing according to an exponential distribution with rate λ. We wish to compute the probability γ that the system fails before ε. For such a trivial problem, we know that $\gamma = 1 - e^{-\lambda\varepsilon}$. Assume that we want to estimate this number using IS, and that we still sample from an exponential density, but with a different rate $\widetilde{\lambda}$. Our IS estimator is the random variable $X = 1_{[0,T]}L$ with L the likelihood ratio. The second moment of this estimator is

$$\widetilde{\mathbb{E}}[X^2] = \int_0^{\varepsilon} \left(\frac{\lambda e^{-\lambda y}}{\widetilde{\lambda} e^{-\widetilde{\lambda} y}}\right)^2 \widetilde{\lambda} e^{-\widetilde{\lambda} y} dy = \frac{\lambda^2}{\widetilde{\lambda}(2\lambda - \widetilde{\lambda})}(1 - e^{-(2\lambda-\widetilde{\lambda})\varepsilon}).$$

The relative error $z_\delta\sigma/\gamma$ is bounded if and only if $\widetilde{\mathbb{E}}[X^2]/\gamma^2$ is bounded as $\varepsilon \to 0$. It can easily be seen that, if $\widetilde{\lambda} = 1/\varepsilon$,

$$\frac{\widetilde{\mathbb{E}}[X^2]}{\gamma^2} = \frac{\lambda^2(1 - e^{-(2\lambda-\widetilde{\lambda})\varepsilon})}{\widetilde{\lambda}(2\lambda - \widetilde{\lambda})(1 - e^{-\lambda\varepsilon})^2} \longrightarrow e - 1 \quad \text{as } \varepsilon \to 0.$$

So, RE remains bounded as $\varepsilon \to 0$.

BRE has often been found difficult to verify in practice. For this reason, people often use logarithmic efficiency, also called asymptotic optimality.

Definition 2. *An unbiased estimator $\widehat{\gamma}_n$ of γ is said to be logarithmic efficient with respect to rarity parameter ε if*

$$\lim_{\varepsilon \to 0} \frac{\ln \mathbb{E}[\widehat{\gamma}_n^2]}{\ln \gamma} = 2.$$

Note that the quantity under limit is always positive and less than or equal to 2. This is because $\mathrm{Var}(\widehat{\gamma}_n) \geq 0$, so $\mathbb{E}[\widehat{\gamma}_n^2] \geq \gamma^2$ and then $\ln \mathbb{E}[\widehat{\gamma}_n^2] \geq 2 \ln \gamma$.

Basically, this property means that the second moment and the square of the mean go to zero at the same *exponential* rate. Asymptotic optimality has been widely used in queuing applications, for the IS class of simulation methods (see Chapter 5).

It can be proved that asymptotic optimality is a necessary but not sufficient condition for BRE. Indeed, if the relative error corresponding to estimator $\widehat{\gamma}_n$ of γ is bounded, then there is some $\kappa > 0$ such that $E[\widehat{\gamma}^2] \leq \kappa^2\gamma^2$, that is, $\ln E[\widehat{\gamma}_n^2] \leq \ln \kappa^2 + 2\ln\gamma$, leading to $\lim_{\varepsilon\to 0} \ln E[\widehat{\gamma}_n^2]/\ln\gamma \geq 2$. Since this ratio is always less than 2, we get the limit 2.

On the other hand, there are plenty of examples for which logarithmic efficiency is verified and not BRE, just by having the same exponential decreasing rate for the second moment and square expectation, but with an additional (polynomial) multiplicative component for the second moment, vanishing for logarithmic efficiency, but not for relative error. Other more practical examples, from queuing analysis and large-deviations theory, can be found in [12]. A simpler basic example is provided in [7], just by looking at an estimator for which $\gamma = e^{-\eta/\varepsilon}$ with $\eta > 0$, but for which the variance is $Q(1/\varepsilon)e^{-2\eta/\varepsilon}$ with Q a polynomial.

Extensions of logarithmic efficiency and BRE were introduced in [7] to higher moments than just the second, to make sure that they are well estimated too. For example, this also allows the variance of the empirical variance to be controlled. A preliminary work on this was [17], where BRE for the empirical variance was studied. In Section 4.4, we further investigate the asymptotic coverage of the confidence interval as $\varepsilon \to 0$.

4.3 Efficiency (or work-normalized variance) analysis

Throughout the above analysis, we have been looking at estimators for which the (relative) variance is as small as possible for a fixed sample size. On the other hand, this improved precision might be attained at the cost of employing a more complex algorithm, which can lead to increased computation time. This variation might also depend on the rarity parameter ε. Similarly, some methods can have an average computation cost decreasing with ε. This trade-off between accuracy and computational complexity has therefore to be taken into account with when analyzing rare event simulators.

The principle is then to combine variance and computation time. In [5], the efficiency is defined as being inversely proportional to the product of the sampling variance and the amount of labor required to obtain this estimate. Formally:

Definition 3. *The efficiency of an estimator $\widehat{\gamma}_n$ based on a sample of size n, with variance σ_n^2 and obtained, on average, in a computation time t_n, is* $1/(\sigma_n^2 t_n)$.

If the estimate is obtained from n independent replications each of variance σ^2 and with sampling average time t, then $\sigma_n^2 = \sigma^2/n$ and $t_n/n \to t$ as $n \to \infty$. Thus, if $n \gg 1$, the efficiency of $\widehat{\gamma}$ is approximately $1/(\sigma^2 t)$. This means that $\sigma_n^2 t_n$

can be also seen as a work-normalized variance. It also allows two estimators to be compared for a given computation budget c: if t and t' are the mean times required to generate one independent replication of X and X' when computing $\widehat{\gamma}_n$ and $\widehat{\gamma}'_{n'}$, the number of replications will be respectively $n = c/t$ and $n' = c/t'$. Thus the best estimator is $\widehat{\gamma}_n$ if $\sigma^2(X)t < \sigma^2(X')t'$, that is, if its efficiency is larger.

This definition is generalized in [4] by looking more precisely at the variance obtained with a budget c, taking into account the random generation time.

Based on this principle, the so-called bounded relative efficiency has been defined in [2]:

Definition 4. *Let $\widehat{\gamma}_n$ be an estimator of γ built using n replications and σ_n^2 its variance. Let t_n be the average simulation time to get those n replications. The relative efficiency of $\widehat{\gamma}_n$ is given by*

$$\mathrm{REff} = \frac{\gamma^2}{\sigma_n^2 t_n}.$$

We will say that $\widehat{\gamma}_n$ has bounded relative efficiency with respect to rarity parameter ε, if there exists a constant $d > 0$ such that REff *is minored by d for all ε.*

This basically means that the normalized relative variance $\sigma_n^2 t_n/\gamma^2$ is upper-bounded whatever the rarity, and is therefore a work-normalized version of the bounded relative error property.

In [2], an illustration of the need for such a definition is provided for the reliability analysis of a network (see Chapter 7 below), where the relative error is unbounded but the method is still efficient as $\varepsilon \to 0$, just due to the fact that the *average* computation time per run decreases to 0 at a proper rate. Sufficient conditions for this are also provided.

Similarly, the work-normalized logarithmic efficiency was defined in [3] to deal with the efficiency of splitting estimators.

Definition 5. *The unbiased estimator $\widehat{\gamma}_n$ of γ has work-normalized logarithmic efficiency if*

$$\lim_{\varepsilon \to 0} \frac{\ln t_n + \ln E[\widehat{\gamma}_n^2]}{\ln \gamma} = 2.$$

Note nonetheless that those definitions of relative efficiency and work-normalized logarithmic efficiency are good for comparing the relative merits of two estimators, but are far from perfect definitions. Indeed, there are some flaws in the above definitions. Computing times are usually random, so looking at a fixed computing budget c might be misleading: the number of replications is roughly c/t, but we would need this number to be *uniformly bounded* to make sure that we can bound the error whatever ε. At least, it would be of interest to consider the second moment of the computation time in the definition. This would lead to what could be the valid definition of work-normalized relative error, that is, the relative error for a computing budget c is bounded as $\varepsilon \to 0$. The above definitions, even if informative, are unfortunately more restrictive.

4.4 Another key issue: confidence interval coverage/reliability

Hitherto we have been dealing with the relative error uniformly in ε (or its weaker work-normalized version), but always based on the idea that the coverage of the confidence interval produced by the central limit theorem is always valid. Making sure that the coverage of the confidence interval is uniformly bounded in ε is of interest too.

Similarly, we have highlighted that, because it is the estimated (rather than the exact) variance that is actually used in the confidence interval computation, we may end up with the simple case of an interval (0, 0) because no occurrence of the rare event is detected, but in any case, as illustrated by Section 4.4.1, with an interval for which relative error seems bounded while it is not, and which does not include the exact value. This unpleasant observation highlights the need to design diagnostic procedures in order to point out if we are in this situation and is the focus of Section 4.5. But first, Section 4.4.2 looks at a property asserting the confidence interval coverage validity, while Section 4.4.3 reviews the coverage function representing the actual coverage in terms of the nominal.

4.4.1 Reliability issue of the observed confidence interval

Consider again the illustrative Example 1, with γ estimated by means of $\widehat{\gamma}_n^{\mathrm{IS}}$, where we fix the number n of samples, $n = 10^4$, using the same pseudo-random number generator, and varying ε from 10^{-2} down to 0. Table 4.1 gives, for different values of ε, $2\varepsilon^2$ (the equivalent of γ), $\widehat{\gamma}_n^{\mathrm{IS}}$, the IS estimator, and the 95% confidence interval obtained, together with the *estimated* variance $\widehat{\sigma}_n$. The estimated value becomes bad as $\varepsilon \to 0$: observe that $\widehat{\gamma}_n^{\mathrm{IS}}$ seems to be close to the expected value for $\varepsilon \geq 2 \times 10^{-4}$, and that the confidence interval seems suitable too, but, between 2×10^{-4} and 1×10^{-4}, as ε decays, the results are far from expectations and $2\varepsilon^2$ is not included in the confidence interval anymore. Actually, in this estimation, some paths important for the estimation of γ and of

Table 4.1 Equivalent $2\varepsilon^2$ of γ, IS estimation $\widehat{\gamma}_n^{\mathrm{IS}}$ of γ, confidence interval and estimated relative error for Example 1 using the failure biasing scheme with $q = 0.8$, for a fixed sample size $n = 10^4$ and different values of ε

ε	$2\varepsilon^2$	$\widehat{\gamma}_n^{\mathrm{IS}}$	Confidence Interval	Est. RE
1e-02	2e-04	2.03e-04	(1.811e-04, 2.249e-04)	1.08e-01
1e-03	2e-06	2.37e-06	(1.561e-06, 3.186e-06)	3.42e-01
2e-04	8e-08	6.48e-08	(1.579e-08, 1.138e-07)	7.56e-01
1e-04	2e-08	9.95e-09	(9.801e-09, 1.010e-08)	1.48e-02
1e-06	2e-12	9.95e-13	(9.798e-13, 1.009e-12)	1.48e-02
1e-08	2e-16	9.95e-17	(9.798e-17, 1.009e-16)	1.48e-02

$\mathrm{Var}(\widehat{\gamma}_n^{\mathrm{IS}})$ (paths whose probability is $\Theta(\varepsilon^2)$ under the original measure) are still rare under the IS measure, leading to wrong estimations.

Let us look at this in some detail. Assume that n is fixed and $\varepsilon \to 0$. At some point, ε will be so small that transitions in $\Theta(\varepsilon)$ (see Figure 4.3) are not sampled anymore (probabilistically speaking). Everything happens as if we were working on the model depicted in Figure 4.4. Let us denote by $\mathcal{P}'_F$ the subset of $\mathcal{P}_F$ whose paths belong to this last chain. The expectation of our estimator will now be, on average,

$$\widehat{\gamma}'_n = \sum_{\pi \in \mathcal{P}'_F} p(\pi) \approx \varepsilon^2$$

and, concerning the variance, we will get, also on average,

$$\frac{1}{n}\left[\sum_{\pi \in \mathcal{P}'_F} \frac{p^2(\pi)}{\widetilde{p}(\pi)} - (\widehat{\gamma}'_n)^2\right] \approx \frac{1-q^2}{nq^2}\varepsilon^4.$$

This leads to a (mean) *observed* RE given by

$$\mathrm{RE} \approx \frac{1.96\sqrt{1-q^2}}{q\sqrt{n}},$$

which is independent of ε. The reader can check that these formulas are coherent with the numerical values observed for $\varepsilon \leq 10^{-4}$. So, this is a case where we know that the relative error of the IS technique used is not bounded when rarity increases, but where we numerically observe exactly the contrary. These problems are much harder to detect than the $(0, 0)$ interval case.

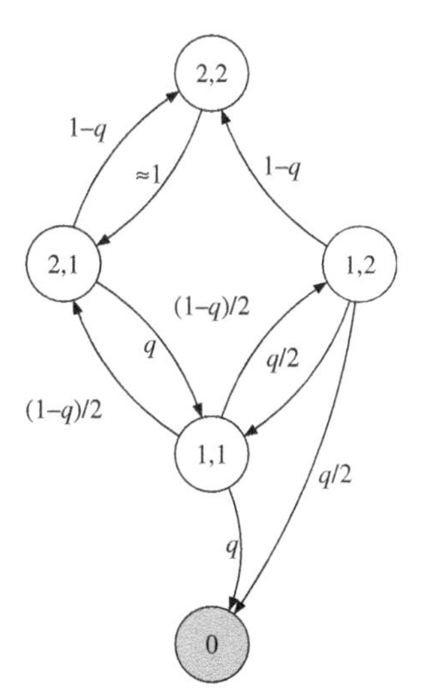

Figure 4.4 Model effectively 'seen' by the IS simulator when transitions in $\Theta(\varepsilon)$ are not observed during n trajectories of the chain.

The question therefore is: what is the validity of the proposed confidence interval? The techniques presented in previous chapters (IS and splitting) consist of different ways to speed up the rare event occurrence, but dealing with the confidence interval coverage might still be an issue.

Consider now the classical $M/M/1/B$ model, where we wish to evaluate $\gamma = \mathbb{P}(\text{reaching } B \text{ before } 0 \mid N(0) = 1)$ (this is Example 3 in Chapter 2), $N(t)$ being the number of customers at time t. More formally:

Example 3. Consider the discrete-time absorbing Markov chain X given in Figure 4.5 and define $\gamma = \mathbb{P}(X(\infty) = B \mid X(0) = 1)$. Observe that this is equal to $\mathbb{P}(\text{reaching } B \text{ before } 0 \mid N(0) = 1)$ in the $M/M/1/B$ queue with arrival rate λ and service rate μ, if $p = \lambda/(\lambda + \mu)$.

This is an elementary example in probability theory, and we know the answer: $\gamma = (r^{-1} - 1)/(r^{-B} - 1)$ if $r = \mu/\lambda = (1 - p)/p \neq 1$ (if $\lambda = \mu$, that is, if $p = 1/2$, then $\gamma = 1/B$). Suppose that we want to estimate γ using the standard simulator. In this example, rarity comes from the combination of values of the parameters p and B, the latter controlling the size of the model, a different situation than in previous example. A typical line of analysis here involves fixing p, varying B, and controlling rarity through $\varepsilon = 1/B$.

For instance, suppose that $p = 0.4$ and $B = 40$. The probability p is not very small, but combined with the size of the chain, we get $\gamma \approx 4.5 \times 10^{-8}$. Suppose we try an IS scheme by simply changing the probability p into some $\widetilde{p} > 1/2$, for instance, $\widetilde{p} = 0.9$, and that we simulate $n = 10^5$ paths of the chain. A standard implementation of this gave the approximate estimate 6.5×10^{-10} and estimated RE $\approx 40\%$. Without knowing the exact value, it is difficult to detect that there is a problem. If we refer to the previous ideas, we can imagine the user increasing B (i.e., increasing rarity), and looking at the behavior of the relative error. In Table 4.2 we provide some numerical results obtained by keeping everything fixed except B, which we increase.

The user may think that the RE looks bounded (while being pretty large), but observe that the exact value is never included in the observed confidence interval. We can suspect the same problem as before, even if the numerical behavior is not exactly the same. Looking again at the case of $B = 40$, it seems reasonable to try increasing the sample size. Keeping everything fixed except the sample size $n = 10^6$, we get an estimate of 1.62×10^{-9} with a relative error $\approx 39\%$. Again, we can suspect the same phenomenon as for the previous example.

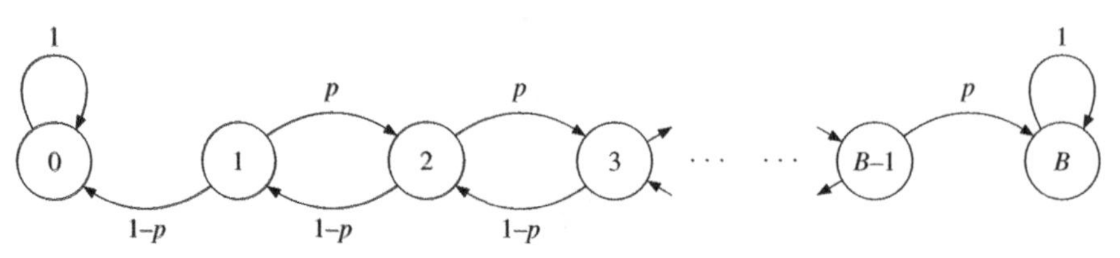

Figure 4.5 Discrete-time Markov chain X associated with the $M/M/1/B$ model, used to compute $\gamma = \mathbb{P}(X(\infty) = B \mid X(0) = 1)$.

Table 4.2 Estimating γ in the $M/M/1/B$ model, with $p = 0.4$, using $n = 10^5$ samples and the failure biasing change of measure with $\widetilde{p} = 0.9$, for different values of the buffer size B. The table gives the exact value of γ, its IS estimate and the estimated RE

B	γ	$\widehat{\gamma}^{\text{IS}}$	Est. RE
40	4.52e-08	6.50e-10	40%
50	7.84e-10	2.46e-12	80%
60	1.36e-11	2.34e-14	120%
70	2.36e-13	1.11e-17	45%
100	1.23e-18	2.21e-24	102%

We observe that in the family of IS methods where the new measure is state-independent (see Chapter 2), the best change of measure for this queue is known: it involves swapping the arrival and the service rate, or equivalently, using $\widetilde{p} = 1 - p$ in discrete time [9]. If we do so, we can check that things go smoothly, and that the estimators behave correctly (no anomaly in the behavior of the RE, nor on the observed likelihood ratio).

The aim of the rest this chapter is to discuss the following questions. How can we define a good estimator? Can it be good whatever the rarity? Can we detect in practice whether an estimate is good or not?

4.4.2 Normal approximation

In [15, 16], the bounded normal approximation (BNA) property is defined, asserting that the Gaussian approximation on which the confidence interval, and thus the confidence interval coverage, is based remains uniformly bounded as ε tends to 0. It finds its roots in the Berry–Esseen theorem which states that if ϱ is the third absolute moment of each of the n independently and identically distributed copies X_i of random variable X (with σ^2 its variance), Φ the standard normal distribution, $\widehat{\gamma}_n = n^{-1} \sum_{i=1}^{n} X_i$, $\widehat{\sigma}_n^2 = n^{-1} \sum_{i=1}^{n} (X_i - \widehat{\gamma}_n)^2$ and F_n the distribution of the centered and normalized sum $(\widehat{\gamma}_n - \gamma)/\widehat{\sigma}_n$, then there exists an absolute constant $a > 0$ such that, for each x and n,

$$|F_n(x) - \Phi(x)| \leq \frac{a\varrho}{\sigma^3 \sqrt{n}}.$$

Definition 6. *We say that $\widehat{\gamma}_n$ satisfies the bounded normal approximation property if ϱ/σ^3 remains bounded as $\varepsilon \to 0$.*

When this property is satisfied, only a fixed number of iterations are required to obtain a confidence interval having a fixed error no matter the level of rarity.

We could also look at a stricter condition, by making sure that the variance satisfies BRE. This is a stricter condition than BNA because it means looking at the fourth moment divided by the square of the variance, and, from the Jensen inequality, BRE for the variance implies BNA [17].

In [15], an example is given where BRE is satisfied, but not BNA, so the coverage of the confidence interval is not validated. BRE is therefore not sufficient alone to guarantee the robustness of a rare event estimator.

Note that BNA is a *sufficient* condition for coverage certification, and not a necessary one [15]. For instance, there exist more general versions of the Berry–Esseen bound (see [10]) for which the moment of order $2 + \delta$ is used (with $\delta > 0$) instead of the third moment, being then less restrictive. Note nonetheless that this is at the expense of the convergence rate to the Gaussian distribution, $O(n^{-\delta/2})$ instead of $O(n^{-1/2})$. A generalized version of BNA property could then be as follows:

Definition 7. *We say that* $\widehat{\gamma}_n$ *satisfies bounded normal approximation if there exists* $\delta > 0$ *such that* $E[|X - \gamma|^{2+\delta}]/\sigma^{2+\delta}$ *remains bounded as* $\varepsilon \to 0$.

4.4.3 Coverage function

In order to more directly investigate the actual coverage of confidence intervals for small values of ε when the number of replications is fixed, we can look at the so-called *coverage function* defined by L.W. Schruben in [13]. Define

$$R(\eta, \mathbb{X}) = \left(\widehat{\gamma}_n - c_\eta \frac{\widehat{\sigma}_n}{\sqrt{n}}, \widehat{\gamma}_n + c_\eta \frac{\widehat{\sigma}_n}{\sqrt{n}}\right)$$

as the confidence interval at confidence level η obtained using data

$$\mathbb{X} = (X_i)_{1 \le i \le n}$$

(i.e., $c_\eta = \Phi^{-1}((1 + \eta)/2)$). Under normality assumptions, it is easy to show that $\mathbb{P}[\gamma \in R(\eta, \mathbb{X})] = \eta$. Now define the random variable

$$\eta^* = \inf\{\eta \in [0, 1] : \gamma \in R(\eta, \mathbb{X})\}.$$

η^* should be uniformly distributed, that is,

$$F_{\eta^*}(\eta) = \mathbb{P}[\eta^* \le \eta] = \eta.$$

Not satisfying normal assumptions leads to two potential sources of error:

- $F_{\eta^*}(\eta) < \eta$ may lead to wrong conclusions (lower coverage),
- while if $F_{\eta^*}(\eta) > \eta$ the method is not efficient because a smaller sample size could have been used to get the desired coverage.

In order to investigate the actual coverage function, one can consider independent blocks of data $\mathbb{X} = (X_i)_{1 \le i \le n}$, producing independent realizations of η^*, from which its empirical distribution can be deduced. Reproducing it for different values of ε and looking at deviations from the uniform distribution illustrates the robustness of the estimator. This will be helpful below when discussing possible diagnostic-oriented approaches.

4.5 Diagnostics ideas

This section discusses the issue of detecting potential problems associated with the reliability of rare event confidence intervals. We will review three ideas to deal with these problems from a diagnostic point of view. First, we will see that using the fact that the expectation of the likelihood ratio equals unity in an importance sampling situation, a relevant idea a priori, is actually not of value when dealing with rare events. Second, we look at the possible numerical anomalies that can occur when looking at the behavior of the relative error as the system becomes rarer. A last diagnostic possibility is to make use of the covering function, that is, to look at how far the empirical coverage function is from the uniform.

4.5.1 Checking the value of the expected likelihood ratio

How should a test concerning the reliability of the confidence interval be constructed? A first thought would be to look at properties of the likelihood ratio when dealing with IS. Consider the expected value of a random variable X under probability measure $\mathbb{P}$. IS generates an unbiased estimator by using an IS measure $\widetilde{\mathbb{P}}$ with $d\widetilde{\mathbb{P}} \neq 0$ when $X d\mathbb{P} \neq 0$. Indeed, we then have $\widetilde{\mathbb{E}}[XL] = \mathbb{E}[X] = \gamma$ with $L = d\mathbb{P}/d\widetilde{\mathbb{P}}$ the likelihood ratio (see Chapter 2). We can then easily see that, with the more stringent condition that $d\widetilde{\mathbb{P}} \neq 0$ when $d\mathbb{P} \neq 0$, the expected value of the likelihood ratio is exactly 1. We will assume that this condition is satisfied for the remainder of this subsection, but remark that it is not true in general since we can construct unbiased IS estimates of γ for which $d\widetilde{\mathbb{P}} = 0$ when $X = 0$, such as the zero-variance change of measure.

This observation on the expected value of L could be thought to be a basis for designing a diagnostic: at the same time as we perform the computations needed to construct $\widehat{\gamma}_n^{\mathrm{IS}}$ and the associated confidence interval, we do the same for estimating $\widetilde{\mathbb{E}}[L]$. If the confidence interval obtained does not contain the exact value 1 under the condition that $d\widetilde{\mathbb{P}} \neq 0$ when $d\mathbb{P} \neq 0$, one has to exercise caution.

Why does this diagnostic not work in general? Let X be the the indicator function of a rare set A, that is, $\gamma = \mathbb{E}[\mathbb{1}(A)] = \widetilde{\mathbb{E}}[L\mathbb{1}(A)]$. Then, defining A^c as the complementary set of A and from the expected value of the likelihood ratio, we get

$$1 = \widetilde{\mathbb{E}}[L\mathbb{1}(A)] + \widetilde{\mathbb{E}}[L\mathbb{1}(A^c)] = \gamma + \widetilde{\mathbb{E}}[L\mathbb{1}(A^c)].$$

In order to use a test based on $\widetilde{\mathbb{E}}[L] = 1$, the variance of L has to be small enough so that we do not encounter the aforementioned problems where its variance is underestimated because the second moment has large values with small probability (so that those cases are not reached for a small to moderate sample size n) under $\widetilde{\mathbb{P}}$, and small vales with high probability. Therefore, $\widetilde{\mathrm{Var}}[L\mathbb{1}(A^c)]/n$ has to be small. This is unfortunately not the case in general because the IS scheme is designed to have a small variance for random variable $L\mathbb{1}(A)$, not for L. We indeed have $L \ll 1$, very small on A to be as close as possible to the value of γ for reducing the variance of the estimator, but $L \gg 1$ is likely to happen at some values in A^c.

The next example illustrates this problem of a properly designed IS scheme for which such a test is not going to work well.

Example 4. Consider a random walk $S_n = X_1 + \cdots + X_n$ on the integers or on the reals, starting from 0, where the X_i are independently and identically distributed with cumulative distribution function F. We wish to estimate the probability γ of reaching a level $b > 0$ before a level $-k < 0$. It is assumed that the random walk has a negative drift, meaning that the probability of going up, $X_i > 0$, is smaller that that of going down, $X_i < 0$, leading to a small value of γ. A class of IS measures, called *exponential twisting*, makes use of large deviations (see Chapter 5 for more details on the application of large-deviations theory to random walks). The exponentially twisted IS measure involves replacing dF by

$$d\widetilde{F} = \frac{e^{\theta x}}{M(\theta)} dF(x)$$

with $M(\theta) = \mathbb{E}[e^{\theta X_1}]$, the moment generation function of the X_i. It is known that there exists a θ^* for which $M(\theta^*) = 1$, and that this IS scheme yields logarithmic efficiency. Let us now investigate more closely the behavior of the likelihood ratio. On the paths for which b is reached before $-k$, we have $L \approx e^{-\theta^* b}$, while $L \approx e^{\theta^* k}$ on paths for which $-k$ is reached before, with probability of the order of $e^{-\theta^* k}$.

Now, if the sample size $n \ll e^{\theta^* k}$, we will therefore end up with an estimation of $\widetilde{\mathbb{E}}[L] \approx e^{-\theta^* b}$ because $-k$ is unlikely to be reached, with a small sample variance too. Worse, to get an estimate around 1 as expected, we need $n \gg e^{\theta^* k}$, which can take a longer time if $k > b$ than in the case of crude Monte Carlo, for which n has to be larger than $e^{\theta^* b}$ on the average.

Another interesting remark arises from looking at Example 1. In that case, estimating the expected value of the likelihood ratio always provides a confidence interval for this expectation that includes 1. For instance, using the same numerical values as in Table 4.1, varying ε in the same way, we get for the mean likelihood ratio under the IS measure almost the same confidence interval (0.99, 1.07). A test is therefore not able to detect the difficulty of estimating γ for that kind of example. It is actually the opposite problem than in previous example: it does not provide a warning even if it should, while for the random walk example, it provides an irrelevant warning.

4.5.2 Observed relative error behavior

In Section 4.4.1 we discussed the fact that, in some cases, the simulation technique can degrade when rarity increases, but the numerical values coming from the simulation run hides this phenomenon, leading the user to accept incorrect results. We illustrated this by means of an example where rarity is parameterized by ε, and where in spite of the fact that RE is unbounded as $\varepsilon \to 0$, we will *necessarily* observe that RE suddenly becomes essentially constant, that is, independent of ε. Of course, this is not a systematic fact appearing in these contexts, but it simply underlines the necessity of being careful if we observe this type of behavior.

Specifically, as a diagnostic rule, the idea is to simulate (with small sample sizes) the network for different values of ε larger than in the original problem, that is, to simulate much less rare events, with a small and fixed sample size, before running the 'real' simulation if things seem to go well. What is the incorrect behavior we try to detect? We look to see whether the estimated relative variance seems first to increase, then suddenly drops and stays fixed. This is due to the fact that important events (or paths, depending on the context) in terms of contribution to the variance (and to the estimation itself), are not sampled anymore. This trend of regular growth and sudden drop is likely to be a good hint of rare event problems.

An illustration of this was provided by Example 1, Table 4.1. If we use a sample size $n = 1000$, ten times smaller than that used in Table 4.1, we observe the same phenomenon, always coherent with the formulas given in Section 4.4.1. We observed the same behavior with different configurations.

This type of phenomenon does not appear in the case of the $M/M/1/B$ model presented in Example 3. Increasing B (see Table 4.2), we observe fluctuations of the relative error, but no trend similar to that exhibited before. The diagnostic can hardly be conclusive in this model, as it was in the first example. This illustrates that the tests of the section are traditional *rejection* tests.

Let us now consider another example [1]:

Example 5. Consider the discrete-time Markov chain X given in Figure 4.6 and define $\gamma = \mathbb{E}(K^{-1} \sum_{k=1}^{K} 1(X(k) = 1) \mid X(0) = 1)$.

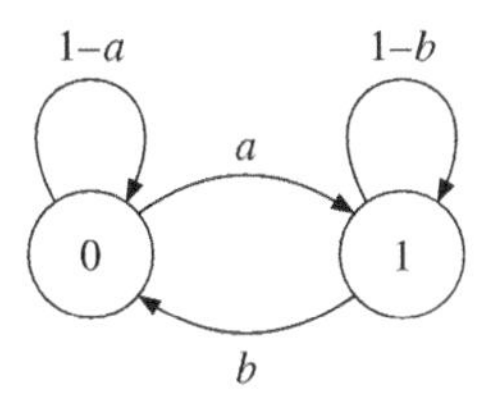

Figure 4.6 A two-state Markov chain. We look at the average fraction of interval $\{1, 2, \ldots, K\}$ where the chain is in state 1, starting at state 0 at time 0.

The exact value of γ is

$$\gamma = \frac{a}{a+b}\left[1-(1-a-b)\frac{1-(1-a-b)^K}{K(a+b)}\right].$$

Assume nevertheless that we use IS to estimate γ and let us consider the cases where

$$P = \begin{pmatrix} 0.99 & 0.01 \\ 0.1 & 0.9 \end{pmatrix}, \quad \widetilde{P} = \begin{pmatrix} 0.4 & 0.6 \\ 0.5 & 0.5 \end{pmatrix}.$$

We look for the value of γ when $K = 30$. We know that the exact answer is $\gamma \approx 6.713^{-2}$, and, of course, this is easy to estimate with the crude estimator. We used the proposed IS scheme for $n = 10^5$ samples, changing the seed of the pseudo-random number generator. We got the results shown in Table 4.3. We can observe here that over these six runs, the relative error fluctuates without a clear trend, but in five of the six cases, the exact value is outside the confidence interval (the case where the exact value is in the confidence interval is for seed).

If we increase the value of K, increasing the possible number of paths, we get the results given in Table 4.4. The RE exhibits no trend again, but we know that the estimations are horribly bad, and that the exact value is never inside the obtained confidence intervals.

In conclusion, for this test, involving checking the behavior of the relative error as a function of rarity, for small sample sizes, we observe good results when rarity is associated with transitions and the state space has a fixed topology, and no clear indications when rarity comes from the increasing length of good paths, as in the $M/M/1/B$ case.

Table 4.3 Estimating γ (whose value is 6.713^{-2}) in the two-state Markov chain of Example 5, with $a = 0.01$, $b = 0.1$, $\widetilde{a} = 0.6$, $\widetilde{b} = 0.5$, $K = 30$, for different seeds (using drand48( ) under Unix), for $n = 10^5$ samples

seed	314159	31415	3141	314	31	3
$\widehat{\gamma}_n^{\text{IS}}$	1.949e-04	1.583e-04	1.282e-04	2.405e-01	6.089e-05	1.021e-04
RE	9.636e-01	8.637e-01	1.667e00	1.958e00	1.263e00	9.686e-01

Table 4.4 Estimating γ in the two-state Markov chain of Example 5, with $a = 0.01$, $b = 0.1$, $\widetilde{a} = 0.6$, $\widetilde{b} = 0.5$, for different values of K, using $n = 10^5$ samples and the same seed (272, with drand48( ) under Unix)

K	30	50	70	90
γ	6.713e-02	7.624e-02	8.040e-02	8.274e-02
$\widehat{\gamma}_n^{\text{IS}}$	4.099e-05	1.748e-10	4.104e-14	2.554e-23
RE	8.723e-01	1.189e00	1.937e00	1.433e00

4.5.3 Diagnostic based on the coverage function

A last diagnostic possibility is to make use of Schruben's coverage function. The algorithm can be described as follows, as hinted in the description of the coverage function. Befored starting to run the (real) simulation, consider smaller sample sizes n and k values of ε, the rarity parameter, $\{\varepsilon_j;\ 1 \leq j \leq k\}$ with $\varepsilon_1 > \cdots > \varepsilon_k$. For each value ε_j, m independent blocks of data $\mathbb{X} = (X_i(\varepsilon))_{1 \leq i \leq n}$ are then used, giving independent realizations of η^*. From those m realizations, the empirical distribution of η^* can be obtained and compared with the uniform distribution. Then one can see if there is a trend: if the empirical distribution gets farther from the uniform as ε_j decreases, the current estimator can be considered as non-robust (unreliable), and a better one should be chosen. Otherwise, the estimator is not rejected by the test.

An important remark is that, in order to apply this diagnostic, the exact value (or at least an equivalent as $\varepsilon \to 0$) has to be known for computing η^*. As a consequence, the diagnostic can only be used for small instances of the problem. For example, when estimating the probability in an $M/M/1$ queue that the occupancy exceeds a value B (with B large), the exact value can be estimated for smaller values of B, and a trend can be derived. The same applies when dealing with a Markov chain on a small state space, but looking at long simulation times T (such as in Example 5 above), by looking at smaller values of T. The case of large Markov chains where rarity comes from rare transitions is more difficult. But one can try to construct a smaller instance of the model, with similar topology or properties (we do not care about the result being the same) and for which the exact value is known, and look to see whether the coverage function does not deviate as critical transition probabilities decrease. Our three examples describe those three situations and are detailed now.

Figure 4.7 displays the coverage function for the $M/M/1$ queue, looking at the probability that B is reached before returning to 0. This is done for sample sizes $n = 1000$ and repeated $k = 500$ times in order to get the empricial distribution function (smoothed thanks to interpolation). In the numerical experiments, $p = 0.3$ and we chose $\widetilde{p} = 0.5$ (not the optimal value, but to illustrate the behavior). It can be seen that as B increases, the coverage function gets worse and worse, so the estimator is not good here.

Look now at the case of the 2×2 matrix of Example 5, with transition matrices

$$P = \begin{pmatrix} 0.2 & 0.8 \\ 0.2 & 0.8 \end{pmatrix}, \quad \widetilde{P} = \begin{pmatrix} 0.5 & 0.5 \\ 0.5 & 0.5 \end{pmatrix}.$$

Again, we take $n = 1000$ and $k = 500$. From Figure 4.8, it can be checked that as the length K of the simulation path increases, the coverage function gets worse and worse, illustrating the bad estimation.

We close our numerical illustrations with Example 1. Figure 4.9 displays the empirical coverage function for different values of ε, still with $n = 1000$ and $k = 500$. Again, as ε decreases, the coverage function gets farther from the uniform, denoting an undesirable behavior.

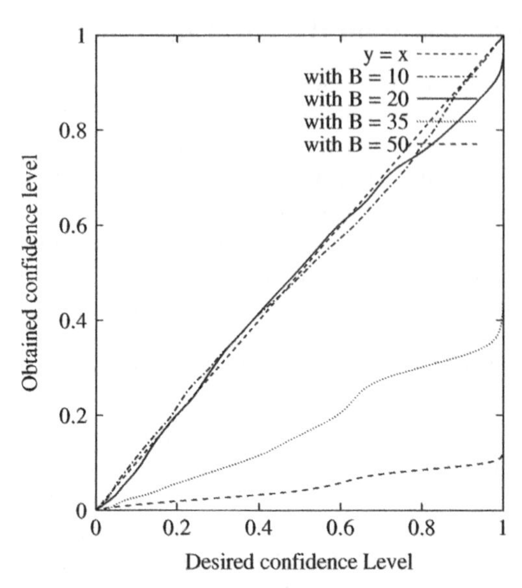

Figure 4.7 Coverage function for the simulation of the $M/M/1$ queue when looking at the probability of exceeding threshold B.

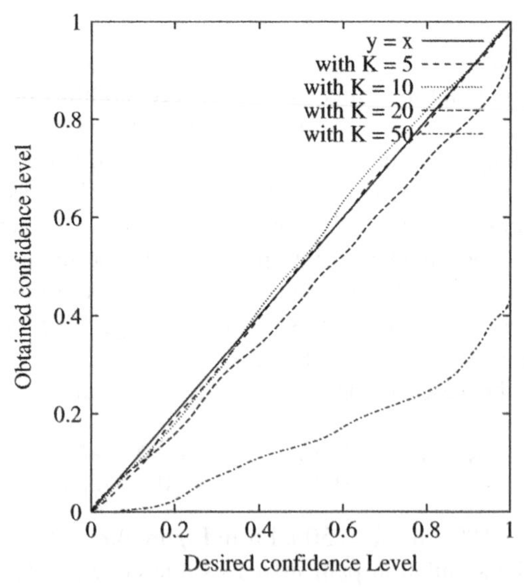

Figure 4.8 Coverage function for the simulation of a two-state Markov chain, as the length K of simulation increases.

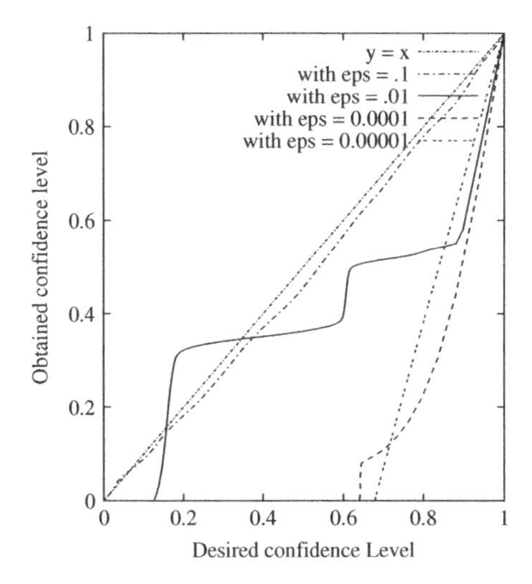

Figure 4.9 Coverage function for Example 1 and various values of ε.

4.6 Conclusions

We discussed the robustness properties (i.e., relative error behavior as the probability of the event goes to zero) we must require an estimator to satisfy when dealing with rare events in a simulation. Together with an overview of these properties and their relations, this chapter also underlined less known problems the practitioner may encounter in this area, concerning the reliability of the confidence interval. The typical situation is a numerical evaluation that can be taken as correctly done, while actually the output of the simulation procedure is completely off target. One of the aims of this chapter is to discuss possible ways of coping with this situation, and to suggest lines of research to derive rules that can be used as diagnostic methods mainly leading to a 'warning' signal along the lines of 'the results of the simulation are suspicious, take care'. But what if such a signal is received? The best advice is to try a different method, or a different parameterization of the technique used.

We concentrated our examples on importance sampling procedures, since this is the most used technique for rare event analysis, and also because it is the one most studied. Observe that the problems underlined here are related to rarity, not just to importance sampling. Also, the rules for detecting problems proposed in this chapter are valid for acceleration methods other than IS-based ones (except obviously for the use of the expected likelihood ratio).

References

[1] S. Andradóttir, D. P. Heyman, and T. J. Ott. On the choice of alternative measures in importance sampling with Markov chains. *Operations Research*, **43**(3): 509–519, 1995.

[2] H. Cancela, G. Rubino, and B. Tuffin. New measures of robustness in rare event simulation. In M. E. Kuhl, N. M. Steiger, F. B. Armstrong, and J. A. Joines, eds, *Proceedings of the 2005 Winter Simulation Conference*, pp. 519–527, 2005.

[3] P. Glasserman, P. Heidelberger, P. Shahabuddin, and T. Zajic. Multilevel splitting for estimating rare event probabilities. *Operations Research*, **47**(4): 585–600, 1999.

[4] P. W. Glynn and W. Whitt. The asymptotic efficiency of simulation estimators. *Operations Research*, **40**: 505–520, 1992.

[5] J. M. Hammersley and D. C. Handscomb. *Monte Carlo Methods*. Methuen, London, 1964.

[6] P. Heidelberger. Fast simulation of rare events in queueing and reliability models. *ACM Transactions on Modeling and Computer Simulation*, **5**(1): 43–85, 1995.

[7] P. L'Ecuyer, J. Blanchet, B. Tuffin, and P. W. Glynn. Asymptotic robustness of estimators in rare-event simulation. Technical Report 6281, INRIA, September 2007.

[8] M. K. Nakayama. General conditions for bounded relative error in simulations of highly reliable Markovian systems. *Advances in Applied Probability*, **28**: 687–727, 1996.

[9] S. Parekh and J. Walrand. Quick simulation of rare events in networks. *IEEE Transactions on Automatic Control*, **34**: 54–66, 1989.

[10] V. V. Petrov. *Limit Theorems in Probability Theory*. Oxford University Press, Oxford, 1995.

[11] T. D. Ross. Accurate confidence intervals for binomial proportion and Poisson rate estimation. *Computers in Biology and Medicine*, **33**(3): 509–531, 2003.

[12] J. S. Sadowsky. On the optimality and stability of exponential twisting in Monte Carlo estimation. *IEEE Transactions on Information Theory*, **IT-39**: 119–128, 1993.

[13] L. W. Schruben. A coverage function for interval estimators of simulation response. *Management Science*, **26**(1): 18–27, 1980.

[14] P. Shahabuddin. Importance sampling for the simulation of highly reliable Markovian systems. *Management Science*, **40**(3): 333–352, 1994.

[15] B. Tuffin. Bounded normal approximation in simulations of highly reliable Markovian systems. *Journal of Applied Probability*, **36**(4): 974–986, 1999.

[16] B. Tuffin. On numerical problems in simulations of highly reliable Markovian systems. In *Proceedings of the 1st International Conference on Quantitative Evaluation of SysTems (QEST)*, pp. 156–164. IEEE Computer Society Press, Los Alamitos, CA, 2004.

[17] B. Tuffin, W. Sandmann, and P. L'Ecuyer. Robustness properties in simulations of highly reliable systems. In *Proceedings of RESIM 2006*, University of Bamberg, Germany, October 2006.

Part II
APPLICATIONS

5

Rare event simulation for queues

José Blanchet and Michel Mandjes

This chapter describes state-of-the-art techniques in rare event simulation for queuing systems, the rare events under consideration being overflow probabilities, probabilities of extremely long delays, etc. We first consider a number of generic examples (and counterexamples) that are very useful in the queuing context. Then we systematically assess importance sampling for the cases of light-tailed input (where large-deviations arguments play a crucial role) and heavy-tailed input (where the change of measure is typically state-dependent). Other issues dealt with are: results under the many-sources scaling, estimation of the tail of the sojourn time distribution for processor-sharing queues, tandem and intree networks, and loss networks.

5.1 Introduction

In the theory of rare event simulation, queuing networks play a pivotal role. Queues are arguably the most widely used concept in applied probability, owing to their generic structure – there are applications in inventory, logistics, supply chains, communications networking, call centers, etc., while related models are heavily used in risk and insurance theory.

The estimation of rare event probabilities has become a topic of great importance in queuing theory. The main motivation for studying these rare events lies

Rare Event Simulation using Monte Carlo Methods Edited by G. Rubino and B. Tuffin

in the fact that the events under consideration relate to situations whose occurrence can be extremely costly for companies or (network) operators to face. Typical examples are: the claims an insurance company is faced with exceed its capital, extremely long delays in call centers that frustrate customers, or a performance collapse in a communication network. The common aspect in these examples is that the rare event under consideration can be rephrased as the event that the workload in an appropriately chosen queuing system exceeds a given high threshold, say B.

Rare events can be analyzed in various ways. When the underlying model is Markovian, numerical solution techniques can be used to compute the rare event probability of interest, say the overflow probability $\pi(B)$. The drawback of this approach is that this can be rather time-consuming, as the state space of the Markov chain tends to be large. Also, solution techniques of this type just give a numerical output, and do not provide any insight into the impact of the system parameters (such as the arrival rate, service rate, or service requirements) on $\pi(B)$. This motivates the search for *asymptotics*, that is, (explicit) functions $\varphi(\cdot)$ such that $\pi(B)/\varphi(B) \to 1$ as $B \to \infty$. These asymptotics can be derived in several ways, for instance by applying large-deviations theory [30]. A problem, however, is that we often lack error bounds; in other words, we do not know a priori from what B on the approximation $\pi(B) \approx \varphi(B)$ is accurate. This explains why one often resorts to a simulation-based analysis. An important complication is that, using naive simulation, the event under consideration occurs infrequently; consequently, it is time-consuming to obtain reliable estimates. Therefore, techniques have had to be developed to speed up the simulation. The present chapter is about simulation techniques for estimating small overflow probabilities. We remark that excellent textbook treatments of this subject are by Asmussen and Glynn [7, Ch. VI] and Bucklew [21, Ch. 11].

The most frequently used technique is *importance sampling* (IS); see Hammersley and Handscomb [41] for an early article, and also Hopmans and Kleijnen [44] and Bratley, Fox, and Schrage [19]. In IS one simulates the model under an alternative probability measure, and translates the simulation output back to the original measure by multiplying with a so-called likelihood ratio (in this chapter often abbreviated to just 'likelihood'). Crucial is the choice of the alternative measure, or, more particularly, one would like to find the measure that provides us with minimum variance. The systematic use of large-deviations theory, in the design of efficient IS algorithms for rare event estimation, started with an influential paper by Siegmund [70] in the 1970s. In many situations [25, 43, 68] this approach led to the identification of a new measure that is optimal according to some specified criterion. Large-deviations arguments also underlie the powerful heuristics of Parekh and Walrand [63], focusing on the estimation of the buffer overflow probability in rather general queuing networks. In the 1990s, however, substantial attention was paid to the limitations of this family of results: counterexamples by Glasserman *et al.* [38, 39] indicated that in specific important cases, the large-deviations based measure performs badly, and can even lead to infinite variance. Various approaches were proposed to circumvent these difficulties.

Another important aspect is that the above large-deviations based framework predominantly relates to queues with *light-tailed* input, which (roughly) means that $\pi(B)$ decays exponentially in B. The rare event of buffer overflow is due to many subsequent 'slightly rare' events ('conspiracy'). In many important queuing systems, however, overflow occurs in a fundamentally different way, namely, because of a single rare event ('catastrophe'). This is for instance the case in the classical $M/G/1$ queue with subexponential service requirements (including the Pareto distribution, as well as a subclass of the Weibull distribution); as a consequence $\pi(B)$ decays subexponentially as well, for instance as B^{α} or $\exp(-\sqrt{B})$. As a result of this crucially different behavior, new rare event simulation techniques have had to be developed for these queues with heavy-tailed input.

On an abstract level, this chapter focuses on the estimation of a probability p_n that goes to 0 as the 'rarity parameter' n goes to ∞. We mainly focus on applying IS: as described earlier in this monograph, one samples from a distribution $\mathbb{Q}$, different from the actual distribution $\mathbb{P}$, and weights the simulation output by the likelihood L, to be interpreted as a Radon–Nikodým derivative $d\mathbb{P}/d\mathbb{Q}$ to recover unbiasedness. A fundamental equality is that for any $\mathbb{Q}$, provided that mild regularity conditions hold, in obvious notation,

$$p_n = \mathbb{E}_n^{\mathbb{Q}}(LI),$$

where I is the indicator function of the event of interest, and the subscript n is added to emphasize the dependence on the rarity parameter. The number of runs needed to obtain an estimate with predefined accuracy (defined as the width of the confidence interval, divided by the estimate) is in general roughly proportional to the variance of the outcome of a single experiment. As a consequence, the quality of the estimator is strongly determined by $\mathbb{E}_n^{\mathbb{Q}}(L^2 I)$; recall that $\mathrm{Var}_n^{\mathbb{Q}}(LI) = \mathbb{E}_n^{\mathbb{Q}}(L^2 I) - p_n^2$. These considerations have led to the following optimality criterion.

Definition 1. *An alternative measure* $\mathbb{Q}$ *is said to be* asymptotically optimal *with respect to scaling parameter n if*

$$\lim_{n\to\infty} \frac{\log \mathbb{E}_n^{\mathbb{Q}}(L^2 I)}{\log \mathbb{E}_n^{\mathbb{Q}}(LI)} = 2.$$

Observe that, because of Jensen's inequality, the above limit is always less than or equal to 2; hence it remains to prove that it is also larger than or equal to 2.

In the light-tailed setting, it is often possible to prove that

$$\lim_{n\to\infty} \frac{1}{n} \log \mathbb{E}_n^{\mathbb{Q}}(LI) = \lim_{n\to\infty} \frac{1}{n} \log p_n = -\vartheta^{\star},$$

for some $\vartheta^\star > 0$. Then it remains to prove that

$$\lim_{n\to\infty} \frac{1}{n} \log \mathbb{E}_n^{\mathbb{Q}}(L^2 I) \leq -2\vartheta^\star.$$

If this is indeed true, then asymptotic optimality means, in this light-tailed setting, that the number of runs required grows at most subexponentially in n. A stronger notion of optimality is given in the following definition; for more background on such optimality notions, see [7, p. 159].

Definition 2. *An alternative measure* $\mathbb{Q}$ *is said to have* bounded relative error *with respect to scaling parameter n if*

$$\limsup_{n\to\infty} \frac{\mathrm{Var}_n^{\mathbb{Q}}(LI)}{p_n^2} < \infty.$$

Needless to say that the survey we present in this chapter is far from complete. The larger part of the chapter is on the light-tailed regime, in line with the literature of, say, the last 25 years. It is noted, however, that attention has recently shifted to the heavy-tailed regime. Also, we hardly pay attention to the use of cross-entropy techniques for identifying suitable IS measures, as this issue is taken care of by other chapters in the book. Finally, we mention that we do not address the use of *splitting techniques*, see for instance [7, Section V.5], and an interesting novel paper [29].

This chapter is organized as follows. In Section 5.2 we highlight a number of standard problems (namely that of a sample mean of independently and identically distributed (i.i.d.) random variables attaining a rare value, and that of a random walk with negative drift exceeding a large threshold) for which a (large-deviations based) change of measure was found that is asymptotically optimal. However, we also show that if the problem is changed slightly, such an alternative measure may perform badly; a few remedies are described. In Section 5.3 we apply the theory of Section 5.2 to queues with light-tailed input. Often we are able to explicitly bound the likelihood, conditional on the fact that the event of interest does indeed occur, which leads to explicit bounds on the variance of the estimator. Section 5.4 focuses on queues with heavy-tailed input. These have to be handled in a completely different manner; they are essentially based on the idea that the rare event happens due to a single random variable attaining an extreme value. Section 5.5 does not consider rarity because of a large buffer threshold that is supposed to be exceeded, but rather rarity because of the number of inputs growing large (and the system parameters scaled accordingly). Then, even if the input is heavy-tailed, we have exponential decay, such that the techniques of Section 5.2 can be applied again. The chapter concludes with a section on networks of queues: we briefly discuss fluid networks, Jackson networks, and loss networks.

5.2 Heuristics and caveats

In this section we consider a number of generic rare event estimation problems that play a crucial role in the setting of queuing networks. The last subsection points out that the underlying heuristics should be handled with care.

5.2.1 Sample means

A classical problem in probability concerns the distribution of *sample means*. With $(X_m)_{m\in\mathbb{N}}$ being a sequence of i.i.d. random variables, the nth partial sum is defined as $S_n := \sum_{m=1}^{n} X_m$, and the sample average is given by S_n/n. Due to laws of large numbers, we have that, under rather mild conditions, $S_n/n \to \mu := \mathbb{E}X_1$ almost surely, as $n \to \infty$. Also, the deviations in the neighborhood of the mean are described by the central limit theorem: this says that, as $n \to \infty$, $(S_n - n\mu)/(\sqrt{n}\mathrm{Var}X_1)$ converges to $\mathcal{N}(0, 1)$. But what can be said about tail probabilities of the type

$$p_n(a) := \mathbb{P}\left(\frac{S_n}{n} \geq a\right),$$

for some $a > \mu$? It is clear that this probability goes to zero for large n, but can we estimate it? We assume throughout that we are in the 'light-tailed regime', that is, the increments X_m are such that their moment generating function (mgf) $M(\vartheta) := \mathbb{E}\exp(\vartheta X_1)$ exists in a neighborhood of the origin (and hence all moments are finite).

A key result in this respect relates to an asymptotically exact asymptotic relation for $p_n(a)$. Using a change of measure, we can determine the asymptotics of $p_n(a)$ as follows. First consider the elementary formula,

$$\mathbb{P}(A) = \int_A L(\omega)d\mathbb{Q}(\omega), \quad \text{with } L(\omega) := \frac{d\mathbb{P}}{d\mathbb{Q}}(\omega),$$

provided that the Radon–Nikodým derivative is well defined. This entails that $p_n(A) = \mathbb{E}_n^{\mathbb{Q}}(LI)$, where $\mathbb{Q}$ is some alternative probability measure, L is the relative likelihood of $\mathbb{P}$ with respect to $\mathbb{Q}$, and I is the indicator function of the event $\{S_n/n \geq a\}$. The idea is to choose $\mathbb{Q}$ such that the event under consideration is not rare anymore, that is, we select a $\mathbb{Q}$ such that $\mathbb{E}^{\mathbb{Q}}X_1 = a$. Also, we choose $\mathbb{Q}$ such that our alternative model is still a random walk, but the increments are 'exponentially twisted'. More precisely: if $f(\cdot)$ is the density of the X_m under $\mathbb{P}$, then the density under $\mathbb{Q}$ belongs to the family

$$f_\vartheta(x) = f(x) \cdot \frac{e^{\vartheta x}}{M(\vartheta)}.$$

It is readily verified that, in order to make sure that $\mathbb{E}^{\mathbb{Q}}X_1 = a$, we should take $\vartheta := \vartheta(a)$, with $M'(\vartheta(a))/M(\vartheta(a)) = a$ (which we assume to exist); one can prove that $a > \mu$ entails that $\vartheta(a) > 0$.

Suppose that this change of measure is performed. Then we obtain

$$p_n(a) = \mathbb{E}_n^{\mathbb{Q}}(LI) = \mathbb{E}_n^{\mathbb{Q}}\left(\left(\frac{M(\vartheta(a))}{e^{\vartheta(a)X_1}} \cdots \frac{M(\vartheta(a))}{e^{\vartheta(a)X_n}}\right) \times I\right)$$
$$= (M(\vartheta(a)))^n \mathbb{E}_n^{\mathbb{Q}}\left(e^{-\vartheta(a)S_n} \times I\right),$$

where we have added the subscript n to stress dependence on n. Interestingly, we find the following corollary: as on $\{I = 1\}$ we have that $S_n \geq na$, this identity directly yields the celebrated *Chernoff bound* $p_n(a) \leq (M(\vartheta(a)) \cdot e^{-\vartheta(a)a})^n$, or, equivalently,

$$p_n(a) \leq e^{-nI(a)},$$

where $I(a) := \sup_\vartheta(\vartheta a - \log M(\vartheta)) = \vartheta(a)a - \log M(\vartheta(a))$ is the Legendre–Fenchel transform of $\log M(\vartheta)$. Let us now further analyze $\mathbb{E}_n^{\mathbb{Q}}\left(e^{-\vartheta(a)S_n} \times I\right)$. Notice that under the measure $\mathbb{Q}$ the event $\{S_n/n \geq a\}$ is not rare anymore; we are therefore essentially in a central limit theorem setting. In other words, one may approximate the distribution of S_n under $\mathbb{Q}$ by a $\mathcal{N}(na, nv)$ random variable, where

$$v := \mathrm{Var}^{\mathbb{Q}}(X_1) = \left.\frac{d^2}{d\vartheta^2}\log M(\vartheta)\right|_{\vartheta=\vartheta(a)}.$$

We thus obtain the following expression (n large) for $\mathbb{E}_n^{\mathbb{Q}}\left(e^{-\vartheta(a)S_n} \times I\right)$:

$$\int_{na}^{\infty} e^{-\vartheta(a)x} d\mathbb{Q}(S_n \leq x) \approx \int_{na}^{\infty} e^{-\vartheta(a)x} \cdot \frac{1}{\sqrt{2\pi nv}} \exp\left(-\frac{(x-na)^2}{2nv}\right) dx$$

$$\stackrel{\text{(i)}}{=} \int_0^{\infty} e^{-n\vartheta(a)a} \cdot e^{-n\vartheta(a)z\sqrt{v}} \left(\frac{n\sqrt{v}}{\sqrt{2\pi nv}}\right) e^{-\frac{1}{2}nz^2} dz$$

$$= e^{-n\vartheta(a)a}\left(\sqrt{\frac{n}{2\pi}}\right) \cdot e^{\frac{1}{2}n\vartheta^2(a)v} \int_0^{\infty} e^{-\frac{1}{2}n(\vartheta(a)\sqrt{v}+z)^2} dz$$

$$\stackrel{\text{(ii)}}{=} e^{-n\vartheta(a)a}\left(\sqrt{\frac{n}{2\pi}}\right) \cdot e^{\frac{1}{2}n\vartheta^2(a)v} \int_{\vartheta(a)\sqrt{nv}}^{\infty} e^{-\frac{1}{2}w^2} \frac{1}{\sqrt{n}} dw \stackrel{\text{(iii)}}{\approx} \frac{e^{-n\vartheta(a)a}}{\sqrt{2\pi nv}\vartheta(a)}.$$

where (i) is due to the transformation $x = n(a + z\sqrt{v})$, (ii) due to $w = \sqrt{n}(\vartheta(a)\sqrt{v} + z)$, whereas (iii) relies on $x \cdot \exp(\frac{1}{2}x^2)\int_x^{\infty} \exp(-\frac{1}{2}y^2)dy \to 1$ for $x \to \infty$ [71]. The above reasoning leads to the following result, first found by Bahadur and Rao [10].

Proposition 1. *$p_n(a)/\bar{p}_n(a) \to 1$ as $n \to \infty$, with*

$$\bar{p}_n(a) := e^{-nI(a)} \frac{1}{\sqrt{2\pi vn}\vartheta(a)}.$$

The only non-rigorous step in the above derivation (i.e., approximating the distribution of S_n by $\mathcal{N}(na, nv)$) can be made formal by using Berry-Esseen bounds, see [30, Section 3.7].

Thus we have found an asymptotically correct expression for $p_n(a)$, but, lacking precise error bounds, it is of course questionable whether the approximation $\bar{p}_n(a)$ is any good. Moreover, one might be interested in higher degrees of accuracy. For that reason, we may opt to estimate $p_n(a)$ by simulation. We may do so by performing IS with the X_m exponentially twisted with parameter $\vartheta(a)$. The variance of the estimator is proportional to $\text{Var}_n^{\mathbb{Q}}(LI) = \mathbb{E}_n^{\mathbb{Q}}(L^2 I) - p_n^2(a)$. With a reasoning analogous to the above

$$\frac{\mathbb{E}_n^{\mathbb{Q}}(L^2 I)}{e^{-2nI(a)}/(2\sqrt{2\pi v n}\vartheta(a))} \to 1 \quad \text{as } n \to \infty,$$

and as $p_n^2(a)$ is roughly of the form $\exp(-2nI(a))/n$ (i.e., decaying faster than $\mathbb{E}_n^{\mathbb{Q}}(L^2 I)$), we have that $\text{Var}_n^{\mathbb{Q}}(LI)$ is asymptotically equal to $\mathbb{E}_n^{\mathbb{Q}}(L^2 I)$. It also means that the number of experiments needed to obtain an estimate with predefined precision (in the sense that the ratio between the width of the confidence interval and the estimate should be below a given number) is roughly proportional to $\sqrt{n}$, where it would have been roughly proportional to $\sqrt{n}e^{-nI(a)}$ under the original measure–hence, a substantial variance reduction is achieved. (As an aside we mention that, bearing in mind that the length of each run is proportional in n, we obtain that under $\mathbb{Q}$ the simulation effort grows as $n\sqrt{n}$, for a given relative error.)

Proposition 2. *Let $\mathbb{Q}$ correspond to the X_m sampled from the exponentially twisted distribution with parameter $\vartheta(a)$. Then $\mathbb{Q}$ is asymptotically optimal for estimating $p_n(a)$, as $n \to \infty$.*

Remark 1. With the Bahadur–Rao result it follows immediately that it makes sense to perform IS with the $\vartheta(a)$-twisted distribution. To see this, recall that it is optimal to mimic as much as possible the zero-variance estimator. Now realize that, for an $m \in \{1, \ldots, n\}$,

$$\lim_{n\to\infty} \mathbb{P}(X_m = x \mid S_n \geq na) = f(x) \lim_{n\to\infty} \frac{\mathbb{P}(S_{n-1} \geq na - x)}{\mathbb{P}(S_n \geq na)}$$

$$= f(x) \frac{\lim_{n\to\infty}(e^{-nI(a)}\sqrt{2\pi n}\vartheta(a))}{\exp\left(-(n-1)I\left(\frac{na-x}{n-1}\right)\right)\sqrt{2\pi(n-1)}\vartheta\left(\frac{na-x}{n-1}\right)}$$

$$= f(x)e^{I(a)+(x-a)I'(a)} = f(x) \cdot \frac{e^{\vartheta(a)x}}{M(\vartheta(a))},$$

where the equality $I'(a) = \vartheta(a)$ is used (see [20, Ch. IV, Exercise 5]).

5.2.2 Supremum of a random walk

A second relevant example again concerns a random walk $(S_n)_{n\in\mathbb{N}}$ with $S_n := \sum_{m=1}^{n} X_m$. Assuming $\mathbb{E}X_1 < 0$, we consider the estimation of the rare event probability

$$q_b := \mathbb{P}\left(\sup_{n\in\mathbb{N}} S_n \geq b\right),$$

for b large. Again it is assumed that the X_m correspond to a light-tailed distribution; in particular, we assume the existence of a $\vartheta^\star > 0$ solving $\log M(\vartheta^\star) = 0$. Suppose that the measure $\mathbb{Q}$ corresponds to an exponential twist with parameter $\vartheta^\star$. As before, we have the following fundamental equality, with $N(b) := \inf\{n : S_n \geq b\}$:

$$q_b = \mathbb{E}_b^{\mathbb{Q}}(LI), \quad \text{where } L := e^{-\vartheta^\star S_{N(b)}}, I := 1_{\{N(b)<\infty\}}.$$

It is readily verified that $\mathbb{E}^{\mathbb{Q}}X_1 = M'(\vartheta^\star) > 0$ so that $N(b) < \infty$ with (under $\mathbb{Q}$) probability 1; hence $q_b = \mathbb{E}_b^{\mathbb{Q}}(e^{-\vartheta^\star S_{N(b)}})$. We immediately conclude that $q_b \leq e^{-\vartheta^\star b}$. Also, with $S_{N(b)} - b$ converging under $\mathbb{Q}$ to some non-negative random variable Z [3], assuming $M'(\vartheta^\star) < \infty$,

$$\lim_{b\to\infty} q_b/\bar{q}_b = 1, \quad \text{with } \bar{q}_b = e^{-\vartheta^\star b}\mathbb{E}^{\mathbb{Q}}e^{-\vartheta^\star Z}.$$

It is also easily checked that $\mathrm{Var}_b^{\mathbb{Q}}(LI)$ behaves as $e^{-2\vartheta^\star b}\mathrm{Var}^{\mathbb{Q}}(e^{-\vartheta^\star Z})$, so that in this case the number of runs needed to obtain a predefined precision (asymptotically) *does not grow with* b. We obtain the following result (cf. [70]).

Proposition 3. *Let $\mathbb{Q}$ correspond to the X_m sampled from the exponentially twisted distribution with parameter $\vartheta^\star$. Then $\mathbb{Q}$ has bounded relative error for estimating q_b, as $b \to \infty$.*

Remark 2. As before, we can argue that the exponential twist is a natural choice, as it mimics the zero-variance change of measure. To see this, observe that, for an $m \in \{1, \ldots, N(b)\}$,

$$\lim_{b\to\infty} \mathbb{P}\left(X_m = x \,\middle|\, \sup_{n\in\mathbb{N}} S_n \geq b\right) \approx f(x) \lim_{b\to\infty} \frac{q_{b-x}}{q_b} = f(x)e^{\vartheta^\star x}.$$

There is an alternative, partly heuristic though intuitively appealing, approach to find good IS distributions. This relies on the theory of *large deviations*, and in particular the concept of 'cost' that a random variable behaves as its exponentially twisted version. In the context of the random walk, this cost function is the Legendre–Fenchel transform $I(\cdot)$ that was introduced in Section 5.2.1: the cost incurred for the increments X_m behaving as their exponentially twisted counterpart (with parameter $\vartheta(a)$, such that the twisted mean is a) during T time units is $TI(a)$. Hence, in order to find the most likely ('cheapest') way for the random walk to reach level b, we have to solve the variational problem [49]

$$b \cdot \min_{a>0} \frac{I(a)}{a};$$

realize that b/a is (approximately) the time the change of measure should be active. This minimization shows that there is an interesting trade-off: if a is relatively small (just slightly larger than 0), then the cost per unit of time is relatively low, but the rare behavior has to be maintained over a long period of time; if, on the other hand, a is large, then it is extremely rare behavior, but the time required is small. The slope of the most likely path, say $a^\star$, finds the optimal trade-off between these two extremes. Interestingly, $I(a^\star)/a^\star$ equals $\vartheta^\star$; to see this, note that, as follows from differentiation with respect to a, $a^\star$ solves $I'(a)a = I(a)$, or, again using $I'(a) = \vartheta(a)$, $\log M(\vartheta(a)) = 0$.

A commonly used heuristic is to devise IS algorithms such that under the new measure $\mathbb{Q}$ an exponentially twisted version of the underlying process is used, where the twisting is such that on average the most likely path is followed.

5.2.3 Caveats

The above two examples suggest that knowledge of the most likely way the rare event is reached is sufficient to come up with an efficient IS procedure. One must be very careful, however. Below we will give an example where, at first glance, one would not expect any difference with the first class of models introduced above, but where there is in fact asymptotic optimality only under some additional condition (cf. [39]).

Suppose we wish to estimate

$$p_n(a,b) := \mathbb{P}\left(\frac{S_n}{n} \geq a \text{ or } \frac{S_n}{n} \leq b\right),$$

for $b < \mathbb{E}X_1 < a$. Assume without loss of generality that $I(a) < I(b)$, so the 'most likely point' in $(-\infty, b] \cup [a, \infty)$ is a, in that we have that $p_n(a)/p_n(a,b) \to 1$ as $n \to \infty$. In light of the above-mentioned large-deviations heuristic, it may seem to make sense to twist with $\vartheta(a)$. The question is: what are the variance properties of the resulting estimator? To this end, we have to analyze

$$\mathbb{E}_n^{\mathbb{Q}}(L^2 I) = \left(\mathbb{E}_n^{\mathbb{Q}}(e^{-2\vartheta(a)S_n} 1_{\{S_n \geq na\}}) + \mathbb{E}_n^{\mathbb{Q}}(e^{-2\vartheta(a)S_n} 1_{\{S_n \leq nb\}})\right) \times (M(\vartheta(a)))^{2n}.$$

We restrict ourselves to the exponential terms (in n); the polynomial terms (such as $1/\sqrt{n}$) do not have any impact here, as can be seen immediately. Relying on the arguments used in our proof of the Bahadur–Rao result, it is found that

$$\lim_{n\to\infty} \frac{1}{n} \log \left(\mathbb{E}_n^{\mathbb{Q}}(e^{-2\vartheta(a)S_n} 1_{\{S_n \geq na\}}) \times (M(\vartheta(a)))^{2n}\right) = -2I(a),$$

as might be expected. The second term is more difficult to handle, though. First observe that the overshoot over level nb (or, more precisely, the undershoot below level nb) will be modest, and therefore

$$\lim_{n\to\infty}\frac{1}{n}\log\left(\mathbb{E}_n^{\mathbb{Q}}(e^{-2\vartheta(a)S_n}1_{\{S_n\le nb\}})\times(M(\vartheta(a)))^{2n}\right)$$
$$=-2\vartheta(a)b+\lim_{n\to\infty}\frac{1}{n}\log\mathbb{Q}\left(\frac{S_n}{n}\le b\right)+2\log M(\vartheta(a)).$$

To find the decay rate of $\mathbb{Q}(S_n/n\le b)$, let $J(\cdot)$ be the Legendre–Fenchel transform under $\mathbb{Q}$, that is

$$J(b)=\sup_{\vartheta}\left(\vartheta b-\log\mathbb{E}^{\mathbb{Q}}e^{\vartheta X_1}\right)$$
$$=\sup_{\vartheta}(\vartheta b-\log M(\vartheta+\vartheta(a))+\log M(\vartheta(a)))$$
$$=-\vartheta(a)b+\sup_{\vartheta}((\vartheta+\vartheta(a))b-\log M(\vartheta+\vartheta(a)))+\log M(\vartheta(a))$$
$$=-\vartheta(a)b+I(b)+\log M(\vartheta(a)).$$

We obtain that

$$\lim_{n\to\infty}\frac{1}{n}\log\left(\mathbb{E}_n^{\mathbb{Q}}(e^{-2\vartheta(a)S_n}1_{\{S_n\le nb\}})\times(M(\vartheta(a)))^{2n}\right)$$
$$=-\vartheta(a)b-I(b)+\log M(\vartheta(a)).$$

Above we saw that, in order for the number of runs to grow subexponentially in n, $\mathrm{Var}_n^{\mathbb{Q}}(LI)$ should decrease (on an exponential scale) as fast as $p_n^2(a,b)$. In other words, we have this behavior if $2I(a)\le\vartheta(a)b+I(b)-\log M(\vartheta(a))$, or

$$I(a)+\vartheta(a)a\le I(b)+\vartheta(a)b.$$

The first question is whether under all circumstances, that is, for all $b<\mathbb{E}X_1<a$, this condition is met. This is clearly not the case. To study this more carefully, fix a, and find b for which the condition is satisfied. Observe that (i) the condition is met with equality at $b=a$; (ii) $I(b)+\vartheta(a)b$ goes to ∞ for $b\to-\infty$; (iii) the equation is not met at $b=\mu$, as $\vartheta(a)(a-b)>0$ and $I(\mu)=0$; and (iv) $I(b)+\vartheta(a)b$ is convex (as any Legendre–Fenchel transform $I(\cdot)$ is convex), so there are at most two points at which the condition is met with equality. Conclude that there is precisely one other intersection, attained for some $b^\star<\mathbb{E}X_1$. For all $b\le b^\star$, the number of runs needed grows subexponentially.

Example 1. Let the X_m be i.i.d. samples from an exponential distribution with mean 1. We wish to estimate $p_n(2,b)$ for some $b<1$. We have that $\theta(a)=1-1/a$ and $I(a)=a-1-\log a$. The inequality to be verified is therefore

$$3-\log 2\le\frac{3}{2}b-\log b,\quad\text{or}\quad\log b\le\frac{3}{2}b-3+\log 2.$$

A standard numerical search procedure yields that the condition is fulfilled when $0 < b < 0.119$.

Heuristically put, the problem of twisting with $\vartheta(a)$ is the following. Even in IS, we may have that $S_n \leq nb$, and if this happens the associated likelihood is huge. When b is smaller than the threshold $b^\star$ this happens so rarely that the performance of the estimator is still good, but for $b \in (b^\star, \mu)$ the variance blows up. In particular, in the latter case the number of runs needed to achieve a certain predefined precision is still exponential. For related examples, and more underlying theory, see [24, 38, 67, 68].

The phenomenon encountered in this setting plays a role in many situations. It shows that knowledge of the most likely way the rare event is reached is not sufficient for finding a fast IS algorithm. Essentially three remedies have been proposed:

- *Partitioning*. In the example above we can split the rare event of interest:

$$p_n(a,b) := \mathbb{P}\left(\frac{S_n}{n} \geq a\right) + \mathbb{P}\left(\frac{S_n}{n} \leq b\right).$$

  Both halves of the split can be efficiently simulated by using the procedure described in Section 5.2.1. This partitioning technique works when the rare event can be split into a finite number of subevents [18, 32].

- *Adaptive change of measure* (or *state-dependent change of measure*). The main problem arising in the above example is that the rare event can be reached through a path that is 'far away' from the most likely path, leading to a large likelihood. This effect can be avoided by updating the change of measure during the simulation run. Consider the example mentioned above. First X_1 is sampled from an exponentially twisted distribution with parameter $\vartheta_1 := \vartheta(a)$ (recall our assumption that $I(a) < I(b)$). Suppose that X_1 had value x_1; then in fact $\sum_{m=2}^n X_m$ should be larger than $na - x_1$ or smaller than $nb - x_1$. If $x_1 + (n-1)\mu \geq na$ or $x_1 + (n-1)\mu \leq nb$ then even 'average behavior' would lead to the rare event, and we draw X_2 from the original distribution. If not, then we sample X_2 using the twist $\vartheta_2 := \vartheta((na - x_1)/(n-1))$ if

$$I\left(\frac{na - x_1}{n-1}\right) < I\left(\frac{nb - x_1}{n-1}\right)$$

  (i.e., it is 'easier' for S_n to reach $[na, \infty)$ than $(-\infty, nb]$); otherwise we use the twist $\vartheta_2 := \vartheta((nb - x_1)/(n-1))$. The sampling procedure continues along these lines: X_3 is drawn using knowledge of x_1 and x_2, etc. In other words, during the simulation run the most likely point of entering $(-\infty, nb] \cup [na, \infty)$ may switch from na to nb. In this way the likelihood is better controlled, thus leading to a procedure with substantially better performance properties [35].

- *Random change of measure*. In this approach, we flip a coin, and the outcome decides from which twisted distribution we should sample. Let p be strictly between 0 and 1, and let our measure $\mathbb{Q}$ be such that we use with probability p (or $1-p$) exponential twisting with parameter $\vartheta(a) > 0$ (or $\vartheta(b) < 0$). The likelihood equals

$$L = \left(pe^{\vartheta(a)S_n} M(\vartheta(a))^n + (1-p)e^{\vartheta(b)S_n} M(\vartheta(b))^n\right)^{-1}.$$

  Now it is readily verified that

$$\mathbb{E}_n^{\mathbb{Q}}(L^2 I) \leq \mathbb{E}_n^{\mathbb{Q}}\left(\left(\frac{1_{\{S_n \geq na\}}}{pe^{\vartheta(a)S_n} M(\vartheta(a))}\right)^2 + \left(\frac{1_{\{S_n \leq nb\}}}{(1-p)e^{\vartheta(b)S_n} M(\vartheta(b))}\right)^2\right).$$

  It is immediate that the previous display is majorized by

$$\frac{1}{p^2}e^{-2nI(a)} + \frac{1}{(1-p)^2}e^{-2nI(b)},$$

  which has decay rate $-2I(a)$, so that the procedure is indeed asymptotically optimal. This random twist method was proposed by Sadowsky and Bucklew [68].

5.3 Queues: the light-tailed case

In this section we will show how the ideas presented in Section 5.2.2 can be used to sample large-deviations probabilities in queuing systems with light-tailed input. We start by addressing this issue for rare events related to long delays or a large workload; then we shift our attention to probabilities related to the number of customers in the queue. We conclude by treating a number of special subjects: queuing systems with Markov-modulated input, and queues operating under service disciplines that are more sophisticated than just first-in-first-out (such as processor sharing).

5.3.1 Long delays, large workload

Consider the class of so-called $GI/G/1$ queues: customers arrive according to some arrival process (with i.i.d. interarrival times $(A_m)_{m\in\mathbb{N}}$, distributed as some random variable A with mgf $\alpha(\vartheta) := \mathbb{E}\exp(\vartheta A)$) at some service resource; let them bring along i.i.d. service requirements $(B_m)_{m\in\mathbb{N}}$ distributed as some random variable B with mgf $\beta(\vartheta) := \mathbb{E}\exp(\vartheta B)$, and let the system be emptied at a constant rate of, say, 1. We assume that the input is light-tailed, that is, there is a $\vartheta > 0$ such that $\beta(\vartheta) < \infty$.

It is well known that waiting time D_n of the nth customer, not including the service time, has the same distribution as $M_n := \sup_{m=0,\ldots,n} S_m$, where $S_m := \sum_{k=1}^m X_k$ and $X_k := B_{k-1} - A_{k-1}$, and in particular, the steady-state waiting time

D is distributed as $M_\infty = \sup_{m\in\mathbb{N}} S_m$; here it is tacitly assumed that $\mathbb{E}X_1 < 0$. Importantly, this means that, in order to estimate the probability that $D > x$, we can use the algorithm presented in Section 5.2.2; $\vartheta^\star$ is the positive solution to the equation $\alpha(-\vartheta)\beta(\vartheta) = 1$.

Example 2. Consider the $M/M/1$ queue with exponential interarrival times (of mean $1/\lambda$) and exponential service times (of mean $1/\mu$), where $\varrho := \lambda/\mu < 1$). It is easily computed that $\vartheta^\star = \mu - \lambda > 0$. The twisted distribution of the increments X_m has moment generating function

$$\left(\frac{\lambda}{\lambda+\vartheta+\vartheta^\star}\bigg/\frac{\lambda}{\lambda+\vartheta^\star}\right)\times\left(\frac{\mu}{\mu-(\vartheta+\vartheta^\star)}\bigg/\frac{\mu}{\mu-\vartheta^\star}\right)=\frac{\mu}{\mu+\vartheta}\cdot\frac{\lambda}{\lambda-\vartheta}.$$

This means that in IS one simulates the interarrival times from an exp(μ) distribution and the service requirements from an exp(λ) distribution, that is, the roles of λ and μ are interchanged. This, inherently unstable, queuing system has apparently optimal variance reduction properties.

The tail probabilities of the steady-state workload W can be simulated similarly. With $A(s,t)$ denoting the amount of work generated in the interval $[s,t)$, Reich's formula states that W has the same distribution as $\sup_{t\geq 0}(A(-t,0)-t)$. As opposed to processes of the type S_n, this is not a discrete-time random walk, but such a process can be embedded. To this end, let us restrict ourselves for convenience' sake to the class of $M/G/1$ queues. Then, with T_n the epoch of the nth arrival *after* 0 (use reversibility!),

$$\sup_{t\geq 0}(A(-t,0)-t) \stackrel{\mathcal{D}}{=} \sup_{n\in\mathbb{N}}(A(0,T_n)-T_n) \stackrel{\mathcal{D}}{=} \sup_{n\in\mathbb{N}}\sum_{m=1}^{n} X_m,$$

with $X_m := B_m - (T_m - T_{m-1})$. The equation $M(\vartheta^\star) = 1$ becomes $\lambda(\beta(\vartheta^\star) - 1) = \vartheta^\star$. Then the procedure described above can be applied.

5.3.2 Large number of customers

We saw above that efficient simulation of long delays on one hand, and a large workload on the other hand, can be done with the *same* change of measure. Interestingly, to achieve a large number of customers this change of measure performs excellently, too. In the setting of a $GI/G/1$ queue operating under the first-come-first-serve discipline, denote by $\zeta(K)$ the probability that within a busy period the number of customers exceeds K; later we extend the analysis to the steady-state probability $\pi(K)$ of having K or more customers in the system. In [54] it was proven that

$$\lim_{K\to\infty}\frac{1}{K}\log\zeta(K) = \lim_{K\to\infty}\frac{1}{K}\log\pi(K) = -\log\beta(\vartheta^\star).$$

We now show that the importance sampling algorithm proposed above does indeed lead to a subexponential number of runs needed to obtain an estimate with predefined precision. To this end, define

$$T := \inf\left\{n : \sum_{m=1}^{n} A_i > \sum_{m=1}^{n} B_i\right\}, \quad T(K) := \inf\left\{n > K : \sum_{m=1}^{n} A_i < \sum_{m=1}^{n-K} B_i\right\}.$$

Clearly, $\zeta(K) = \mathbb{P}(T(K) < T)$. Also, $\zeta(K) \leq \mathbb{E}^{\mathbb{Q}}(L \mid I = 1)$, with I being the indicator function of $\{T(K) < T\}$. Consider L on $\{I = 1\}$. Let C be the number of customers whose service has started before the overflow; observe that we have scheduled $T(K) - 1$ interarrival times. Hence

$$L = \prod_{m=1}^{T(K)-1} \left(e^{\vartheta^\star A_m}\alpha(-\vartheta^\star)\right) \prod_{m=1}^{C} \left(e^{-\vartheta^\star B_m}\beta(\vartheta^\star)\right),$$

which can be simplified to

$$\exp\left(\vartheta^\star \sum_{m=1}^{T(K)-1} A_m - \vartheta^\star \sum_{m=1}^{C} B_m\right) \times (\beta(\vartheta^\star))^{C-T(K)+1}.$$

The difference between the number of customers who entered the system and those who have been served (the backlog) exceeds (at an overflow) K, so C cannot be larger than $T(K) - K$. Apart from that, we have $\sum_{m=1}^{T(K)-1} A_i < \sum_{m=1}^{C} B_m$. We conclude that, under $\mathbb{Q}$,

$$LI \leq (\beta(\vartheta^\star))^{1-K}.$$

This also yields $\mathbb{E}_K^{\mathbb{Q}}(L^2 I) \leq (\beta(\vartheta^\star))^{2-2K}$. An immediate consequence is the following.

Proposition 4. *Let $\mathbb{Q}$ correspond to the A_m (B_m) sampled from the exponentially twisted distribution with parameter $-\vartheta^\star$ ($\vartheta^\star$). Then $\mathbb{Q}$ is asymptotically optimal for estimating $\zeta(K)$, as $K \to \infty$.*

Recall that $\pi(K)$ denotes the steady-state probability that there are more than K customers in the queue. Using renewal arguments, it is clear that $\pi(K)$ equals the ratio of the mean number of customers seeing K or more customers upon arrival during a busy cycle, say $\mathbb{E}N(K)$, and the mean number of customers arriving during a busy cycle, say $\mathbb{E}N$. The denominator does not involve a rare event, and its estimation is therefore standard. The numerator, however, corresponds to a rare event, particularly for K large. Observe that one cannot estimate $\mathbb{E}N(K)$ by sampling under $\mathbb{Q}$ (as defined above) all the time, as there is a positive probability that under $\mathbb{Q}$ the busy cycle does not end. This has led to the idea of 'measure-specific dynamic importance sampling' [40]. This concept means that one simulates the queue under $\mathbb{Q}$ (and updates the likelihood) until either the busy cycle has ended or the number of customers has exceeded K. In the latter case one 'turns off' the IS: the remainder of the busy cycle is simulated under the original measure $\mathbb{P}$ (the fact that the system is stable entails that the busy cycle will end under $\mathbb{P}$); evidently, in this second part of the busy cycle the likelihood is not updated.

As an aside, we mention that this type of IS can be used very well to estimate the probability that, in the context of a $GI/G/1$ queue, the length of the busy period, say P, exceeds some large value x. Let us, for convenience' sake, restrict ourselves to the $M/G/1$ queue; the arrival rate of jobs is λ. Using the large-deviations heuristics introduced in Section 5.2.2 one can guess a good alternative measure $\mathbb{Q}$, as follows. The heuristic indicates that the optimal path is such that one generates traffic at a constant rate for a while. If one generates traffic at a rate a during a time interval of length s, and with μ denoting the drift $\lambda\,\mathbb{E}B-1$, the decay rate $\vartheta^\star$ solves the variational problem $\min_{a\rangle 0} I(a)s$, where s is such that $as+\mu(x-s)=0$ (because then the busy period does indeed end at x; hence $s=-\mu x/(a-\mu)$). Here $I(a)$ is the Legendre–Fenchel transform corresponding to the (net) amount of work generated in one unit of time; it is readily verified that the corresponding mgf equals

$$M(\vartheta)=e^{-\vartheta}\sum_{k=0}^{\infty}e^{-\lambda}\frac{\lambda^k}{k!}(\beta(\vartheta))^k=\exp(-\vartheta-\lambda+\lambda\beta(\vartheta)).$$

We thus find that the decay rate should equal

$$(-\mu)\cdot x\cdot\inf_{a\geq 0}\frac{I(a)}{a-\mu}$$

(recall that $-\mu$ is a positive number). It can be checked that the optimum is attained for $a^\star=0$ (this follows immediately from the convexity of $I(\cdot)$ and $I(\mu)=0$); in other words, the optimal path is such that the queue is 'almost empty' all the time between 0 and x. It can be checked that twisting the distribution of the interarrival times with $-\vartheta(a^\star)$ and the distribution of the job sizes with $\vartheta(a^\star)$ (which is a change of measure that converts the model into a system with load 1; check!) indeed yields an asymptotically optimal procedure. Details are omitted here, but see also Section 5.3.4 on sojourn times in the processor-sharing queue.

5.3.3 Queues with Markov-modulated input

The $GI/G/1$ queues described above do not readily lend themselves to modeling queues with correlated input. A model that does incorporate this feature is the queue with so-called Markov-modulated fluid input. The simplest model of this kind is as follows. Consider an irreducible continuous-time Markov process $X(\cdot)$, living on a finite state space $\{1,\ldots,d\}$, with generator matrix $\Lambda=(\lambda_{ij})_{i,j=1}^{d}$ and π denoting its (unique) steady-state distribution. Then traffic is generated at a constant rate $r_i \geq 0$ if the Markov process is in state i, and this traffic is transmitted into a queue. The queue is emptied at a constant service rate C; to ensure stability it is assumed that the mean input rate is smaller than the service rate: $\sum_{i=1}^{d} \pi_i r_i < C$.

There are algorithms for finding the steady-state buffer content distribution of this system [2, 50], but these involve the solution of an eigensystem and, in addition, a linear system. As the corresponding numerical computations can be time-consuming and error-prone (particularly the linear system), one may want to resort to estimating overflow probabilities relying on IS.

First recall that the steady-state workload is distributed as $\sup_{t\geq 0}(A(-t,0) - Ct)$, with

$$A(s,t) = \int_s^t r_{X(u)} du,$$

or $\sup_{t\geq 0} \bar{A}(-t,0)$, with $\bar{A}(s,t) := A(s,t) - C \cdot (t-s)$. We are interested in

$$\gamma(B) := \mathbb{P}\left(\sup_{t\geq 0} \bar{A}(-t,0) \geq B\right).$$

As the process is assumed to be in steady state at time 0, we can let it start in state i with probability π_i. To get a good idea of a suitable change of measure, let us first consider the values of $A(-t,0)$ at the epochs when the Markov process (started at time 0 in state i, but *looking backwards in time*) enters this state i again. With T_m the epoch of the mth such visit, we let X_m denote $\bar{A}(T_m, T_{m-1})$. These increments X_m are i.i.d.; the theory of Section 5.2.2 suggests that we have to solve the equation $\mathbb{E}\exp(\vartheta X_1) = 1$. As we are looking backwards in time, we have to work with the time-reversed generator $\bar{\Lambda} = (\bar{\lambda}_{ij})_{i,j=1}^{d}$, where $\bar{\lambda}_{ij} := \lambda_{ji}\pi_j/\pi_i$ for $i \neq j$ (and $\bar{\lambda}_{ii} = \lambda_{ii}$; verify that this does lead to a proper generator matrix!). In general, we cannot find a closed-form expression for $\mathbb{E}\exp(\vartheta X_1)$, but we can write, with $\lambda_i := -\lambda_{ii}$,

$$\mathbb{E}e^{\vartheta X_1} = \frac{\sum_{j\neq i} \bar{\lambda}_{ij} x_j}{\lambda_i - (r_i - C)\vartheta}$$

where $x_j = x_j(\vartheta)$ is the mgf of the net amount of fluid generated starting in state j until absorption in state i. Also, bearing in mind that, as we want to find $\vartheta^\star$, we can a priori impose that $x_i = 1$,

$$x_j = \frac{\bar{\lambda}_{ji} + \sum_{k\neq i, k\neq j} \bar{\lambda}_{jk} x_k}{\lambda_j - (r_j - C)\vartheta}.$$

Now we have to find the $\vartheta^\star$ such that $\mathbb{E}\exp(\vartheta^\star X_1) = 1$. It is readily verified that the above equations can be rewritten as follows: with $x_i \equiv 1$, we get for $j = 1, \ldots, d$,

$$-(r_i - C)\vartheta^\star x_j = \sum_{k=1}^{d} \bar{\lambda}_{jk} x_k.$$

With $R := \text{diag}\{r\}$, and assuming that $r_i \neq C$ for all $i = 1, \ldots, d$, this set of equations can be rewritten as the eigensystem

$$-\vartheta^\star x = (R - CI)^{-1}\bar{\Lambda}x.$$

Since the x_j represent mgfs, it is immediate that they are positive; this could also be seen by using the Perron–Frobenius theorem. It is also clear that the value of $\vartheta^\star$ does not depend on the specific state i.

Now that we know how to find the twist parameter $\vartheta^\star$, we can study the following change of measure. Let $\mathbb{Q}$ correspond to a Markov-modulated fluid process in which the generator $M = (\mu_{ij})_{i,j=1}^d$ is used rather than $\bar{\Lambda}$, where $\mu_{ij} := \bar{\lambda}_{ij}x_j/x_i$. Then

$$\mu_i := -\mu_{ii} = \sum_{j\neq i}\mu_{ij} = \sum_{j\neq i}\bar{\lambda}_{ij}\frac{x_j}{x_i} = \lambda_i - (r_i - C)\vartheta^\star$$

(where it is noted that the last equation has the intuitive explanation that, under this new measure, the process spends per visit more time in states i with $r_i > C$ and less time in states with $r_i < C$; it can even be verified that under $\mathbb{Q}$ the drift becomes positive). Now consider the likelihood L of a path, generated under $\mathbb{Q}$, that is such that $\sup_{t\geq 0}\bar{A}(-t,0) > B$. Let J_m be the state of the Markov process after the mth jump, T_m the time spent there, and N the number of jumps until the process exceeds B. Then it is easily verified that the likelihood is given by

$$\begin{aligned}
L &= \frac{\bar{\lambda}_{I_0I_1}\cdots\bar{\lambda}_{J_{N-1}J_N}}{\mu_{I_0I_1}\cdots\mu_{J_{N-1}J_N}}\cdot\frac{\lambda_{J_N}}{\mu_{J_N}}\cdot\exp\left(-\sum_{m=0}^{N}(\bar{\lambda}_{J_m}-\mu_{J_m})T_m\right)\\
&= \frac{x_{J_0}}{x_{J_N}}\frac{\lambda_{J_N}}{\mu_{J_N}}\cdot\exp\left(-\sum_{m=0}^{N}(\bar{\lambda}_{J_m}-\mu_{J_m})T_m\right)\\
&\leq k\exp\left(-\sum_{m=0}^{N}(\bar{\lambda}_{J_m}-\mu_{J_m})T_m\right),
\end{aligned}$$

where

$$k := \max_{i,j}\frac{x_i}{x_j}\frac{\lambda_j}{\mu_j} < \infty;$$

notice that our specific choice for $\mathbb{Q}$ is such that the likelihood has a nice and manageable form, as almost all x_i-terms vanish (cf. [9, 57]). Observe that on $\{I = 1\}$ (i.e., in overflow cycles) we have that $L \leq k\exp(-\vartheta^\star B)$, so that $\gamma(B) \leq k\exp(-\vartheta^\star B)$. In fact, it can be proven that $(1/B)\log\gamma(B) \to -\vartheta^\star$ as $B \to \infty$, so that asymptotic optimality directly follows.

Proposition 5. *Let $\mathbb{Q}$ correspond to the $X(\cdot)$ generated under the transition rates μ_{ij}. Then $\mathbb{Q}$ is asymptotically optimal for estimating $\gamma(B)$, as $B \to \infty$.*

Also in this context large-deviations heuristics can be used to find the change of measure $\mathbb{Q}$. The cost per unit of time of behaving as the exponentially twisted version (such that the 'new' input rate becomes a) is

$$I(a) = \sup_{\vartheta}(\vartheta a - N(\vartheta)), \quad \text{where } N(\vartheta) := \lim_{t\to\infty} \frac{1}{t} \log e^{\vartheta A(t)};$$

interestingly $N(\vartheta)$ can be alternatively characterized as $\vartheta C(\vartheta)$, with $C(\vartheta)$ being the largest (real) eigenvalue of $R + \Lambda/\vartheta$ [49]. Then the 'optimal slope' $a^\star$ follows from minimizing $I(a)/(a - C)$ over all $a > C$ (and, in addition, $\vartheta^\star = I(a^\star)/(a^\star - C)$). As shown in [57], yet another way of characterizing the decay rate as a variational problem is by using the concept of relative entropy of M with respect to $\bar{\Lambda}$. Informally, by analogy with the i.i.d. case, the cost of $X(\cdot)$ behaving as a Markov process with generator matrix M for one unit of time, with ϱ being the invariant measure of M, is given by a form of Kullback–Leibler distance:

$$I(M \mid \bar{\Lambda}) := \sum_{i=1}^{d} \varrho_i \sum_{j\neq i} \mu_{ij} \log \frac{\mu_{ij}}{\bar{\lambda}_{ij}} + \sum_{i=1}^{d} \varrho_i (\mu_{ii} - \bar{\lambda}_{ii}).$$

To find the rates μ_{ij} that should be used in the IS, one has to minimize

$$b \cdot \frac{I(M \mid \bar{\Lambda})}{\sum_{i=1}^{d} \varrho_i r_i - C}$$

over all generator matrices M such that $\sum_{i=1}^{d} \varrho_i r_i > C$. It can be verified that this leads to the same change of measure $\mathbb{Q}$ as derived above.

5.3.4 Sojourn times in processor-sharing queues

In the previous subsection, the rare event under consideration related to the queue attaining an unusually high level, in terms of either workload or the number of customers. Importance sampling, however, can also be used to analyze the efficacy of scheduling disciplines that are more sophisticated than just first-come-first-serve. Let us, for example, consider the estimation of tail probabilities of the sojourn times in a queue that is operating under the processor-sharing discipline; for ease of exposition we look specifically at the case of Poisson input, that is, the $M/G/1$-PS queue.

Let jobs arrive according to a Poisson process with rate λ, and let the jobs be independently and identically distributed as a random variable B with mgf $\beta(\vartheta)$ (again assumed to exist for some positive ϑ); also $\varrho := \lambda\beta'(0) < 1$. It has been proven [59] that the probability that an arbitrary sojourn time exceeds x, say $\delta(x)$, obeys, under weak conditions on the tail of the job-size distribution,

$$\lim_{x\to\infty} \frac{1}{x} \log \delta(x) = -\vartheta^\star - \lambda + \lambda\beta(\vartheta^\star),$$

where $\vartheta^\star$ solves $\lambda\beta'(\vartheta^\star) = 1$. Interestingly, this decay rate coincides with the decay rate of the probability that a busy period exceeds x. In view of this, it makes sense to consider IS under the measure $\mathbb{Q}$, in which λ is replaced by $\lambda^\star := \lambda\beta(\vartheta^\star)$, and that the jobs have mgf $\beta(\vartheta + \vartheta^\star)/\beta(\vartheta^\star)$ rather than $\beta(\vartheta)$; notice here that the load under $\mathbb{Q}$ equals

$$\varrho^{\mathbb{Q}} := \lambda^\star \mathbb{E}^{\mathbb{Q}} B = \lambda\beta(\vartheta^\star) \cdot \frac{\beta'(\vartheta^\star)}{\beta(\vartheta^\star)} = 1.$$

The sojourn time of a 'tagged' job (say that it arrives at time 0, seeing the system in equilibrium) exceeding x, is the result of three factors:

(i) the jobs already present at time zero (both their number, say Q_0, and their residual sizes, say $\bar{B}_1, \ldots, \bar{B}_{Q_0}$; it is a well-known result that under the original distribution Q_0 has a geometric distribution with success probability ϱ, whereas the $\bar{B}_i$ have the so-called residual life distribution of the B_i);

(ii) the size B_0 of the tagged job itself;

(iii) the jobs arriving between 0 and x.

Suppose that we indeed change the arrival rate and distribution of the jobs arriving in $(0, x)$ as described above, and that we twist the distribution of Q_0 with ϑ_{Q_0}, the residual jobs present at time 0 with $\vartheta_{\bar{B}}$, and the tagged job with ϑ_{B_0}. With $W := \sum_{n=1}^{Q_0} \bar{B}_n$ denoting the workload at time 0, it can be seen that the likelihood L is given by $L_1L_2L_3$, where L_1 is the contribution due to the jobs present at time 0, L_2 due to the tagged jobs, and L_3 due to the job arriving at time 0:

$$L_1 := \left(e^{-\vartheta_{Q_0}}\mathbb{E}e^{\vartheta_{Q_0}Q_0}\right) \cdot \left(e^{-\vartheta_{\bar{B}}W}\mathbb{E}e^{\vartheta_{\bar{B}}\bar{B}}\right),$$

$$L_2 := e^{-\vartheta_{B_0}}\beta(\vartheta_{B_0}),$$

$$L_3 := \left(e^{(\lambda^\star-\lambda)x}\left(\frac{\lambda}{\lambda^\star}\right)^{N(x)}\right)\left(e^{-\vartheta^\star A(x)}\left(\beta(\vartheta^\star)\right)^{N(x)}\right) = e^{(\lambda^\star-\lambda)x-\vartheta^\star A(x)},$$

where $A(x)$ ($N(x)$) is the amount of work (number of customers) arriving in $(0, x)$.

The comparison with the busy period asymptotics suggests that we may take $\vartheta_{Q_0} = \vartheta_{\bar{B}} = \vartheta_{B_0} = 0$. Using the inequality $W + B_0 + A(x) > x$ under $\{I = 1\}$ (a necessary condition is that the busy period has not ended), it is not hard to see that then

$$\begin{aligned}\mathbb{E}^{\mathbb{Q}}(L^2 I) &\le e^{-2(\lambda-\lambda^\star)x-2\vartheta^\star A(x)}\mathbb{E}^{\mathbb{Q}}e^{2\vartheta^\star(B_0+W)} \\ &= e^{-2(\lambda-\lambda^\star)x-2\vartheta^\star A(x)}\beta(2\vartheta^\star)\mathbb{E}^{\mathbb{Q}}e^{2\vartheta^\star W},\end{aligned}$$

but neither $\beta(2\vartheta^\star)$ nor $\mathbb{E}^{\mathbb{Q}}e^{2\vartheta^\star W}$ is necessarily finite. So, interestingly, one needs to twist Q_0, the $\bar{B}_n$, and B_0. Consider, for instance, the choice $\vartheta_{Q_0} = \log\beta(\vartheta^\star)$ and $\vartheta_{\bar{B}} = \vartheta_{B_0} = \vartheta^\star$; then, on $\{I = 1\}$, for some finite constant k,

$$L \le ke^{-(\lambda-\lambda^\star)x-\vartheta^\star A(x)},$$

so that we have asymptotic optimality. For instance in the $M/M/1$ case (with mean job size μ^{-1}), we have that $\lambda^\star = \sqrt{\lambda\mu}$, that the jobs arriving in $(0, x)$ are also exponentially distributed with mean $(\sqrt{\lambda\mu})^{-1}$, so that the system has load 1; in addition, however, it turns out that we need to sample both the $\bar{B}_n$ and B_0 from an exponential distribution with mean $(\sqrt{\lambda\mu})^{-1}$, and Q_0 from a geometric distribution with success probability $\sqrt{\varrho}$ rather than ϱ.

5.4 Queues: the heavy-tailed case

As indicated in Section 5.1, in queues with heavy-tailed input overflow is often essentially due to a *single* rare event, and, as a consequence, the algorithms presented in the previous section do not apply. In this section we review algorithms that have been designed for heavy-tailed queues. The first subsection is devoted to so-called 'conditional Monte Carlo estimators', whereas the second subsection focuses on IS-based procedures.

Let us start by gathering together some useful definitions related to heavy-tailed systems; a general reference in this respect is [36]. A non-negative random variable X is said to be *subexponential* if, for two independent copies X_1 and X_2 of X, we have $\mathbb{P}(X_1 + X_2 > x) \sim 2\mathbb{P}(X > x)$ as $x \to \infty$, where $f(x) \sim g(x)$ means $f(x)/g(x) \to 1$ as $x \to \infty$.

An important special subclass of subexponential random variables is given by those whose right tail is *regularly varying*. A random variable X, with tail distribution $\overline{F}(x) := \mathbb{P}(X > x) = 1 - F(x)$, has a regularly varying right tail with index $\alpha > 0$ if $\overline{F}(\beta x)/\overline{F}(x) \to \beta^{-\alpha}$ as $x \to \infty$ for all $\beta > 0$. In addition to regularly varying distributions, the class of subexponential distributions includes Weibull distributions with 'shape parameter' $\gamma \in (0, 1)$; the latter distributions are roughly such that $\overline{F}(x) \sim c\exp(-\beta x^\gamma)$ for c, $\beta > 0$, as $x \to \infty$. Other popular special cases are log-normal, log-gamma, and t-distributions (where it is noted that the last two models are particular cases of regular variation).

The subexponential property provides insight into how large deviations tend to occur in heavy-tailed models. In particular, suppose that $S_m = X_1 + \ldots + X_m$ where the X_i are i.i.d. non-negative subexponential random variables. It follows by induction and simple manipulations that $\mathbb{P}(S_m > x) \sim m\overline{F}(x) \sim \mathbb{P}\left(\max_{1\le j\le m} X_j > x\right)$ as $x \to \infty$. As a consequence we have the following property.

Proposition 6. *As $b \to \infty$,*

$$\mathbb{P}\left(\max_{1\le j\le m} X_j > b \,\middle|\, S_m > b\right) \longrightarrow 1.$$

The previous proposition illustrates the so-called 'catastrophe' principle, which, as mentioned in Section 5.1, governs the extreme behavior of heavy-tailed systems. Informally, in this setting it says that large deviations are caused by extremes in just a single component: the sum is large because one increment is large. In previous sections we saw that a convenient change of measure for IS is suggested by studying the asymptotic conditional distribution of the underlying process given the rare event of interest. Now, consider applying the idea behind such conditional description to a simple example. Suppose that we are interested in estimating $\mathbb{P}(S_m > b)$ efficiently as $b \to \infty$. Proposition 6 indicates that an asymptotic conditional description of the X_j, $j = 1, \ldots, m$, given $S_m > b$, assigns zero mass to sample paths for which all the random variables are less than b. As a consequence, the natural asymptotic description of X_j given $S_m > b$ is singular with respect to the nominal (original) distribution and therefore, contrary to the light-tailed case, a direct importance sampling approach is not feasible. This feature was observed by Asmussen *et al.* [6], who provide an extended discussion of the difficulties that are inherent in the design of efficient simulation estimators for heavy-tailed systems.

Throughout the rest of the section we shall provide a concise overview of the techniques that are applicable to rare event simulation for heavy-tailed models. We shall focus on analyzing the waiting time in the $M/G/1$ queue; extensions are briefly discussed at the end of this section. The Pollaczek–Khinchine representation (see [4, p. 237] yields that the steady-state waiting time, D, in the $M/G/1$ queue with load ϱ can be written as a random geometric sum. More precisely, let M be a geometric random variable with success parameter $1 - \varrho$, that is, $\mathbb{P}(M = m) = (1 - \varrho)\varrho^m$, and let X_j, $j = 1, 2, \ldots$, be a sequence of i.i.d. positive random variables independent of M such that

$$\mathbb{P}(X_1 > x) = \int_x^\infty \frac{\mathbb{P}(B > s)}{\mathbb{E}B} ds,$$

where B represents, as before, a generic service time. Then we have

$$\mathbb{P}(D > b) = \mathbb{P}(S_M > b).$$

Consequently, if X_1 is subexponential (which is the case if B is), then $\mathbb{P}(D > b) \sim \mathbb{E}M \cdot \mathbb{P}(X_1 > b)$ as $b \to \infty$ (see [36]). The development of efficient simulation estimators for the tail of D has traditionally focused (as we will do here) on estimating $\mathbb{P}(S_m > b)$ efficiently as $b \to \infty$ (for *fixed* m). The reason is that, in this heavy-tailed setting, the event $\{S_M > b\}$ is essentially caused by extreme behavior of the increments X_j, and not so much by 'unusual behavior' of the geometric random variable M.

5.4.1 Conditional Monte Carlo estimators

As indicated before, developing efficient IS estimators for the tail of S_m is not straightforward. An alternative idea is based on 'conditional Monte Carlo', and was first studied by Asmussen and Binswanger [5]; see also [6]. Note that we evidently have direct access to the distribution of $X_{(m)} := \max_{1 \le j \le m} X_j$. Using the catastrophe principle, it seems appropriate to estimate $\mathbb{P}(S_m > b)$ by conditioning on the first $m - 1$ order statistics $X_{(1)}, \ldots, X_{(m-1)}$ (i.e., integrating out the most relevant contribution, namely $X_{(m)}$). Based on this idea, [5] provided the following result which is the first instance of a provably efficient estimator in a heavy-tailed setting. Throughout the rest of the section we use the notation $x \vee y := \max(x, y)$.

Proposition 7. *Assume that X_1 has a density and a regularly varying right tail with index $\alpha > 0$. Define $S_{(m-1)} := X_{(1)} + \ldots + X_{(m-1)}$ and set*

$$Z_0(b) := \mathbb{P}\left(S_m > b \mid X_{(1)}, \ldots, X_{(m-1)}\right) = \frac{\overline{F}\left(\left(b - S_{(m-1)}\right) \vee X_{(m-1)}\right)}{\overline{F}\left(X_{(m-1)}\right)}.$$

Then $Z_0(b)$ is an asymptotically optimal estimator for $\mathbb{P}(S_m > b)$ as $b \to \infty$.

Proof. We here sketch the proof of the result; for more details, see the Appendix to [6]. Note that, with $\mathbb{E}(X; E) := \mathbb{E}(X \cdot 1_{\{E\}})$,

$$\mathbb{E}\left(Z_0(b)^2\right) = \mathbb{E}\left(Z_0(b)^2 ; X_{(m-1)} \leq b/2\right) + \mathbb{E}\left(Z_0(b)^2 ; X_{(m-1)} > b/2\right).$$

Observe that

$$\mathbb{E}\left(Z_0(b)^2 ; X_{(m-1)} > b/2\right) \leq \mathbb{P}\left(X_{(m-1)} > b/2\right),$$

where the right-hand side is $O(\overline{F}(b)^2)$, that is, roughly proportional to $b^{-2\alpha}$, as $b \to \infty$, as two (out of m) components have to be larger than $b/2$.

Now, let us denote the density of X_1 by $f(\cdot)$. It is evident from

$$\mathbb{P}(X_{(m-1)} \leq x) = m(F(x))^{m-1}\overline{F}(x) + (F(x))^m$$

that the density of $X_{(m-1)}$, which we denote by $f_{(m-1)}(\cdot)$, satisfies

$$f_{(m-1)}(x) = m(m-1)F(x)^{m-2}\overline{F}(x) f(x).$$

In addition,

$$\mathbb{E}\left(Z_0(b)^2 ; X_{(m-1)} \leq \frac{b}{2}\right) = \int_0^{b/m} \mathbb{E}\left(Z_0(b)^2 ; X_{(m-1)} \in dx\right) + \int_{b/m}^{b/2} \mathbb{E}\left(Z_0(b)^2 ; X_{(m-1)} \in dx\right).$$

Consider the first of these integrals. Observe that on the event $X_{(m-1)} < b/m$ we have that $S_{(m-1)} < (m-1)b/m$ and therefore

$$\begin{aligned}\int_0^{b/m} \mathbb{E}\left(Z_0(b)^2 ; X_{(m-1)} \in dx\right) &\leq \int_0^{b/m} \frac{m^2\overline{F}(b/m)^2 f(x)}{\overline{F}(x)} dx \\ &= \left[-m^2\overline{F}(b/m)^2 \cdot \log \overline{F}(x)\right]_0^{b/m} \\ &= O\left(\overline{F}(b)^2 \log b\right).\end{aligned}$$

The second integral (i.e., the one ranging from b/m to $b/2$) is handled as indicated above. Asymptotic optimality then follows as a result of the previous estimates, in conjunction with the fact that the probability of interest essentially vanishes as $\overline{F}(b)$. □

Regular variation has been used extensively in the previous result, so it may not be surprising that the estimator $Z_0(b)$ fails to be asymptotically optimal for other subexponential distributions (e.g., Weibull). Improved Monte Carlo estimators have been recently proposed by Asmussen and Kroese [8]. In particular, they observe that in order to reduce uncertainty, one can also consider the index corresponding to the largest jump and note that

$$\begin{aligned}\mathbb{P}(S_m > b) &= \sum_{j=1}^{m} \mathbb{E}\left(\mathbb{P}\left(S_m > x, X_{(m)} = X_j \middle| X_1, \ldots, X_{j-1}, X_{j+1}, \ldots, X_m\right)\right) \\ &= m \cdot \mathbb{E}\left(\mathbb{P}\left(S_m > x, X_{(m)} = X_m \middle| X_1, \ldots, X_{m-1}\right)\right).\end{aligned}$$

Therefore, a natural conditional Monte Carlo estimator that one can consider is

$$\begin{aligned}Z_1(b) &:= m \cdot \mathbb{P}\left(S_m > x, X_{(m)} = X_m \middle| X_1, \ldots, X_{m-1}\right) \\ &= m \cdot \overline{F}\left((b - S_{m-1}) \vee \max(X_j : j \leq m-1)\right).\end{aligned}$$

In [8] the following result is proven (the analysis is similar to that given in the proof of Proposition 7 and therefore the details are omitted).

Proposition 8. *If X_1 has a density and a regularly varying tail, then the estimator $Z_1(b)$ for $\mathbb{P}(S_m > b)$ has bounded relative error as $b \to \infty$. In addition, if X_1 has Weibull-type tails with index $\gamma \in (0, 0.58)$, then $Z_1(b)$ is an asymptotically optimal estimator for $\mathbb{P}(S_m > b)$ as $b \to \infty$.*

The previous proposition indicates that $Z_1(b)$ can only be guaranteed to be efficient if γ is sufficiently small. In the design of efficient estimators in the heavy-tailed setting it often occurs that models with less heavy tails than, say, regularly varying tend to require more and more information of all the components (not only the largest one); for $\gamma \uparrow 1$ it will resemble the case in which the X_j are exponentially distributed, in that *all* components matter. It is partly due to this feature that the majority of the efficient estimators developed for heavy-tailed systems assume special characteristics (such as regular variation or Weibull-type tails). Notable exceptions are the algorithms presented in [12, 15], which apply to general subexponential distributions. Additional conditional Monte Carlo algorithms have been proposed for the *transient* distribution of an $M/G/1$ queue in [66]; the proposed estimator is proved to have bounded relative error for regularly varying input.

5.4.2 Importance sampling estimators

As mentioned before, [6] discusses the difficulties of applying IS to heavy-tailed problems – the main problems are summarized by the singularity issue indicated above. Nevertheless, [6] also studies ideas that give rise to provably efficient algorithms. For instance, it is noted that if the X_j have a density and regularly varying tails with index $\alpha > 0$, then an asymptotically optimal importance sampling scheme for estimating $P(S_m > b)$ is obtained by sampling the X_j in an i.i.d. fashion according to the tail distribution $\tilde{\mathbb{P}}(X_j > x) = 1/\log(e+x)$ for $x \geq 0$. Asymptotic optimality of this estimator easily follows by applying Karamata's theorem (see [7, p. 176]). This IS selection biases the increments to induce very heavy-tailed distributions and therefore 'oversamples' paths for which *several* (i.e., not only the maximum) components contribute to the occurrence of the rare event. However, this procedure, although asymptotically optimal, does not seem to perform well in practice [7, p. 176].

Another IS approach was suggested by Juneja and Shahabuddin [46], and is based on applying exponential tilting type ideas via the hazard rate corresponding to the X_j. A basic observation behind this hazard rate tilting approach is the fact that if the X_j have a positive density, then $\mathbb{P}(X_j > t) = \exp(-\Lambda(t))$, where $\Lambda(t) := \int_0^t \lambda(s)\,ds$ and $\lambda(\cdot)$ is the hazard rate of X_j. In particular, if T is exponentially distributed with mean 1, then $\Lambda^{-1}(T)$ is a copy of X_j and therefore, for appropriate θ, we can define hazard rate tilting densities $f_\theta(\cdot)$ via

$$f_\theta(x) = \frac{\exp(\theta\Lambda(x)) f(x)}{\mathbb{E}\exp(\theta\Lambda(X_j))} = \exp(\theta\Lambda(x)) f(x)(1-\theta).$$

The corresponding hazard-rate IS estimator for $\mathbb{P}(S_m > b)$ takes the form

$$Z_2(b) := \frac{\exp\left(-\sum_{j=1}^m \theta\Lambda(X_j)\right)}{(1-\theta)^m} \cdot 1_{\{S_m > b\}}.$$

Using $\mathbb{E}_\theta(\cdot)$ to denote the expectation operator assuming that the X_j are i.i.d. with density $f_\theta(\cdot)$, we obtain

$$\mathbb{E}_\theta\left(Z_2(b)^2\right) = (1-\theta)^{-2m}\,\mathbb{E}_\theta\left(\exp\left(-2\sum_{j=1}^m \theta\Lambda(X_j)\right); S_m > b\right).$$

Assuming that $\Lambda(\cdot)$ is a concave function (as is the case for Pareto and Weibull random variables), we obtain that if $S_m > b$ then

$$\sum_{j=1}^m \Lambda(X_j) \geq \Lambda\left(\sum_{j=1}^m X_j\right) = \Lambda(S_m) \geq \Lambda(b) \tag{5.1}$$

and therefore, if $\theta \geq 0$, we obtain

$$\mathbb{E}_\theta\left(Z_2(b)^2\right) \leq (1-\theta)^{-2m}\exp(-2\theta\Lambda(b)).$$

Taking $\theta = 1 - \eta/\Lambda(b)$ for some $\eta > 0$ yields the following result.

Proposition 9. *If the X_j, $j = 1, \ldots, m$, have a concave cumulative hazard rate function $\Lambda(\cdot)$, then $Z_2(b)$ is an asymptotically optimal estimator for $\mathbb{P}(S_m > b)$ as $b \to \infty$.*

The previous hazard rate tilting strategy has been improved for the case in which the number of increments follows a geometric distribution (as is the case in the $M/G/1$ setting; see the remarks above). Such approaches involve suitable translation of the function $\Lambda(\cdot)$ applied when tilting. This adjustment allows us to apply the concavity argument given in (5.1) at a more convenient location relative to the rare event $\{S_m > b\}$ (see [47] and references therein). The idea of hazard-rate tilting has inspired further study in the field as it tries to develop rare event simulation methodology through a structure that resembles that of light-tailed input systems (via exponential twisting; see Section 5.3). Nevertheless, virtually all estimators that take advantage of this idea utilize the random walk structure substantially and some sort of subadditivity argument on $\Lambda(\cdot)$, as we did in (5.1).

We conclude this subsection with the discussion of recent state-dependent IS algorithms. Dupuis *et al.* [33] proposed a change of measure based on mixture densities that captures the 'catastrophic' behavior typical in heavy-tailed large deviations. A modification to the mixture in [33], analyzed in [14], in the more general context of a $G/G/1$ queue, can be described as follows. Given that $S_{k-1} = s$ for $1 \leq k < m$, then the next increment is sampled according to the density

$$f_k(x|s) = \frac{p_k(s) f(x)}{\overline{F}(a(b-s))} 1_{\{x > a(b-s)\}} + \frac{(1 - p_k(s)) f(x)}{F(a(b-s))} 1_{\{x \leq a(b-s)\}}, \tag{5.2}$$

where $p_k(s) \in (0, 1)$ is selected appropriately and $a \in (0, 1)$. If $k = m$ (i.e., $S_{m-1} = s$), then the increment is sampled according to the law of X_1 given that $X_1 > b - s$. Dupuis *et al.* [33] obtained a limiting control problem that allows optimal selection of the p_k. Bounded relative error of their estimator is obtained using a weak-convergence analysis. The parameter $a \in (0, 1)$ is important in order to incorporate the contribution of sample paths for which more than one large jump is required to achieve the rare event.

An improved proof technique introduced by Blanchet *et al.* [14], based on Lyapunov inequalities, simplifies the analysis in [33] and can be used to design efficient simulation estimators in more general multidimensional settings; see also [16]. In [12] it is noted that if one can find positive functions $g_k(s)$, for $k = 1, \ldots, m$, such that

$$g_{k-1}(s) \geq \mathbb{E}\left(g_k(s + X_k) \frac{f(X_k)}{f_k(X_k \mid s)} \right) \tag{5.3}$$

and $g_m(s) = 1_{\{s > b\}}$, then $g_0(0) \geq \tilde{\mathbb{E}}_0 Z_3(b)^2$, where $\tilde{\mathbb{P}}_0(\cdot)$ denotes the probability measure induced by the changes of measure explained in (5.2) given $S_0 = 0$

and

$$Z_3(b) := \frac{f(X_1)}{f_1(X_1|\, S_0)} \times \frac{f(X_2)}{f_2(X_2|\, S_1)} \times \cdots \times \frac{f(X_m)}{f_m(X_m|\, S_{m-1})}.$$

The solution to inequality (5.3) can be considered as a Lyapunov function. Blanchet *et al.* [14] describe, in the context of the $G/G/1$ queue, how to select the $g_k(s)$ approximately of order $O(\mathbb{P}^2(S_{m-k} > b - s))$ in order to achieve bounded relative error. Using Proposition 6 it is then natural to propose as a candidate Lyapunov function

$$g_k(s) = \min\left(1, c_k \overline{F}(a(b-s))^2\right)$$

for some constants $c_k > 0$. The parameter $a \in (0, 1)$ does not change the asymptotic behavior of g_k as $b \to \infty$, but is introduced for mathematical convenience. The parameters $p_k(s)$ and c_k are selected to satisfy inequality (5.3) which is equivalent to showing that $J_1 + J_2 \leq 1$, where

$$J_1 := \frac{\mathbb{E}(g_k(s+X); X > a(b-s))\, \overline{F}(a(b-s))}{g_{k-1}(s)\, p_k(s)},$$

$$J_2 := \frac{\mathbb{E}(g_k(s+X); X \leq a(b-s))\, F(a(b-s))}{g_{k-1}(s)(1 - p_k(s))}.$$

Note that $J_1 \leq \overline{F}(a(b-s))^2 / (g_{k-1}(s)\, p_k(s))$. Since each of the increments can cause the rare event, we let $p_k := 1/(m-k)$. If $g_{k-1}(s) < 1$, then conclude that

$$J_1 \leq \frac{\overline{F}(a(b-s))^2}{c_{k-1}\overline{F}(a(b-s))^2\, p_k} \leq \frac{1}{c_{k-1} p_k}.$$

On the other hand, we have that

$$J_2 \leq \frac{\mathbb{E}\left(g_k(s) + g_k'(s+\xi)X; X \leq a(b-s)\right) F(a(b-s))}{(1 - p_k)\, g_{k-1}(s)}$$

where $\xi \in (0, a(b-s))$. Therefore, we obtain (assuming the existence of a regularly varying density $f(\cdot)$) that if $g_{k-1}(s) < 1$, then it is possible to find $\kappa \in (0, \infty)$ such that

$$J_2 \leq \frac{c_k}{c_{k-1}(1 - p_k)} + \frac{\kappa c_k}{c_{k-1}(b - s + 1)(1 - p_k)}.$$

The previous bound requires that $\mathbb{E}X_1 < \infty$ and also that $a \in (0, 1)$ in order to facilitate a suitable dominated convergence argument. Note that if we select $p_k := 1/(m-k)$ and $c_k := c(m-k)$, then we obtain that

$$J_1 + J_2 \leq \frac{1}{c(1 + 1/(m-k))} + \frac{1}{1 - 1/(m-k)^2}\left(1 + \frac{\kappa}{b - s + 1}\right).$$

Clearly, the previous quantity can be set to be smaller than 1 by choosing $c > 0$ large enough and $g_k(s) < 1$ (which implies that $b - s > 0$ is sufficiently large). The algorithm then proceeds by applying the importance sampler (5.2) for the kth increment only if $g_k(s) < 1$ and sample from the nominal distribution if $g_k(s) = 1$. The previous analysis yields the following result.

Proposition 10. *If X_1 is regularly varying with a density and $\mathbb{E}X_1 < \infty$, then the estimator $Z_3(b)$ for $\mathbb{P}(S_m > b)$ satisfies*

$$\tilde{\mathbb{E}}_0 Z_3(b) \leq \min\left(cm\overline{F}(b)^2, 1\right),$$

and therefore it has bounded relative error, as $b \to \infty$.

As a final remark we mention that the use of Lyapunov inequalities seems to be the method of choice for heavy-tailed settings as it is the only approach which has been extended to more general settings, including heavy-tailed multi-server queues [13], multidimensional regularly varying random walks [16], random walks with subexponential increments [12], and large-deviations probabilities that involve path-dependent events [17].

5.5 Queues under the many-sources scaling

As we saw in the previous section, the presence of heavy tails is a serious complication that forces us to use relatively advanced IS schemes (as opposed to the rather straightforward exponential twists that can be applied for light tails).

One important remark needs to be made: the above complication particularly relates to the *large buffer regime*, that is, the regime in which rarity is due to the large threshold that needs to be exceeded. There are, however, other asymptotic regimes that are interesting to consider, one of them being the so-called many-sources regime. In this regime n i.i.d. sources feed into a queuing resource that is emptied at a constant rate nc. The probability p_n that the steady-state buffer content, say Q_n, exceeds level nb decreases as a function of n. More specifically, p_n decays, under mild conditions, *exponentially* in n, even when the individual sources have heavy-tailed and/or long-range dependent characteristics. As a consequence, in this asymptotic regime, elements of the theory for light tails become applicable again. In this section we demonstrate a number of effective IS procedures.

We restrict ourselves to the discrete-time case. A particularly useful result, due to Likhanov and Mazumdar [53], is the following. Consider n sources with stationary increments, and let $A_i(j_1, j_2)$, for natural numbers j_1, j_2 such that $j_1 < j_2$, be the traffic generated by the ith source, in slots $j_1 + 1, \ldots, j_2$; we denote by $A(\cdot, \cdot)$ a generic source. Because of the stationary increments $A(j_1 + k, j_2 + k)$ has the same distribution as $A(j_1, j_2)$ for any $k \in \mathbb{Z}$. We abbreviate $A(j) := A(0, j)$. We assume that $M_k(\vartheta) := \mathbb{E}\exp(\vartheta A(k)) < \infty$ for some positive ϑ. With

$I_k := \sup_\vartheta(\vartheta(b + ck) - \log M_k(\vartheta))$, it is shown in [53] that, under the (mild) condition that $\liminf_{k\to\infty} I_k/\log k > 0$, p_n decays exponentially:

$$\lim_{n\to\infty} \frac{1}{n} \log p_n = -\inf_{k\in\mathbb{N}} I_k.$$

Let $k^\star$ be the infimizer in the right-hand side of the previous display (assumed to be unique), and $\vartheta(k)$ be the optimizing ϑ in the definition of I_k.

In this section we will specialize to two different (and rather generic) classes of sources: *Gaussian sources* and *on–off sources*. As is readily verified, both are reversible in time, and hence p_n can be rewritten as

$$\mathbb{P}\left(\exists k \in \mathbb{N} : \sum_{i=1}^{n} A_i(k) \geq nb + nck\right).$$

For Gaussian sources, $A_i(k)$ has a normal distribution with mean μk (where $\mu < c$ can be set to 0 without loss of generality) and variance $v(k)$; for example, in the case of fractional Brownian motion $v(k) = k^{2H}$, with $H \in (0, 1)]$ denoting the so-called Hurst parameter [72].

On–off sources generate traffic at a constant rate, say 1, when on, and are silent when off. The on-periods are distributed as a random variable A with finite mean μ_A, whereas the off-periods are distributed as S with mean S; clearly the system is stable when $\pi := \mathbb{E}A/(\mathbb{E}A + \mathbb{E}S) < c$. At time 0 any source is in its on-state with probability π, and off with probability $1 - \pi$. When we observe that it is on at time 0, the remaining on-time $A^\star$ has the well-known residual distribution $\mathbb{P}(A^\star = k) = (\mathbb{E}A)^{-1}\mathbb{P}(A \geq k)$; the residual off-time is distributed likewise.

A conceptual difficulty with this framework is that we should monitor the process $\sum_{i=1}^{n} A_i(k)$ for all $k \in \mathbb{N}$ in order to see whether overflow has been reached or not. If one allows, however, a (controlled) error of at most ε, that is,

$$\mathbb{P}\left(\exists k \in \{T+1, T+2, \ldots\} : \sum_{i=1}^{n} A_i(k) \geq nb + nck\right) \Big/ p_n < \varepsilon, \qquad (5.4)$$

then one can approximate p_n by $p_n^T := \mathbb{P}(\sup_{k\in\{1,\ldots,T\}} \sum_{i=1}^{n} A_i(k) \geq nb + nck)$. In [18, 32] it is shown how to find such a T; for later purposes we mention here that $T = o(n)$. It is clear that $T \geq k^\star$, interpreting $k^\star$ as the epoch with the most probability mass.

Interestingly, the most likely path to overflow (cf. Section 5.2.2) can also be explicitly determined, and, as before, this may give us a handle on devising asymptotically optimal algorithms. Without giving the precise definition of 'most likely path' here, we mention that Wischik [73] showed that this path equals

$$f(j) = \frac{\mathbb{E}A(j)e^{\vartheta(k^\star)A(k^\star)}}{\mathbb{E}e^{\vartheta(k^\star)A(k^\star)}},$$

where $f(j)$ corresponds to the amount of work generated by a single source during the slots $1, \ldots, j$; notice that indeed $f(k^\star) = b + ck^\star$. It is not clear a priori, though, how to find a measure $\mathbb{Q}$ such that the sources on average follow this path f. For on–off sources such a change of measure is rather involved, and given explicitly in [18]. For Gaussian sources [32] the distribution of $A(1), \ldots, A(T)$ under the new measure $\mathbb{Q}$ is again multivariate normal, with the mean per source (evidently) equal to $f(1), \ldots, f(T)$, whereas the covariance between $A(j_1)$ and $A(j_2)$ (without loss of generality assuming $j_1 \leq j_2$) *remains*, as under the original measure,

$$\Gamma(j_1, j_2) := \mathrm{Cov}(A(j_1), A(j_2)) = \frac{1}{2}\left(v(j_2) + v(j_1) - v(j_2 - j_1)\right).$$

Notice that we now see an example of a 'non-constant' (i.e., state-independent) change of measure, as the most likely path to overflow does not correspond (necessarily) to a straight line; see [25] for an early reference that uses this idea.

The above alternative measure $\mathbb{Q}$ requires a procedure to sample from a multivariate normal distribution. As we assume Gaussian processes with stationary increments, this multivariate normal distribution has a special form (the correlation between two slots only depends on the distance between these two slots, rather than their positions). Due to this property, traces of fixed length, say T, can be sampled relatively fast [26]; in the special case of fractional Brownian motion, the effort required to sample a trace of length T is roughly proportional to $T \log T$. For more reflections on these methods, see [32]. Interesting (and fast, but not exact) algorithms are given in [62, 64].

Under the above change of measure, it is readily verified that if *the first time* that $\sum_{i=1}^{n} A_i(k)$ exceeds $nb + nck$ is exactly at $k^\star$, then the likelihood is bounded by $\exp(-nI_{k^\star})$ [18, 32], suggesting that this procedure may be asymptotically optimal. If the first exceedance, however, takes place at a different epoch, this bound does not (always) apply, and hence this change of measure is *not* (necessarily) asymptotically optimal.

Now note that the problem described above in fact falls in the setting of Section 5.2.3: the target set is a (finite) union of events (T rather than 2); we can decompose p_n^T as $\sum_{j=1}^{T} p_{n,j}^T$, with $p_{n,j}^T$ the probability that at epoch j overflow occurs for the first time:

$$p_{n,j}^T := \mathbb{P}\left(\forall k \in \{1, \ldots, j-1\} : \sum_{i=1}^{n} A_i(k) < nb + nck; \sum_{i=1}^{n} A_i(j) \geq nb + ncj\right);$$

the corresponding events are disjoint. In this way we can use the partitioning method described in Section 5.2.3 to obtain an efficient IS procedure. We then estimate the $p_{n,j}^T$ separately, for $j = 1, \ldots, T$, but, recalling that T grows (for a fixed error ε in (5.4)) sublinearly in n, the procedure remains asymptotically optimal. A similar optimality result can also be achieved by performing an adaptive twist (see [32, 35]). A third approach yielding asymptotic optimality is by the random twist method proposed by Sadowsky and Bucklew [68].

We conclude this section with a few biographical notes. For overviews on methods to generate Gaussian traces (and in particular fractional Brownian motion), see [31, 62]. Some related results on fast simulation of queues with Gaussian input have been reported in [60, 45]. Michna [60] focuses on fractional Brownian motion input under the large-buffer scaling, but does not consider asymptotic efficiency of his simulation scheme (in fact, one may check that his estimator is asymptotically inefficient). Huang *et al.* [45] also work in the large-buffer asymptotic regime, and present a constant-mean change of measure; this consequently does not correspond to the (curved) most likely path, and is therefore asymptotically inefficient in the many-sources regime. We also mention an interesting work [11] which uses recent insights into certain Gaussian martingales, and a novel 'bridge' approach [37].

5.6 Networks

In this section we focus on the extension of the theory to networks of queues. An important conclusion is that networks with fixed service rates should be handled substantially differently from those with random service times. First we consider a class of fluid queues with constant service rate, so-called *intree* networks, and indicate that an asymptotically optimal IS procedure can be found by applying rather straightforward heuristics. In situations with random service times, such as Jackson networks, these heuristics do not apply, and more sophisticated methods have to be used instead.

5.6.1 Fluid networks

In the light-tailed setting, the results for a single queue have been extended to a rather broad class of networks, so-called intree networks (see [22, 23]). These are acyclic networks of queues (each having a constant service rate), fed by fluid input, of which the output goes either to one next queue, or leaves the system; a typical example is a tandem system with Markov-modulated fluid input (see Section 5.3.3).

Let us for convenience' sake restrict ourselves to considering the event of buffer overflow in the second queue of a two-node tandem queue, where the service rate of the first (second) queue is C_1 (C_2). Evidently we should assume $C_1 > C_2$, because otherwise the first queue remains empty all the time. As stated above, we consider light-tailed input, but let us for the moment restrict ourselves even more, and suppose that the input is Markov fluid with parameters $\Lambda = (\lambda_{ij})_{i,j=1}^{d}$ and $r = (r_1, \ldots, r_d)$. We identified in Section 5.3.3 a function $I(a)$ that represented the cost per unit of time when generating traffic at a rate of roughly a per unit of time.

A crucial observation is that it does not make sense to use a rate a that is bigger than C_1: this is 'more expensive' than transmitting at C_1 and does not help to fill the second queue. Using this idea, the decay rate should read

$$\inf_{C_2 \le a \le C_1} \frac{I(a)}{a - C_2};$$

realize that $1/(a - C_2)$ can be thought of as the time the alternative measure is active. As before, the optimizing a, say $a^\star$, corresponds to an exponential twist of the Markov fluid source. The heuristics are then as follows. If $a^\star \in (C_2, C_1)$ the first queue is essentially transparent, as, in order to cause overflow in the second queue, the first queue hardly plays a role in shaping the traffic stream. If, on the other hand, $a^\star = C_1$ the output rate of the first queue is 'throttled', that is, there is a significant shaping effect of the first queue. The resulting importance sampling scheme is asymptotically optimal [22, 23].

The above ideas can be made rigorous by using sample-path large deviations, as developed in [22]. The class of source models considered there is significantly broader than just Markov fluid. The dichotomy, with a regime in which the first queue is 'transparent' and a regime in which the first queue really 'shapes' the traffic, was also found in several other studies (see, for instance, [55, 58]). The fact that the network is intree is crucial; see [65] for structural results on networks in which traffic streams are allowed to split.

5.6.2 Jackson networks

Let us now focus on networks with random service times, to see whether the theory of the previous subsection carries over. To be able to draw the parallel with Section 5.6.1, let us focus on the $M/M/1$ tandem queue: jobs arrive according to a Poisson process of rate λ, and have a service time in the first (second) queue that is exponentially distributed with mean μ_1^{-1} (μ_2^{-1}). In this case it is known that that the decay rate of overflow in the second queue is, in obvious notation,

$$\lim_{K\to\infty} \frac{1}{K} \log \mathbb{P}(Q_2 > K) = -\log \varrho_2 = -\log\left(\frac{\lambda}{\mu_2}\right),$$

that is, independent of μ_1.

To find a good alternative measure $\mathbb{Q}$, it is tempting to use the heuristic of Section 5.6.1. The 'cost' of a Poisson process with rate λ behaving like a Poisson process with rate $\bar{\lambda}$ is [69, pp. 14, 20]

$$I_\lambda(\bar{\lambda}) := \lambda - \bar{\lambda} + \bar{\lambda} \log \frac{\bar{\lambda}}{\lambda}$$

which can be interpreted as a Kullback–Leibler distance. Parallelling the argument used above for intree networks, we would have to solve the variational problem

$$\inf \frac{I_\lambda(\bar{\lambda}) + I_{\mu_1}(\bar{\mu}_1) + I_{\mu_2}(\bar{\mu}_2)}{\min\{\bar{\lambda}, \bar{\mu}_1\} - \bar{\mu}_2}, \tag{5.5}$$

over $\bar{\lambda}, \bar{\mu}_1, \bar{\mu}_2$ such that $\min\{\bar{\lambda}, \bar{\mu_1}\} < \bar{\mu}_2$; observe that under the new parameters the output rate of the first queue is essentially $\min\{\bar{\lambda}, \bar{\mu}_1\}$. If $\mu_2 \leq \mu_1$ this optimization program indeed gives the right result. We then find that $\bar{\lambda} = \mu_2$ and $\bar{\mu}_2 = \lambda$. Under these new parameters the first queue remains stable, so that jobs leave the first queue according to a Poisson process of rate μ_2. In other words, we are essentially in the setting of Example 2, where the arrival rate and service rate were interchanged.

If, however, $\mu_1 < \mu_2$, then interchanging λ and μ_2 leads to a situation where jobs leave the first queue at a rate μ_1. In fact, the above optimization program always yields a value larger than the correct value, $-\log \varrho_2$. The reason for this is that the most likely path to overflow in the second queue *is not a straight line!* More precisely, in order for the second queue to exceed K (see [1, 61]), the first queue builds up to, roughly, level $K \cdot (\mu_1 - \mu_2)/(\mu_1 - \lambda)$ (by interchanging λ and μ_1), and after that, the first queue drains while the second builds up (by cyclically interchanging λ, μ_1, and μ_2), thus indeed yielding cost $K \cdot (-\log \varrho_2)$.

State-independent IS distributions, for instance those proposed by Parekh and Walrand [63] for estimating the probability that the *total* network population exceeds K, do not guarantee asymptotic optimality, as further investigated in [27, 38]. Proposals for state-dependent schemes, and proofs of asymptotic optimality, are provided in, for example, [34]. Good state-dependent IS distributions can also be found by applying cross-entropy techniques [28].

5.6.3 Loss networks

A classical model in communications engineering is the *Erlang loss model*: calls arrive at a link according to a Poisson process of rate λ, the link can accommodate at most C calls at the same time, and the call holding times constitute a sequence of i.i.d. non-negative random variables with mean $1/\mu$. With $\nu := \lambda/\mu$ denoting the offered load, the stationary probability $\pi(k)$ of having $k \in \{0, \ldots, C\}$ calls in the system is given by the truncated Poisson distribution:

$$\pi(k) = \frac{\nu^k/k!}{\sum_{\ell=1}^{C}(\nu^\ell/\ell!)};$$

importantly, this distribution is *insensitive*, that is, it depends on the call holding time distribution only through its mean.

Loss networks can be considered as the extension of the Erlang loss model to a setting in which there are multiple types of calls, and in which these calls use circuits on multiple links simultaneously. Suppose that the set of links is given by $\mathcal{J}$, and the set of call types by $\mathcal{R}$. Each call type is characterized by an arrival rate λ_r and mean holding time $1/\mu_r$. With C_j denoting the capacity of link j, and A_{jr} the number of trunks required by a type r call on a type j link, it is clear that the state space is

$$\mathcal{S} := \left\{\vec{k} : \sum_{r \in \mathcal{R}} A_{jr}k_r \leq C_j, \text{for all } j \in \mathcal{J}\right\}.$$

It turns out that the theory for the single link and single class carries over: the distribution of the number of calls is again truncated Poisson [48]: with $\nu_r := \lambda_r/\mu_r$,

$$\pi(\vec{k}) = \frac{\prod_{r\in\mathcal{R}}(\nu_r^{k_r}/k_r!)}{\sum_{\vec{\ell}\in\mathcal{S}}\prod_{r\in\mathcal{R}}(\nu_r^{\ell_r}/\ell_r!)} = \frac{\prod_{r\in\mathcal{R}}((e^{-\nu_r}\nu_r^{k_r})/k_r!)}{\sum_{\vec{\ell}\in\mathcal{S}}\prod_{r\in\mathcal{R}}((e^{-\nu_r}\nu_r^{\ell_r})/\ell_r!)}$$

In particular, the probability $\beta^{(r)}$ of a call of type r being blocked can be expressed in terms of the $\pi(\vec{k})$:

$$\beta^{(r)} = \sum_{k\in\mathcal{S}_r}\pi(\vec{k}) = \frac{\sum_{\vec{k}\in\mathcal{S}_r}\prod_{s\in\mathcal{R}}((e^{-\nu_s}\nu_s^{k_s})/k_s!)}{\sum_{\vec{\ell}\in\mathcal{S}}\prod_{s\in\mathcal{R}}((e^{-\nu_s}\nu_s^{\ell_s})/\ell_s!)},$$

where

$$\mathcal{S}_r := \left\{\vec{k} : \sum_{s\in\mathcal{R}} A_{js}k_s + A_{jr} > C_j, \text{ for some } j \in \mathcal{J}\right\}.$$

Despite the availability of these explicit formulae, numerical evaluation of the blocking probabilities is cumbersome; in particular, the evaluation of the normalizing constant (i.e., the numerator in the previous display) can be time-consuming. An idea could be to resort to simulation. To this end, [42] proposed an acceptance–rejection method: one samples from a multivariate Poisson distribution with mean $\vec{\nu}$ (and independent marginal distributions), and estimates $\beta^{(r)}$ by the fraction of the samples in $\mathcal{S}$ that is also in $\mathcal{S}_r$. If $\nu \in \mathcal{S}$, then the denominator does not correspond to a rare event, so that one has relatively many 'useful' samples; the numerator, however, *does* correspond to a rare event, so that just a small fraction of the useful samples will fall in $\mathcal{S}_r$, and hence it will take a long time to obtain a sufficiently precise estimate. If, on the other hand, $\vec{\nu} \notin \mathcal{S}$, then even the estimation of the denominator will be time-consuming, so that in this case too the acceptance–rejection technique will be slow.

In light of the above reasoning, it seems natural to sample under another Poisson distribution. To find a good candidate, a scaling is applied: the arrival rates λ_r were replaced by $n\lambda_r$, and the link capacities C_j by nC_j. It was shown [56] that the type r blocking probability, which now depends on n, decays exponentially:

$$\lim_{n\to\infty}\frac{1}{n}\log\beta_n^{(r)} = -\inf_{\vec{x}\in\bar{\mathcal{S}}_r}\sum_{s\in\mathscr{R}}\left(x_s\log\frac{x_s}{\nu_s} - x_s + \nu_s\right) + \inf_{\vec{y}\in\bar{\mathcal{S}}}\sum_{s\in\mathscr{R}}\left(y_s\log\frac{y_s}{\nu_s} - y_s + \nu_s\right),$$

where

$$\bar{\mathcal{S}} := \left\{ \vec{y} : \sum_{s \in \mathcal{R}} A_{js} y_s \leq C_j, \text{ for all } j \in \mathcal{J} \right\},$$

$$\bar{\mathcal{S}}_r := \left\{ \vec{x} : \sum_{s \in \mathcal{R}} A_{js} x_s = C_j, \text{ for some } j \in \mathcal{J} \text{ with } A_{jr} > 0 \right\}.$$

With $\vec{x}^\star$ and $\vec{y}^\star$ denoting the optimizers in the above infima, [56] proposes to estimate the numerator and the denominator of the blocking probability separately, the numerator (denominator) by applying IS with a Poisson distribution with mean $\vec{x}^\star$ ($\vec{y}^\star$).

Observe that $x^\star$ can be interpreted as the most likely blocking state, and the $j^\star$ (with $A_{j^\star r} > 0$) for which $\sum_{s \in \mathcal{R}} A_{j^\star s} x_s = C_{j^\star}$ the most likely blocked link (when considering type r calls). It is readily verified that the above IS distribution behaves nicely when blocking indeed occurs at link $j^\star$; if it occurs at another link the likelihood may explode (cf. Section 5.2.3). The remedies mentioned in Section 5.2.3 can be used to overcome this problem; see, for instance, [51] for reflections on the use of the random twist approach of [68] (which is asymptotically optimal). A more advanced decomposition, leading to a very efficient IS distribution, was studied in [52].

References

[1] V. Anantharam, P. Heidelberger, and P. Tsoucas. Analysis of rare events in continuous time marked chains via time reversal and fluid approximation. IBM Research Report, REC 16280, 1990.

[2] D. Anick, D. Mitra, and M. Sondhi. Stochastic theory of a data handling system with multiple resources. *Bell System Technical Journal*, **61**: 1871–1894, 1982.

[3] S. Asmussen. *Ruin Probabilities*. World Scientific, London, 2000.

[4] S. Asmussen. *Applied Probability and Queues*. Springer, New York, 2003.

[5] S. Asmussen and K. Binswanger. Simulation of ruin probabilities for subexponential claims. *ASTIN Bulletin*, **27**: 297–318, 1997.

[6] S. Asmussen, K. Binswanger, and B. Højgaard. Rare events simulation for heavy-tailed distributions. *Bernoulli*, **6**: 303–322, 2000.

[7] S. Asmussen and P. Glynn. *Stochastic Simulation: Algorithms and Analysis*. Springer, New York, 2007.

[8] S. Asmussen and D. Kroese. Improved algorithms for rare event simulation with heavy tails. *Advances in Applied Probability*, **38**: 545–558, 2006.

[9] S. Asmussen and R. Rubinstein. Steady state rare events simulation in queueing models and its complexity properties. In J. Dshalalow, ed., *Advances in Queueing*, pp. 429–461. CRC Press, Boca Raton, FL, 1995.

[10] R. Bahadur and R. R. Rao. On deviations of the sample mean. *Annals of Mathematical Statistics*, **31**: 1015–1027, 1960.

[11] P. Baldi and B. Pacchiarotti. Importance sampling for the ruin problem for general Gaussian processes. Submitted, University Paris VI, Technical Report 875, 2004.

[12] J. Blanchet and P. Glynn. Efficient rare-event simulation for the maximum of a heavy-tailed random walk. *Annals of Applied Probability*, 2008.

[13] J. Blanchet, P. Glynn, and J. C. Liu. Efficient rare event simulation for multiserver queues. Submitted, 2009.

[14] J. Blanchet, P. Glynn, and J. C. Liu. Fluid heuristics, Lyapunov bounds and efficient importance sampling for a heavy-tailed G/G/1 queue. *Queueing Systems*, 2007.

[15] J. Blanchet and C. Li. Efficient rare-event simulation of compound sums. Submitted, 2009.

[16] J. Blanchet and J. C. Liu. Rare event simulation for multidimensional random walks with t-distributed increments. In *Proceedings of the 40th Winter Simulation Conference*, Washington, DC, 2007.

[17] J. Blanchet and J. C. Liu. State-dependent importance sampling for regularly varying random walks. *Advances in Applied Probability*, **40**: 1104–1128, 2008.

[18] N.-K. Boots and M. Mandjes. Fast simulation of a queue fed by a superposition of many (heavy-tailed) sources. *Probability in the Engineering and Informational Sciences*, **16**: 205–232, 2002.

[19] P. Bratley, B. Fox, and L. Schrage. *A Guide to Simulation*, 2nd edition. Springer, New York, 1987.

[20] J. Bucklew. *Large Deviation Techniques in Decision, Simulation and Estimation*. John Wiley & Sons, Inc., New York, 1990.

[21] J. Bucklew. *Introduction to Rare Event Simulation*. Springer, New York, 2004.

[22] C.-S. Chang. Sample path large deviations and intree networks. *Queueing Systems*, **20**: 7–36, 1995.

[23] C.-S. Chang, P. Heidelberger, S. Juneja, and P. Shahabuddin. Effective bandwidth and fast simulation of ATM intree networks. *Performance Evaluation*, **20**: 45–65, 1994.

[24] J. Collamore. Importance sampling techniques for the multidimensional ruin problem for general Markov additive sequences of random vectors. *Annals of Applied Probability*, **12**: 382–421, 2002.

[25] M. Cottrell, J.-C. Fort, and G. Malgouyres. Large deviations and rare events in the study of stochastic algorithms. *IEEE Transactions on Automat. Contr.*, **28**: 907–920, 1983.

[26] R. Davies and D. Harte. Tests for Hurst effect. *Biometrika*, **74**: 95–101, 1987.

[27] P.-T. de Boer. Analysis of state-independent importance-sampling measures for the two-node tandem queue. *ACM Transactions on Modeling and Computer Simulation*, **16**: 225–250, 2006.

[28] P.-T. de Boer, D. Kroese, S. Mannor, and R. Rubinstein. A tutorial on the cross-entropy method. *Annals of Operations Research*, **134**: 19–67, 2005.

[29] T. Dean and P. Dupuis. Splitting for rare event simulation: A large deviation approach to design and analysis. *Stoch. Proc. and Appl.*, **119**: 562–587, 2008.

[30] A. Dembo and O. Zeitouni. *Large Deviations Techniques and Applications*, 2nd edition. Springer, New York, 1998.

[31] T. Dieker and M. Mandjes. On spectral simulation of fractional Brownian motion. *Probability in the Engineering and Informational Sciences*, **17**: 417–434, 2003.

[32] T. Dieker and M. Mandjes. Fast simulation of overflow probabilities in a queue with Gaussian input. *ACM Transactions on Modeling and Computer Simulation*, **16**: 119–151, 2006.

[33] P. Dupuis, K. Leder, and H. Wang. Importance sampling for sums of random variables with regularly varying tails. *ACM Transactions on Modeling and Computer Simulation*, **17**, 2007.

[34] P. Dupuis, A. D. Sezer, and H. Wang. Dynamic importance sampling for queueing networks. *Annals of Applied Probability*, **17**: 1306–1346, 2007.

[35] P. Dupuis and H. Wang. Importance sampling, large deviations, and differential games. *Stochastics and Stochastics Reports*, **76**: 481–508, 2004.

[36] P. Embrechts, C. Klüppelberg, and T. Mikosch. *Modelling Extreme Events with Applications to Insurance and Finance*. Springer, New York, 1997.

[37] S. Giordano, M. Gubinelli, and M. Pagano. Bridge Monte-Carlo: A novel approach to rare events of Gaussian processes. In *Proceedings of the 5th Workshop on Simulation*, pp. 281–286, St Petersburg, 2005.

[38] P. Glasserman and S. Kou. Analysis of an importance sampling estimator for tandem queues. *ACM Transactions on Modeling and Computer Simulation*, **5**: 22–42, 1995.

[39] P. Glasserman and Y. Wang. Counterexamples in importance sampling for large deviations probabilities. *Annals of Applied Probability*, **7**: 731–746, 1997.

[40] A. Goyal, P. Shahabuddin, P. Heidelberger, V. Nicola, and P. Glynn. A unified framework for simulating Markovian models of highly reliable systems. *IEEE Transactions on Computers*, **41**: 36–51, 1992.

[41] J. Hammersley and D. Handscomb. *Monte Carlo Methods*. Methuen, London, 1964.

[42] C. Harvey and C. Hills. Determining grades of service in a network. In *9th International Teletraffic Congress ITC9*, Torremolinos, Spain, 1979.

[43] P. Heidelberger. Fast simulation of rare events in queueing and reliability models. *ACM Transactions on Modeling and Computer Simulation*, **5**: 43–85, 1995.

[44] A. Hopmans and J. Kleijnen. Importance sampling in systems simulation: A practical failure? *Mathematics and and Computers in Simulation*, **21**: 209–220, 1979.

[45] C. Huang, M. Devetsikiotis, I. Lambadaris, and A. Kaye. Fast simulation of queues with long-range dependent traffic. *Communications in Statistics. Stochastic Models* , **15**: 429–460, 1999.

[46] S. Juneja and P. Shahabuddin. Simulating heavy-tailed processes using delayed hazard rate twisting. *ACM Transactions on Modeling and Computer Simulation*, **12**: 94–118, 2002.

[47] S. Juneja and P. Shahabuddin. Rare event simulation techniques: An introduction and recent advances. In S. Henderson and B. Nelson, eds, *Handbook on Simulation*, pp. 291–350. Elsevier, Amsterdam, 2006.

[48] F. Kelly. Loss networks. *Annals of Applied Probability*, **1**: 319–378, 1991.

[49] G. Kesidis, J. Walrand, and C.-S. Chang. Effective bandwidths for multiclass Markov fluids and other ATM sources. *IEEE/ACM Transactions on Networking*, **1**: 424–428, 1993.

[50] L. Kosten. Stochastic theory of data-handling systems with groups of multiple sources. In H. Rudin and W. Bux, eds, *Performance of Computer-Communication Systems*, pp. 321–331. Elsevier, New York, 1984.

[51] P. Lassila and J. Virtamo. Efficient importance sampling for Monte Carlo simulation of loss systems. In *16th International Teletraffic Congress ITC16*, Edinburgh, 1999.

[52] P. Lassila and J. Virtamo. Nearly optimal importance sampling for Monte Carlo simulation of loss systems. *ACM Transactions on Modeling and Computer Simulation*, **10**: 326–347, 2000.

[53] N. Likhanov and R. Mazumdar. Cell loss asymptotics in buffers fed with a large number of independent stationary sources. *Journal of Applied Probability*, **36**: 86–96, 1999.

[54] M. Mandjes. Rare event analysis of batch-arrival queues. *Telecommunication Systems*, **6**: 161–180.

[55] M. Mandjes. Asymptotically optimal importance sampling for tandem queues with Markov fluid input. *AEU International Journal on Electronics and Communications* , **52**: 152–161, 1998.

[56] M. Mandjes. Fast simulation of blocking probabilities in loss networks. *European Journal of Operational Research*, **101**: 393–405, 1998.

[57] M. Mandjes and A. Ridder. Finding the conjugate of Markov fluid processes. *Probability in the Engineering and Informational Sciences*, **9**: 297–315, 1995.

[58] M. Mandjes and M. van Uitert. Sample-path large deviations for tandem and priority queues with Gaussian inputs. *Annals of Applied Probability*, **15**: 1193–1226, 2005.

[59] M. Mandjes and B. Zwart. Large deviations for sojourn times in processor sharing queues. *Queueing Systems*, **52**: 237–250, 2006.

[60] Z. Michna. On tail probabilities and first passage times for fractional Brownian motion. *Mathematical Methods of Operations Research*, **49**: 335–354, 1999.

[61] D. Miretskiy, W. Scheinhardt, and M. Mandjes. Efficient simulation of a tandem queue with server slow-down. *Simulation*, **83**: 751–767, 2007.

[62] I. Norros, P. Mannersalo, and J. L. Wang. Simulation of fractional Brownian motion with conditionalized random midpoint displacement. *Advances in Performance Analysis*, **2**: 77–101, 1999.

[63] S. Parekh and J. Walrand. Quick simulation of rare events in networks. *IEEE Transactions on Automatic Control*, **34**: 54–66.

[64] V. Paxson. Fast, approximate synthesis of fractional Gaussian noise for generating self-similar network traffic. *Computer Communication Review*, **27**: 5–18, 1997.

[65] K. Ramanan and P. Dupuis. Large deviation properties of data streams that share a buffer. *Annals of Applied Probability*, **8**: 1070–1129, 1998.

[66] L. Rojas-Nandayapa and S. Asmussen. Efficient simulation of finite horizon problems in queueing and insurance risk. *Queueing Systems*, **57**: 85–97, 2007.

[67] J. Sadowsky. On Monte Carlo estimation of large deviations probabilities. *Annals of Applied Probability*, **6**: 399–422, 1996.

[68] J. Sadowsky and J. Bucklew. On large deviations theory and asymptotically efficient Monte Carlo estimation. *IEEE Transactions on Information Theory*, **36**: 579–588, 1990.

[69] A. Shwartz and A. Weiss. *Large Deviations for Performance Analysis. Queues, Communications, and Computing*. Chapman & Hall, London, 1995.

[70] D. Siegmund. Importance sampling in the Monte Carlo study of sequential tests. *Annals of Statistics*, **4**: 673–684, 1976.

[71] Wikipedia. http://en.wikipedia.org/wiki/Normal_distribution, 2008.

[72] W. Willinger, M. Taqqu, R. Sherman, and D. Wilson. Self-similarity through high-variability: statistical analysis of Ethernet LAN traffic at the source level. *IEEE/ACM Transactions on Networking*, **5**: 71–86, 1997.

[73] D. Wischik. Sample path large deviations for queues with many inputs. *Annals of Applied Probability*, **11**: 379–404, 2001.

6

Markovian models for dependability analysis

Gerardo Rubino and Bruno Tuffin

6.1 Introduction

This chapter deals with dependability models originating, for instance, in computer science or networking systems. The basic modeling assumption is to consider a multicomponent system (i.e., a multidimensional model) such that each component is subject to failure and possibly to repair. The global system is then considered to be down (typically a rare event in most applications of interest) when it is in some subset of states such that given components are themselves down. The principal dependability measures we are interested in are the *mean time to failure* (MTTF), representing the average time to reach the set of failed states starting from a given point; the unavailability at a given time t, that is, the probability that at time t the model is in one of its failed states; the steady-state unavailability; and the unreliability at time t, which is the probability that the model enters the subset of failed states at or before t. These are not the only dependability measures of interest. There are many others, such as the interval availability at t, a random variable defined as the fraction of $[0, t]$ during which the system is in an operational state. In the chapter, we will discuss only methods designed to evaluate the main metrics in the area.

Rare Event Simulation using Monte Carlo Methods Edited by G. Rubino and B. Tuffin

The applications are numerous. Dependability analysis (which inherits from the classical theory an older concept of reliability analysis) is a critical issue in telecommunications, computer science, and manufacturing, among other fields, [13]. For instance, catastrophic failures in transport systems or in nuclear power plants could lead to human losses. Similarly, the failure of a computer or a network (telecommunication, electricity, etc.) may lead to important monetary losses. These systems are thus designed in such a way that these undesirable events with serious consequences happen rarely (i.e., have very small probabilities). The ability to numerically evaluate the risks associated with their use is therefore a major concern.

Example 1. A typical example is that of a large computing system. This type of model, originally from [11], has been used in most of the papers focusing on highly reliable Markovian systems. Our example consists of two sets of processors, each with two sets of disk controllers and six clusters of disks with four disks per cluster. Data are replicated in each cluster so that one disk can fail without affecting the system. Figure 6.1 describes the system. Failure propagations are possible, as will be seen later. There are two failure modes for each component. The system is considered as operational if all data are accessible from each processor type. This translates as follows: at least one processor, one controller in each set, and three out of four disks in each cluster are operational.

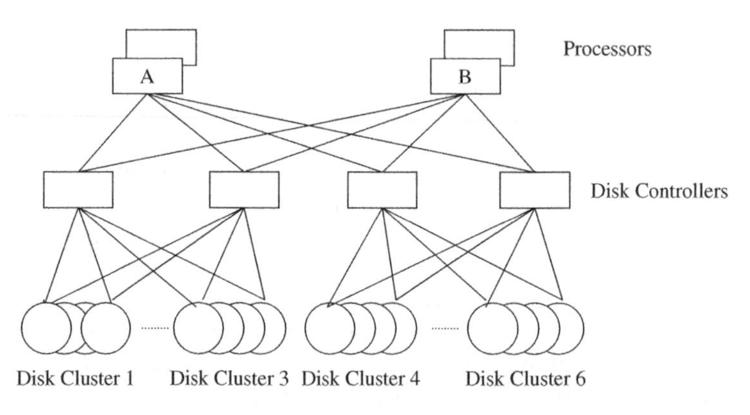

Figure 6.1 Block diagram for Example 15.

These finite Markovian models can in theory be analyzed by means of a rich set of efficient numerical procedures. Moreover, in some cases these techniques are basically insensitive to the rarity phenomenon. In practice things are different: the power of these representations that can capture quite accurately the behavior of complex systems leads very often to huge state spaces, rendering numerical approaches impracticable. Simulation is then the only possible evaluation tool, but the rarity problem becomes the bottleneck of the solution process. The same happens obviously if the model is not Markovian, for instance, if it is a semi-Markov process. The theory is much less rich in this case, and simulation is almost always the method of choice. We will briefly consider this case at the end of the chapter.

Another issue is the fact that some transitions, namely failures, are much rarer than repairs. From a modeling assumption point of view, we will introduce a rarity parameter ε, such that transition rates are decreasing with ε, while repair rates do not depend on it. Then the smaller ε is, the less likely the general system will fail. As extensively explained in previous chapters, standard (naive) simulation is inefficient because the event of interest, the failure of the system, is rare, and special procedures have to be implemented. Splitting and importance sampling (IS) are again the tools at hand. Note that, here, splitting is not relevant when $\varepsilon \to 0$. Indeed, observe that for the type of model we are looking at, it is each individual failure which is rare, and the number of transitions required to reach a failed state is therefore small (otherwise, the probability of the event would be meaningless). Splitting can therefore hardly be efficiently applied because it does not change the probabilities of individual transitions; it would be necessary to decompose each (rare) component failure in sub-events to make those events less rare themselves, which would be cumbersome. We will then focus on IS.

The chapter reviews the main results and techniques obtained in the domain mainly for steady-state but also for transient analysis [3, 5, 7, 9–11, 14, 17, 18, 21, 23, 25, 27, 36, 39–43], in the case of Markovian modeling. We also briefly deal with the non-Markovian case [15, 30, 32–35, 38,], and with sensitivity analysis [26, 29].

The chapter is organized as follows. Section 6.2 describes the mathematical model and its rare event parameterization. Section 6.3 looks at the estimation of steady-state measures, unavailability and the MTTF;[1] it shows how they can be estimated and describes the known IS schemes for obtaining an efficient simulation. Robustness properties, as described in Chapter 4 of this book, are also discussed. Section 6.4 looks in the same way at transient measures, such as the reliability at a given time. Section 6.5 gives a short introduction to and references on sensitivity analysis and non-Markovian models.

The following notation is used throughout the chapter. For a function $f : (0, \infty) \to \mathbb{R}$, we say that $f(\varepsilon) = o(\varepsilon^d)$ if $f(\varepsilon)/\varepsilon^d \to 0$ as $\varepsilon \to 0$; $f(\varepsilon) = O(\varepsilon^d)$ if $|f(\varepsilon)| \le c_1\varepsilon^d$ for some constant $c_1 > 0$ for all ε sufficiently small; $f(\varepsilon) = \underline{O}(\varepsilon^d)$ if $|f(\varepsilon)| \ge c_2\varepsilon^d$ for some constant $c_2 > 0$ for all ε sufficiently small; and $f(\varepsilon) = \Theta(\varepsilon^d)$ if $f(\varepsilon) = \underline{O}(\varepsilon^d)$ and $f(\varepsilon) = O(\varepsilon^d)$.

6.2 Model

We consider a highly reliable Markovian system with c types of components, n_i components of type i, for $i = 1, \ldots, c$, and $n = \sum_{i=1}^{c} n_i$ components in total. Each component is either in a failed state or in an operational state. The model is

[1] Even if this is rather a transient measure, we will see that we express it in a way that requires the estimation of steady-state quantities; the reason is that the time horizon is not finite in this case.

given by a continuous-time Markov chain (CTMC) $(Y(t))_{t\geq 0}$ defined over some state space $\mathcal{S}$. In the simplest version, a state $Y(t) = y \in \mathcal{S}$ is represented by a vector $y = (y^{(1)}, \ldots, y^{(c)})$, where $y^{(i)}$ is the number of *failed* components of type i. But the definition of a state can be more general: we can have several failure modes, and different classes of repairer with different scheduling or priority policies. More information may then be required to fully define a state. In the simplest case, the (finite) state space $\mathcal{S}$ is of cardinality $(n_1 + 1) \cdots (n_c + 1)$, therefore increasing exponentially with the number of component types. It is partitioned into two subsets $\mathcal{U}$ and $\mathcal{D}$, where $\mathcal{U}$ is the set of operational states, and $\mathcal{D}$ the set of failed states. $\mathcal{U}$ and $\mathcal{D}$ are such that from $y \in \mathcal{U}$ $(\mathcal{D})$, any repair (failure) still leads to a state in $\mathcal{U}$ $(\mathcal{D})$. We also define $\mathbf{0} \in \mathcal{U}$, the state in which all the components are operational.

We assume that the times to failure and times to repair of the individual components are independent exponential random variables. The rates are $\lambda_i(y) = o(1)$ for type i component failures when the current state is y, and $\mu(x, y) = \Theta(1)$ for repairs from a state x to a state y, grouped repairs being possible, as well as deferred repairs, even if this case is more complicated and will only be briefly considered here. Repairs do not depend asymptotically on ε when $\varepsilon \to 0$, meaning that they are not rare. The fact that $\lambda_i(y)$ decreases to zero with ε illustrates how rare failures can be. In most of the literature, as introduced in [39], $\lambda_i(y) = a_i(y)\varepsilon^{b_i(y)}$, where $a_i(y)$ and $b_i(y)$ are strictly positive values independent of ε. To deal with full generality, there may be some failure propagation, such that from state x, there is a probability $p_i(x, y)$ (which may depend on ε) that the failure of a type i component directly drives the system to state y, in which there could be additional component failures. The global failure rate from x to y is

$$\lambda(x, y) = \sum_{i=1}^{c} \lambda_i(x) p_i(x, y) = o(1).$$

It is useful to define Γ as the set of pairs $(x, y) \in \mathcal{S}^2$ for which a transition from x to y is possible, that is, such that $\lambda(x, y) > 0$ or $\mu(x, y) > 0$.

Note that when the performance measure of interest does not depend on the jump times of the CTMC $(Y(t))_{t\geq 0}$, it is relevant to rather simulate its canonically embedded discrete-time Markov chain (DTMC) $(X_j)_{j\geq 0}$, defined by $X_j = Y(\xi_j)$ for $j = 0, 1, 2, \ldots$, where $\xi_0 = 0$ and $0 < \xi_1 < \xi_2 < \ldots$ are the jump times of the CTMC. Its transition probability matrix $\mathbf{P}$ is

$$\mathbf{P}(x, y) = \mathbb{P}[X_j = y \mid X_{j-1} = x] = \frac{\lambda(x, y)}{q(x)}$$

if the transition from x to y corresponds to a failure, and

$$\mathbf{P}(x, y) = \frac{\mu(x, y)}{q(x)}$$

if it corresponds to a repair, where

$$q(x) = \sum_{z \in \mathcal{S}} (\lambda(x, z) + \mu(x, z))$$

is the total jump rate out of x, for all x, y in $\mathcal{S}$. By simulating the embedded DTMC instead of its counterpart CTMC, we reduce the variance of the estimator. Indeed, less randomness is included in the model since random jump times are replaced by their expected values (see also [16] for insights on this point); the randomness is now only in defining the sequences of states forming the paths observed during the simulation.

In our analysis, we will assume that the Markov chain is irreducible, that for each state $x \in \mathcal{S} \neq \mathbf{0}$ there exists a repair transition (meaning that a repairer is always working if a component is down), and that, if there is a direct transition from $\mathbf{0}$ to $z \in \mathcal{D}$, then $\mathbb{P}(\mathbf{0}, z) = o(1)$ (otherwise, reaching the failure set is not a rare event).

6.3 Steady-state analysis

6.3.1 Performance measures and simulation

In this section, we deal with the estimation of steady-state measures. In a reliability setting, two main measures of interest are the MTTF, representing the mean time to reach a failed state, and the steady-state unavailability, which is the steady-state probability of being in a failed state. How might we simulate those metrics? We then come to the classical steady-state output analysis of Monte Carlo simulation (see [2]): estimating the variance in order to get a valid confidence interval is a key issue. In this context *regenerative simulation* is an appropriate technique. Indeed, note first that, from a renewal argument [20], for any asymptotic measure of the form

$$r = \lim_{t \to \infty} \mathbb{E}(f(Y(t)))$$

we have

$$r = \frac{\mathbb{E}\left[\int_0^{\tau_x} f(Y_s)ds\right]}{\mathbb{E}\left[\tau_x\right]}, \tag{6.1}$$

for a real function f, where $\mathbb{E}[\cdot]$ is the expectation under the probability distribution $\mathbb{P}$ and τ_x is the return time (with finite expectation) to x given that we started in x. In our simulation context, assuming that we start from state $\mathbf{0}$, the stopping time $\tau_{\mathbf{0}}$ will be used.

Following a similar principle, if $\tau_M = \min(\tau_{\mathcal{D}}, \tau_{\mathbf{0}})$,

$$MTTF = \mathbb{E}[\tau_{\mathcal{D}}] = \frac{\mathbb{E}[\tau_M]}{\mathbb{P}[\tau_{\mathcal{D}} < \tau_{\mathbf{0}}]} \tag{6.2}$$

because, starting from $\mathbf{0}$,

$$\begin{aligned} MTTF &= \mathbb{E}[\tau_M] + \mathbb{E}[MTTF - \tau_M | \tau_{\mathbf{0}} < \tau_{\mathcal{D}}]\mathbb{P}[\tau_{\mathbf{0}} < \tau_{\mathcal{D}}] \\ &= \mathbb{E}[\tau_M] + \mathbb{E}[MTTF - \tau_M | Y(\tau_{\mathbf{0}}) = \mathbf{0}]\mathbb{P}[\tau_{\mathbf{0}} < \tau_{\mathcal{D}}] \\ &= \mathbb{E}[\tau_M] + \mathbb{E}[MTTF][1 - \mathbb{P}(\tau_{\mathcal{D}} < \tau_{\mathbf{0}}]). \end{aligned}$$

Using the discrete-time version, let $T_{\mathbf{0}}$ and $T_{\mathcal{D}}$ be respectively the corresponding return time to $\mathbf{0}$ and time when $\mathcal{D}$ is reached. Sojourn times in given states are then constant, equal to mean values ($1/q(i)$ for state $i \in \mathcal{S}$), so that (6.1) is written

$$r = \frac{\mathbb{P}\left[\sum_{k=0}^{T_{\mathbf{0}}-1} f_1(X_k)\right]}{\mathbb{P}\left[\sum_{k=0}^{T_{\mathbf{0}}-1} (1/q(X_k))\right]}$$

where, for all $i \in \mathcal{S}$, $f_1(i) = f(i)/q(i)$ (with $f(x) = 1_{\mathcal{D}}$ when estimating the unreliability), and

$$MTTF = \frac{\mathbb{E}\left[\sum_{k=0}^{\min(T_{\mathbf{0}}, T_{\mathcal{D}})-1} (1/q(X_k))\right]}{\mathbb{E}\left[1_{(T_{\mathcal{D}} < T_{\mathbf{0}})}\right]}. \quad (6.3)$$

We then simulate regenerative cycles, the ith being $C_i = \left(X_{T_{\mathbf{0}}^{(i)}}, \ldots, X_{T_{\mathbf{0}}^{(i+1)}-1}\right)$, where $T_{\mathbf{0}}^{(i)}$ is ith return to $\mathbf{0}$ and $T_{\mathbf{0}}^{(0)} = 0$.

Estimators of the unavailability u and of the MTTF are

$$\hat{u}_n = \frac{\sum_{i=1}^{n} F_1(C_i)}{\sum_{i=1}^{I} Q(C_i)} \quad (6.4)$$

$$\widehat{MTTF}_n = \frac{\sum_{i=1}^{n} G(C_i)}{\sum_{i=1}^{n} H(C_i)} \quad (6.5)$$

with $F_1(C_i)$ the sum of expected sojourn times in states of $\mathcal{D}$ over cycle C_i, $Q(C_i)$ the sum of expected sojourn times in all states over C_i, $G(C_i)$ the same sum but up to reaching $\mathcal{D}$ or returning to $\mathbf{0}$, and $H(C_i) = 1_{(T_{\mathcal{D}} < T_{\mathbf{0}})}(C_i)$ the indicator function that $\mathcal{D}$ was reached before coming to $\mathbf{0}$ over cycle C_i.

From now on, we derive the results only for the MTTF, but the same can easily be done for the unavailability. How can we get a confidence interval for the estimation? We apply the central limit theorem to the *independent* random variables $G(C_i) - MTTF \times H(C_i)$, with expectation zero and variance

$$\sigma^2 = \sigma^2(G) - 2MTTF\ \mathrm{Cov}(G, H) + (MTTF)^2\sigma^2(H)$$

(here G and H denote the generic random variable defined over a cycle, and $\sigma^2(\cdot)$ and $\mathrm{Cov}(\cdot, \cdot)$ are the variance and covariance under $\mathbb{P}$); we then have

$$\frac{\frac{1}{n}\sum_{i=1}^{n}(G(C_i) - MTTF\, H(C_i))}{\sigma/\sqrt{n}} \to \mathcal{N}(0, 1),$$

where $\mathcal{N}(0, 1)$ is the standard normal distribution, when the number n of cycles tends to infinity. In other words, just by dividing both the numerator and denominator by $\bar{H}_n = n^{-1} \sum_{i=1}^{n} H(C_i)$,

$$\frac{\sqrt{n}(\widehat{MTTF} - MTTF)}{\sigma / \bar{H}_n} \rightarrow \mathcal{N}(0, 1),$$

when the number n of cycles tends to infinity.

A Monte Carlo standard estimation of the MTTF will be inefficient because the denominator is the probability of a rare event (it will be the numerator if we deal with the unavailability). IS is a relevant way to cope with that problem. The first class of IS strategies is called *dynamic importance sampling* (DIS). We will not redefine IS here; for a more precise description, see Chapter 2. Basically, we replace the transition matrix $\mathbf{P}$ by another one $\tilde{\mathbf{P}}$ (with corresponding probability measure $\tilde{\mathbb{P}}$ and expectation $\tilde{\mathbb{E}}$). If the likelihood ratio over a cycle is

$$L(x_0, \ldots, x_T) = \frac{\mathbb{P}\{(X_0, \ldots, X_T) = (x_0, \ldots, x_n)\}}{\tilde{\mathbb{P}}\{(X_0, \ldots, X_T) = (x_0, \ldots, x_n)\}} = \frac{\prod_{i=0}^{T-1} \mathbf{P}(x_i, x_{i+1})}{\prod_{i=0}^{T-1} \tilde{\mathbf{P}}(x_i, x_{i+1})},$$

provided T has finite expectation under $\tilde{\mathbb{P}}$, then for any random variable Z defined over paths,

$$\mathbb{E}[Z] = \tilde{\mathbb{E}}[ZL].$$

A new estimator of the MTTF is then

$$\widehat{MTTF} = \frac{\sum_{i=1}^{n} G(C_i) L_i}{\sum_{i=1}^{n} H(C_i) L_i}$$

where L_i is the likelihood associated with the ith cycle.

Another method, *measure-specific dynamic importance sampling* (MSDIS), giving better results, was introduced in [11]. This involves simulating independently the numerator and denominator of (6.3), using different IS measures $\tilde{\mathbb{P}}_1$ for the numerator and $\tilde{\mathbb{P}}_2$ for the denominator. Indeed, the functions being different, reducing the variance for one does not necessary mean the same for the other. Of the total of n cycles, ξn are used to estimate the numerator, and $(1 - \xi)n$ for the denominator. A new estimator is then

$$\widehat{MTTF} = \frac{\sum_{i=1}^{\xi n} G(C_i^{(1)}) L_i^{(1)} / (\xi n)}{\sum_{i=1}^{(1-\xi)n} H(C_i^{(2)}) L_i^{(2)} / ((1 - \xi)n)}$$

where the $C_i^{(1)}$ and $L_i^{(1)}$ ($C_i^{(2)}$ and $L_i^{(2)}$) are the cycles and likelihood ratios corresponding to IS measure $\tilde{\mathbb{P}}_1$ ($\tilde{\mathbb{P}}_2$).

We then have the following result, using independent cycles (thus, the covariance term does not exist anymore): if

$$\bar{H}_{(1-\xi)n} = \frac{1}{(1-\xi)n} \sum_{i=1}^{(1-\xi)n} H(C_i^{(2)}) L_i^{(2)}$$

is an estimator of $E_{\mathbb{P}_2}(HL^{(2)})$, and if

$$\tilde{\sigma}^2 = \tilde{\sigma}_1^2(GL^{(1)}) + (MTTF)^2 \tilde{\sigma}_2^2(HL^{(2)})$$

where $\tilde{\sigma}_i^2(\cdot)$ is for the variance using $\tilde{\mathbb{P}}_i$ as the underlying probability measure, then

$$\frac{\sqrt{n}(\widehat{MTTF} - MTTF)}{\sigma / \bar{H}_{(1-\xi)n}} \to \mathcal{N}(0, 1).$$

6.3.2 Importance sampling simulation schemes and robustness properties

Many IS simulation schemes have been proposed in the literature. The basic principle is to increase the occurrence of failures. We review such schemes here, dividing them into three categories: first the basic schemes first; then those using some topological information; and finally those directly trying to approach the zero-variance change of measure. In each case, we will discuss the robustness properties as $\varepsilon \to 0$. The properties the literature has looked at are bounded relative error (BRE) and bounded normal approximation (BNA). Recall that BRE means that the relative variance remains bounded as $\varepsilon \to 0$, so that the relative precision of the confidence interval is insensitive to the rarity of the event, and BNA is a sufficient condition to assert that the coverage of the confidence interval will remain valid whatever the rarity. For more precise definitions, see Chapter 4 devoted to robustness properties. Those properties have been discussed at great length for highly reliable Markovian systems [27, 39, 41, 42]. Looking at all sample paths, necessary and sufficient conditions have been obtained. Basically, it is not sufficient that the most likely paths to failure are not rare (i.e., their probability is $\Theta(1)$) under the IS measure; other paths should not be *too rare* either (but not necessarily $\Theta(1)$). A string of properties has also been shown in [41, 42]: BNA implies that paths contributing the most to the variance are $\Theta(1)$ under IS measure, meaning that the variance is asymptotically properly estimated (Chapter 4 illustrates the problems that could occur otherwise), implying BRE, implying in turn that most likely paths to failure are $\Theta(1)$ under IS measure. For all those implications, the reverse assertion is not true in general; counterexamples have been highlighted in [42].

In what follows, since in our model transitions are either failures or repairs, we denote by $\mathcal{F}$ the set of failures and by $\mathcal{R}$ the set of repairs. If $x \neq \mathbf{0}$, we also denote $F_x = \{y : (x, y) \in \mathcal{F}\}$ and $R_x = \{y : (x, y) \in \mathcal{R}\}$, and let $f_x =$

$\sum_{y \in F_x} \mathbb{P}(x, y)$ and $r_x = \sum_{y \in R_x} \mathbb{P}(x, y)$ be the failure and repair probabilities from state x.

Basic schemes

The first proposal, called *failure biasing* (FB), first appeared in [23]. It simply increases the probability of the failure transitions to a fixed value $\alpha \in (0, 1)$; typically, $0.5 \leq \alpha \leq 0.9$. Then the probability of getting a failure is no longer $o(1)$. The transition probabilities are changed as follows:

- $\forall x \in \mathcal{U},\ x \neq \mathbf{0},\ (x, y) \in \mathcal{F}: \quad \tilde{\mathbb{P}}(x, y) = \alpha \dfrac{\mathbb{P}(x, y)}{f_x};$
- $\forall x \in \mathcal{U},\ x \neq \mathbf{0},\ (x, y) \in \mathcal{R}: \quad \tilde{\mathbb{P}}(x, y) = (1 - \alpha) \dfrac{\mathbb{P}(x, y)}{r_x}.$

The $\mathbb{P}(\mathbf{0}, \cdot)$s are not modified (since only a failure can happen from $\mathbf{0}$). Observe that the total probability of failure from x is now equal to α. From states in $\mathcal{D}$, the probabilities are not changed. Similarly, as soon as $\mathcal{D}$ has been reached, we switch back to $\mathbb{P}$. It was shown in [39] that, even if BRE is not satisfied in general by FB, it is the case for so-called *balanced systems*, that is, systems for which, from every state x, each failure transition has a probability of the same order of magnitude in terms of ε. But some (important) paths are still too rare when using FB, because one of its failure transitions in a given state can still have probability $o(1)$ due to a less rare failure under the initial law and which does not lead to 'interesting' states.

Based on this, *balanced failure biasing* (BFB) was suggested [39]; here, for the subset of failure transitions, the conditional individual probabilities, taken proportionally to the initial ones in FB, are replaced by uniform ones. Formally,

- $\forall x \in \mathcal{U},\ (x, y) \in \mathcal{F}: \quad \tilde{\mathbb{P}}(x, y) = \alpha \dfrac{1}{\mathrm{Card}(F_x)};$
- $\forall x \in \mathcal{U},\ x \neq \mathbf{0},\ (x, y) \in \mathcal{R}: \quad \tilde{\mathbb{P}}(x, y) = (1 - \alpha) \dfrac{\mathbb{P}(x, y)}{r_x}.$

From $\mathbf{0}$, we just use $\alpha = 1$. It is then shown in [39] that this scheme satisfies BRE, and in [41] that BNA is also satisfied.

Inverse failure biasing (IFB) [36] is based on the efficient simulation of the $M/M/1$ queue, involving switching arrival and service rates. $\tilde{\mathbb{P}}$ is then chosen as follows:

- If $x = \mathbf{0}, \forall y : \mathbb{P}(\mathbf{0}, y) > 0, \quad \tilde{\mathbb{P}}(\mathbf{0}, y) = \dfrac{1}{\mathrm{Card}(F_{\mathbf{0}})}.$
- $\forall x \in \mathcal{U},\ x \neq \mathbf{0}\ \forall y : (x, y) \in \mathcal{F}, \quad \tilde{\mathbb{P}}(x, y) = \dfrac{r_x}{\mathrm{Card}(F_x)},$

  and if $(x, y) \in \mathcal{R}, \quad \tilde{\mathbb{P}}(x, y) = \dfrac{f_x}{\mathrm{Card}(R_x)}.$

The probability of having a repair is then $o(1)$. Very efficient when most likely paths to failure involve only failures, it performs poorly when those paths involve some repairs [5] (because those paths are $o(1)$ under IFB).

In [3], *simple balanced likelihood ratio* methods are introduced to increase the frequency of component failures, but keeping bounded at the same time the likelihood ratios associated with regenerative cycles. The idea is to define stacks, initialized to empty sets, corresponding to failures with a given order of magnitude in terms of ε. Throughout the simulation of a cycle, likelihood ratios for a component failure are put on top of the corresponding stack, and this value is taken back (and removed) from the stack when there is a component repair which has a failure with the same order of magnitude, in order to cancel the current likelihood ratio. As a consequence, BRE is satisfied. See [3] for more details and a complete description.

In the above IS estimators, the variance comes from the variations of the likelihood ratio (disregarding the fact that we either do or do not hit the rare set). In [21] it is highlighted that reducing the variance of that likelihood ratio can increase the efficiency of the IS estimator if its does not significantly reduce the probability of the rare event, and if it does not require much more work. Such a variance reduction can be obtained using weight windows. Indeed, the likelihood ratio can be viewed as a weight that the simulated chain has accumulated so far. For each state the current weight multiplied by the expected remaining likelihood ratio must be equal to the value of interest. Given some estimation of the expected likelihood ratio from any state, we can decide at each step of the simulation to apply splitting to chains with excessive weights (therefore decreasing those weights), or to apply 'Russian roulette' (increasing the weight if the chain is not killed), if the weight is not included in a window. Another version simulates a fixed number of chains in parallel. The weight windows algorithm is then applied just to keep the number of chains constant, and each weight as close as possible to the expected value. It has been noted in [21] that the savings can be large, but the estimator can also be very poor if the weight windows are wrongly selected. The difficulty is to get a good approximation of the expected likelihood ratio from any state. As a rough approximation, the most likely path, or direct paths, can be considered.

Some refinements were proposed in [17, 18] for the case where we have deferred (or grouped) repairs; that is, when there are states other than $\mathbf{0}$ for which only failures are possible. This induces high probability cycles for which the above methods can lead to very large and even infinite variances. In [17, 18] the probabilities along those cycles are not reduced too much.

How do these IS schemes perform in practice? For space reasons, we limit ourselves to studying Example 1 with BFB and IFB. For an extensive numerical study, depending on the topology, the reader may consult [5]. Assume that failure rates of processors, controllers and disks are 5×10^{-6}, 2×10^{-6}, and 2×10^{-6} respectively, leading to $\gamma = \mathbb{E}\left[1_{(T_D < T_0)}\right] \approx 5.55 \times 10^{-6}$, the difficult component in the estimation of the MTTF. This is not too rare, but allows us to compare with standard simulation (a rarer event would indeed yield a confidence interval (0, 0) because the rare event is unlikely to occur even once). We compare the results for $n = 10^7$ independent replications. Crude Monte Carlo yields a confidence interval $(4.28015 \times 10^{-6}, 6.71985 \times 10^{-6})$, for an empirical variance with value 5.5000×10^{-6}. On the other hand, BFB yields a confidence interval $(5.50120 \times 10^{-6}, 5.58070 \times 10^{-6})$ and an empirical variance 5.8397×10^{-10}, which is an improvement by a factor of about ten thousand. IFB is even more efficient due to the fact that only direct paths to failure are important here; it yields a confidence interval $(5.53473 \times 10^{-6}, 5.55203 \times 10^{-6})$, and an empirical variance of 2.7681×10^{-10}.

Using topological information

The above sampling schemes use only information that is 'local' (with respect to the mathematical model), about direct transitions being failures or repairs. On the other hand, more general topological information could be used, detecting how far we are from the set of failed states.

A first technique, probably the simplest, called *selective failure biasing*, also called Bias2 failure biasing [11], is a refinement of FB. The only difference is that from any $x \in \mathcal{U}$, the set of failure transitions is separated into two subsets: the set of failure transitions corresponding to component types having already at least one failed component of this type, for which a conditional probability is set to α_1; and the set of transitions that fails types of components that are always functioning (with conditional probability $1 - \alpha_1$). If the first (second) set is empty, then $\alpha_1 = 0$ ($\alpha_1 = 1$). In each subset, individual probabilities are (still) taken proportional to the initial ones. The intuition is that pushing more failures of component types already having failed components should get the chain closer to the set of failed states.

Using a similar approach, a *selective failure biasing for 'series-like' systems* (SFBS) can be designed [5]. Assume that the system's structure is close to the situation where the system is functioning if and only if, for each type k component, the number of operational components is greater than or equal to some threshold l_k. A reasonable way to improve selective failure biasing is to make more probable the failures of class-k components when the number of components remaining operational, $n_k - x_k$ if the state is x, is closer to the threshold l_k. The set to which a probability α_1 is assigned in state x is then the set of transitions including type k component failures such that $(n_k - x_k) - l_k$ is minimum.

Still in [5], another version, called *selective failure biasing for 'parallel-like' systems*, behaves in a similar way but is designed to deal with systems working as sets of l_k-out-of-N_k modules in parallel, $1 \le k \le K$. From $x \in \mathcal{U}$, a component type k is said to be *critical* if its number of operational components is larger than l_k. A transition (x, y) is said to be *critical* if $y_k > x_k$ for some critical type k. The method proceeds like SFBS: using in the same way two parameters α and α_1, the principle is to accelerate the critical transitions first, then the non-critical ones, by means of the respective weights $\alpha\beta$ and $\alpha(1 - \beta)$. Again, in each subset, individual probabilities are taken proportionally to the original ones.

Distance-based selected failure biasing (DSFB) [5, 7] was created to deal with systems involving a more general structure, with failure propagation, and with less generic properties. It therefore requires more information about the system, that is, some work to learn the topology of the model. From current state x, we first need to compute for each y such that $(x, y) \in \Gamma$ the distance $d(y)$ to $\mathcal{D}$ defined as $d(y) = \min_{z \in D} \sum_k (y_k - z_k)$, which might be computationally demanding for general failure sets, but could also be very easy to evaluate for some specific models. The set of failure transitions (to which a probability α is still assigned) is decomposed into the set of failure transitions with the ℓth smallest distance to $\mathcal{D}$, which receives the conditional probability $\alpha_1(1-\alpha_1)^{\ell-1}$, except the last one which has conditional probability $(1-\alpha_1)^{\ell}$. Again, in each subset, individual probabilities are taken proportionally to the original ones.

None of the above methods satisfy BRE in general, even it is proved to be the case for balanced systems [5]. For this reason, balanced versions of those IS schemes have been considered in [5, 41], by taking in each subset of *failure* transitions uniform probabilities instead of probabilities proportional to the initial ones. That way, BRE and even the stronger BNA property are satisfied.

Structural information is also used in [3] to improve the efficiency of the simple balanced likelihood ratio, putting more probability on short paths to failure. Looking at the Markov chain as a graph, it identifies from the current state x transitions that are on mincuts, and put a high conditional probability on that set of transitions. Different stacks are then defined, depending on whether transitions are on mincuts or not. The same observations as for the DSFB technique are relevant here: the way the model is specified and the regularities in the chain's structure are critical to the applicability of the method.

Trying to approach the zero-variance change of measure

All the aforementioned IS proposals are designed to reach the set of failed states faster than with a standard simulation, but none directly investigates how the zero-variance change of measure can be approached, or how far the algorithm is from it. The zero-variance change of measure is extensively studied in Chapter 2. To try to be specific, we only deal here with the estimation of $\gamma = \mathbb{P}[T_{\mathcal{D}} < T_{\mathbf{0}}]$, the denominator and critical quantity in the estimation of the MTTF.

Define $\gamma(x) = \mathbb{P}[T_{\mathcal{D}} < T_{\mathbf{0}} | X_0 = x]$. It is then straightforward to see that the (Markov chain) zero-variance change of measure is given by (see Chapter 2)

$$\tilde{\mathbf{P}}(x, y) = \frac{\mathbf{P}(x, y)\gamma(y)}{\sum_z \mathbf{P}(x, z)\gamma(z)} \tag{6.6}$$

(with $\gamma(z)$ replaced by 0 on the right-hand side if $z = \mathbf{0}$). Implementing this change of measure involves he knowledge of the values $\gamma(z)$ for all z, that is, even more than what we are looking for (the number $\gamma(\mathbf{0})$), but it might be of interest to replace $\gamma(x)$ by an approximation in (6.6) in order to be close to the zero-variance estimator. The learning algorithms of [1, 19] require storage of an estimation of $\gamma(x)$ for each x, updated in the course of the simulation. This is realistic if the model is small, but not relevant when we address most interesting models, having at least very large state spaces. Similarly, in [37], the whole IS matrix $\tilde{\mathbf{P}}$ is learned using the cross-entropy technique (again see Chapter 2), which requires the storage of even more information. For this reason, a suggestion for obtaining an approximation of the zero-variance estimator at no cost in terms of storage and computation (and as a consequence which resists a state space size increase) is to use a rough (once and for all) guess of function $\gamma(x)$. In [22], $\gamma(x)$ is replaced by the probability of the most likely path to failure (an underestimate), a problem equivalent to computing the shortest path from a state to a set of states in a graph (this could add some computational burden depending on the structure of the set of failed states, as observed before for procedures such as DSFB or the one in [3]). This way, the method is difficult to use in situations where the distance to the set of failed states is difficult, costly or simply impossible to evaluate. For Example 1, using the same parameters as when testing BFB and IFB, the method yields a confidence interval $(5.54579 \times 10^{-6}, 5.54595 \times 10^{-6})$ and an impressive variance reduction, the empirical variance being 2.5407×10^{-15}, which is two hundred thousand times smaller than for BFB and IFB. This is due to the very small value of ε, and therefore the very good approximation of $\gamma(x)$.

6.4 Transient analysis

6.4.1 Performance metrics and simulation

Steady-state performance metrics and (independent of a finite time horizon) the MTTF are not the the only measures of interest in dependability. In some situations we need to know what happens over a time interval $[0, t]$. This can be captured, for instance, by the unreliability at t (the probability of reaching $\mathcal{D}$ before t), or by the expected interval unavailability on the interval (the proportion of time within $[0, t]$ the system spends in $\mathcal{D}$). We will investigate two cases here: first, when the time horizon t does not depend on ε (i.e., $t = \Theta(1)$); then, the case where t is $\underline{O}(\varepsilon^{-1})$, that is, t increases as the rarity parameter goes to zero. The case where it is $o(1)$ is a special case of $\Theta(1)$, provided repair rates can also be arbitrarily small, just by multiplying (rescaling) all transition rates by t and looking at the performance measure on the time interval $[0, 1]$.

How do we simulate such a process? In contrast to steady-state measures which only require us to simulate the embedded DTMC, here we need to take the distribution of holding times into account. We therefore simulate the CTMC. A standard simulation generates n trajectories of the CTMC on $[0, t]$. It progressively gererates the successive states of the chain from the embedded DTMC (using transition matrix $\mathbf{P}$), and the holding time in each state, up to time t. From the n trajectories, the average value of the point estimates (depending on the performance measure we are looking at) gives the estimator, and the sample variance (and, as a consequence, a confidence interval) can easily be constructed. Here again, since the event of interest is rare, specific procedures need to be applied. For instance, if t is fixed, just obtaining the first failure event before t is a rare event.

6.4.2 Importance sampling simulation schemes and robustness properties for small time horizons

The main technique in the case where the time horizon is small (i.e., $\Theta(1)$) is 'forcing+BFB' [23, 31, 38]. Forcing means that the probability distribution for the time until the first failure from the initial state $\mathbf{0}$ is replaced by the distribution conditional on that failure happening before the time horizon. We therefore force that transition to happen before t. When that transition has happened, the holding times are not changed; they still follow their initial distributions. Observe that if a repairer is active in each of those states and the repair rates are $\Theta(1)$, then no rarity is involved in the holding times. However, a new probability matrix $\tilde{\mathbf{P}}$, based on the BFB procedure, is used for choosing the sequence of successive states. Note that any of the other biasing schemes could be applied in place of BFB, but a balanced scheme is preferable (see below). In this situation, the likelihood ratio is actually the ratio of the holding time densities mutiplied by the likelihood ratio for the embedded DTMC. It has been proved for the unreliability in [31], and for the expected interval unavailability in [38], that forcing+BFB (or forcing + any other balanced scheme) yields BRE for small time horizons.

In [8], a general expression for the zero-variance change of measure applied to the estimation of the unreliability in this transient case is proposed. Denote the time limit by T and define $\gamma(x, t)$ as the probability of reaching the set of failed states before T if the CTMC is in state x and there remain t units of time (i.e., the current time is $T - t$). The zero-variance density of going from x to x', given that the remaining time was t when reaching x, and time to go to x' is δ, is given by

$$g_0(x', t - \delta \mid x, t) = \frac{\gamma(x', t - \delta)}{\gamma(x, t)} \mathbf{P}(x, x') q(x) e^{-q(x)\delta}.$$

Note that it is no longer exponential. Still in [8], an approximation is developed via a power series, replacing $\gamma(x, t)$ by the probability of reaching $\mathcal{D}$ in $[0, t]$ with direct paths to failure. It is shown that this approximation can provide a BRE when the rarity parameter goes to zero.

In [6] another approach is followed to estimate the same unreliability-at-t metric. The specific property of the model that is exploited here is that it captures the behavior of a network, which means that the function specifying when the system is operational is based on the properties of the graph corresponding to the network connections. More specifically, assume the network (system) is operational if and only if there is at least a path connecting the sender of messages and the receiver with all its links working. The basic idea is close to forcing. It consists in writing the reliability at t by conditioning on the event 'at least one component in a path set fails before t'. This changes the distributions of the times to failure of the components in the selected path. The same idea can be followed using a set of disjoint paths between the node sending data and the node receiving it (or, in a more general setting, when more than two nodes are exchanging information, replacing paths by trees). Very large variance reductions and efficiency indexes can be obtained in this way for this type of system. See [6] for details.

6.4.3 Importance sampling simulation schemes and robustness properties for large time horizons

Assume that the failure rate of a type i component is $\lambda_i \varepsilon^{r_i}$ and define r_0 as the minimum of the r_i, $i = 1, \ldots, c$. Also let r be such that $\gamma = \mathbb{P}[\tau_{\mathcal{D}} < \tau_{\mathbf{0}}] = \Theta(\varepsilon^r)$. The unreliability and interval unavailability are both $\Theta(\varepsilon^{r_0+r})$. Now let t be such that $t = \Theta(\varepsilon^{-r_t})$ for some strictly positive r_t. This models arbitrarily large horizon times depending on the rarity parameter at the same time as transition rates. In [31, 38], it is proved that if $r_t \leq r_0$ (i.e., if the probability of reaching $\mathcal{D}$ is smaller than getting the first failure before t), forcing+BFB still yields BRE. On the other hand, it is not the case anymore for $r_t > r_0$, because the variance of the likelihood ratio increases exponentially fast in the number of transitions of the Markov chain. Another simulation scheme is required in that case. An alternative for obtaining robust estimations is to use upper and lower bounds of the unreliability, as explored in [31], or in [38] for the interval unavailability. Those bounds use the property that the number of regenerative cycles between returns to $\mathbf{0}$ until the first system failure follows a geometric distribution with success probability γ. The time to first failure is then approximated by an exponential distribution with rate $\gamma/\mathbb{E}[\tau_{\mathbf{0}}]$. For specific (long) formulas, see [31, 38]. The basic idea is to replace the estimation of the measure of interest by upper and lower bounds requiring the estimation of measures for which the number of simulated events is not a critical issue. For instance, a simple upper bound of the time interval unavailability is $\mathbb{E}[D]q(\mathbf{0})$, where D is the random amount of time during which the system is down. Then it is shown that the *relative error of the bound*, defined as the difference between the upper and the lower bounds divided by twice the measure of interest, tends to zero when $\varepsilon \to 0$, this if and only if $r_t > r_0$.

Note that, to the best of our knowledge, no zero-variance-based estimator has been developed yet in this context. This thus remains an open question.

6.5 Other issues

This section is devoted to the issues of sensitivity analysis and non-Markovian simulation, which space considerations prevent us from developing in much depth.

Sensitivity analysis is an important issue for an architecture designer. It helps identify those parts of the system into which she should direct her efforts to improve the overall system dependability. Sensitivity analysis is done by computing derivatives of the metric considered in terms of the parameters we can play with. With full generality, assuming the performance measure is $\alpha(t,\theta) = \mathbb{E}[V(t,\theta)]$ for some random variable V, for a time horizon t ($t = \infty$ for steady-state measures) and θ the parameter value of interest, and denoting by $\mathbb{P}_\theta$ is the probability measure with that parameter value, let us write

$$\alpha(t,\theta) = \int V(t,\theta)d\mathbb{P}_\theta = \int V(t,\theta)\frac{d\mathbb{P}_\theta}{d\mathbb{P}_{\theta_0}}d\mathbb{P}_{\theta_0} = \mathbb{E}_{\theta_0}[V(t,\theta)L(t,\theta)],$$

with $L(t,\theta) = d\mathbb{P}_\theta/d\mathbb{P}_{\theta_0}$ the likelihood ratio. Under mild conditions,

$$\frac{d}{d\theta}\alpha(t,\theta) = \mathbb{E}_{\theta_0}\left[\frac{d}{d\theta}(V(t,\theta)L(t,\theta))\right].$$

Specific conditions for this to hold, and the performance of the biasing techniques for this estimation, when θ is the repair or failure rate of a component type, are discussed in [26, 27, 29, 30]. It can be shown that, in many situations, the numbers $\alpha(t,\theta)$ and $d\alpha(t,\theta)/d\theta$ have the same logarithmic asymptotics as rarity increases, for fixed θ. This means that efficient techniques for the estimation of α can in general be extended to the estimation of its sensitivities. See [4] for this, and [24, 28] for other related results.

In the case where the process is not driven by a Markov chain because failures or repairs are not exponentially distributed, there are two options. We can simulate the generalized semi-Markov process representing the global state of the system; this basically involves extending the state space by including the remaining clock for each non-exponential distribution. The augmented state then defines a CTMC where again failures can be accelerated [33, 38]. Another possibility is to use the uniformization procedure [15, 32, 34, 35]. To give a specific example of a possible approach in this context, IS schemes such as the ones described in this chapter for Markovian models can be designed by replacing the failure rate λ_i of a type i component, in the Markovian case, by the hazard function associated with class i. Recall that if $f_i(t)$ and $F_i(t)$ are respectively the density and cdf of the time to failure of a class-i component at time t, then the corresponding hazard function value is $h_i(t) = f_i(t)/[1 - F_i(t)]$. The interpretation is that the probability of observing a failure for such a component between t and $t + dt$, given that the component is working at time t, is $\approx h_i(t)dt$ (equal to $\lambda_i dt$ in the Markovian case). Under some assumptions about the hazard functions of these distributions (and the corresponding ones for the repair transitions), it is possible to follow approaches similar to the methods designed in a Markovian setting. See [12] and the references therein for details.

References

[1] I. Ahamed, V.S. Borkar, and S. Juneja. Adaptive importance sampling for Markov chains using stochastic approximation. *Operations Research*, **54**(3): 489–504, 2006.

[2] C. Alexopoulos and S.-H. Kim. Output data analysis for simulations. In E. Yücesan, C.-H. Chen, J.L. Snowdon, and J.M. Charnes, eds, *Proceedings of the 2002 Winter Simulation Conference*, pp. 85–96. IEEE Press, 2002.

[3] C. Alexopoulos and B.C. Shultes. Estimating reliability measures for highly-dependable Markov systems, using balanced likelihoos ratios. *IEEE Transactions on Reliability*, **50**(3): 265–280, 2001.

[4] S. Asmussen and R. Rubinstein. Sensitivity analysis of insurance risk models via simulation. *Management Science*, **45**: 1125–1141, 1999.

[5] H. Cancela, G. Rubino, and B. Tuffin. MTTF estimation by Monte Carlo methods using Markov models. *Monte Carlo Methods and Applications*, **8**(4): 312–341, 2002.

[6] H. Cancela, G. Rubino, and M. Urquhart. Path set based conditioning for transient simulation of highly dependable networks. In *Proceedings of the 7th IFAC Symposium on Cost Oriented Automation*, 2004.

[7] J.A. Carrasco. Failure distance based on simulation of repairable fault tolerant systems. *Proceedings of the 5th International Conference on Modeling Techniques and Tools for Computer Performance Evaluation*, pp. 351–365, 1992.

[8] P.T. De Boer, P. L'Ecuyer, G. Rubino, and B. Tuffin. Estimating the probability of a rare event over a finite horizon. In *Proceedings of the 2007 Winter Simulation Conference*. IEEE Press, 2007.

[9] V. Demers, P. L'Ecuyer, and B. Tuffin. A combination of randomized quasi-Monte Carlo with splitting for rare-event simulation. *In Proceedings of the 2005 European Simulation and Modeling Conference*, pp. 25–32, Ghent, Belgium, 2005. EUROSIS.

[10] A. Goyal, P. Heidelberger, and Shahabuddin P. Measure specific dynamic importance sampling for availability simulations. *In Proceedings of the 1987 Winter Simulation Conference*, pp. 351–357, 1987.

[11] A. Goyal, P. Shahabuddin, P. Heidelberger, V.F. Nicola, and P.W. Glynn. A unified framework for simulating Markovian models of highly reliable systems. *IEEE Transactions on Computers*, **C-41**: 36–51, 1992.

[12] P. Heidelberger. Fast simulation of rare events in queueing and reliability models. In L. Donatiello and R. Nelson, eds, *Performance Evaluation of Computer and Communication Systems*, Lecture Notes in Computer Science 729, pp. 165–202. Springer, New York, 1993.

[13] P. Heidelberger. Fast simulation of rare events in queueing and reliability models. *ACM Transactions on Modeling and Computer Simulation*, **5**(1): 43–85, 1995.

[14] P. Heidelberger, V.F. Nicola, and P. Shahabuddin. Simultaneous and efficient simulation of highly dependable systems with different underlying distributions. In *Proceedings of the 1992 Winter Simulation Conference*, pp. 458–465. IEEE Press, 1992.

[15] P. Heidelberger, P. Shahabuddin, and V.F. Nicola. Bounded relative error in estimating transient measures of highly dependable non-Markovian systems. *ACM Transactions on Modeling and Computer Simulation*, **4**(2): 137–164, 1994.

[16] D.L. Iglehart and R. Schassberger. Discrete time methods for simulating continuous time Markov chains. *Advances in Applied Probability*, **8**: 772–788, 1976.

[17] S. Juneja and P. Shahabuddin. In *Proceedings of the 22nd International Symposium on Fault-Tolerant Computing*, pp. 150–159. IEEE Computer Society Press, 1992.

[18] S. Juneja and P. Shahabuddin. Fast simulation of Markov chains with small transition probabilities. *Management Science*, **47**(4): 547–562, 2001.

[19] S. Juneja and P. Shahabuddin. Rare event simulation techniques: An introduction and recent advances. In S.G. Henderson and B.L. Nelson, eds, *Simulation*, Handbooks in Operations Research and Management Science, pp. 291–350. Elsevier, Amsterdam, 2006.

[20] J. Keilson. *Markov Chain Models – Rarity and Exponentiality*. Springer, New York, 1979.

[21] P. L'Ecuyer and B. Tuffin. Splitting with weight windows to control the likelihood ratio in importance sampling. In *Proceedings of ValueTools 2006: International Conference on Performance Evaluation Methodologies and Tools*, Pisa, Italy, 2006. ACM Publications.

[22] P. L'Ecuyer and B. Tuffin. Effective approximation of zero-variance simulation in a reliability setting. In *Proceedings of the 2007 European Simulation and Modeling Conference*, pp. 48–54, Ghent, Belgium, 2007. EUROSIS.

[23] E.E. Lewis and F. Böhm. Monte Carlo simulation of Markov unreliability models. *Nuclear Engineering and Design*, **77**: 49–62, 1984.

[24] M.K. Nakayama. General conditions for bounded relative error in simulations of highly reliable Markovian systems. Technical Report RC 18993, IBM Research Division, T. J. Watson Research Center, Yorktown Heights, NY, 1993.

[25] M.K. Nakayama. A characterization of the simple failure biasing method for simulations of highly reliable Markovian systems. *ACM Transactions on Modeling and Computer Simulation*, **4**(1): 52–88, 1994.

[26] M.K. Nakayama. Asymptotics of likelihood ratio derivatives estimators in simulations of highly reliable Markovian systems. *Management Science*, **41**(3): 524–554, 1995.

[27] M.K. Nakayama. General conditions for bounded relative error in simulations of highly reliable Markovian systems. *Advances in Applied Probability*, **28**: 687–727, 1996.

[28] M.K. Nakayama. On derivative estimation of the mean time to failure in simulations of highly reliable Markovian systems. *Operations Research*, **46**: 285–290, 1998.

[29] M.K. Nakayama, A. Goyal, and P.W. Glynn. Likelihood ratio sensitivity analysis for Markovian models of highly dependable systems. *Operations Research*, **42**(1): 137–157, 1994.

[30] M.K. Nakayama and P. Shahabuddin. Likelihood ratio derivative estimation for finite-time performance measures in generalized semi-Markov processes. *Management Science*, **44**(10): 1426–1441, 1998.

[31] M.K. Nakayama and P. Shahabuddin. Quick simulation methods for estimating the unreliability of regenerative models of large highly reliable systems. *Probability in the Engineering and Information Sciences*, **18**: 339–368, 2004.

[32] V.F. Nicola, P. Heidelberger, and P. Shahabuddin. Uniformization and exponential transformation: Techniques for fast simulation of highly dependable non-Markovian systems. In *Proceedings of the 22nd International Symposium on Fault-Tolerant Computing*, pp. 130–139. IEEE Computer Society Press, 1992.

[33] V.F. Nicola, M.K. Nakayama, P. Heidelberger, and A. Goyal. Fast simulation of dependability models with general failure, repair and maintenance processes. In *Proceedings of the 20th International Symposium on Fault-Tolerant Computing*, pp. 491–498. IEEE Computer Society Press, 1990.

[34] V.F. Nicola, M.K. Nakayama, P. Heidelberger, and A. Goyal. Fast simulation of highly dependable systems with general failure and repair processes. *IEEE Transactions on Computers*, **42**(12): 1440–1452, December 1993.

[35] V.F. Nicola, P. Shahabuddin, P. Heidelberger, and P.W. Glynn. Fast simulation of steady-state availability in non-Markovian highly dependable systems. In *Proceedings of the 23rd International Symposium on Fault-Tolerant Computing*, pp. 38–47. IEEE Computer Society Press, 1993.

[36] C. Papadopoulos. A new technique for MTTF estimation in highly reliable Markovian systems. *Monte Carlo Methods and Applications*, **4**(2): 95–112, 1998.

[37] A. Ridder. Importance sampling simulations of Markovian reliability systems using cross-entropy. *Annals of Operations Research*, **143**: 119–136, 2005.

[38] P. Shahabuddin. Fast transient simulation of Markovian models of highly dependable systems. *Performance Evaluation*, **20**: 267–286, 1994.

[39] P. Shahabuddin. Importance sampling for the simulation of highly reliable Markovian systems. *Management Science*, **40**(3): 333–352, 1994.

[40] P. Shahabuddin, V.F. Nicola, P. Heidelberger, A. Goyal, and P.W. Glynn. Variance reduction in mean time to failure simulations. In *Proceedings of the 1988 Winter Simulation Conference*, pp. 491–499. IEEE Press, 1988.

[41] B. Tuffin. Bounded normal approximation in simulations of highly reliable Markovian systems. *Journal of Applied Probability*, **36**(4): 974–986, 1999.

[42] B. Tuffin. On numerical problems in simulations of highly reliable Markovian systems. In *Proceedings of the 1st International Conference on Quantitative Evaluation of Systems (QEST)*, pp. 156–164. Los Alamitos, CA: IEEE Computer Society Press, 2004.

[43] B. Tuffin, W. Sandmann, and P. L'Ecuyer. Robustness properties in simulations of highly reliable systems. In *Proceedings of RESIM 2006*, University of Bamberg, Germany, October 2006.

7

Rare event analysis by Monte Carlo techniques in static models

Héctor Cancela, Mohamed El Khadiri and Gerardo Rubino

This chapter discusses Monte Carlo techniques for rare event simulation in the case of *static models*, that is, models in which time is not an explicit variable. The main example and the one that will be used in the chapter is the network reliability analysis problem, where the models are graphs with probabilities associated with their components (with arcs or edges, and/or with nodes). Other typical names in this domain are fault trees, block diagrams, etc. All these models are in general solved using combinatorial techniques, but only for quite small sizes, because their analysis is extremely costly in terms of computational resources. The only methods able to deal with models having arbitrary size are Monte Carlo techniques, but there the main difficulty is with the rare event case, the focus of this chapter. In many areas (e.g., telecommunications, transportation systems, energy productions plants), either the components are very reliable or redundancy schemes are adopted, resulting in extremely reliable systems. This means that a system's failure is (or should be) a rare event.

Rare Event Simulation using Monte Carlo Methods Edited by G. Rubino and B. Tuffin

7.1 Introduction

The most commonly discussed example in the area of static models in dependability analysis is the network reliability problem. This concerns the evaluation of reliability metrics of large classes of multicomponent systems. We will denote by $\mathcal{E}$ the set of components in the system (which will shortly be represented by the set of edges of the undirected graph modeling the system). In general, the structure of such a system is represented by a binary function Φ of $|\mathcal{E}|$ binary variables. The usual convention for the state of a component or for the whole system is that 1 represents the operational state (the device, component or system is operational or *up*) and 0 represents the failed or *down* state. A *state vector* or *system configuration* is a vector $\vec{x} = (x_1, \ldots, x_{|\mathcal{E}|})$ where x_i is a possible state, 0 or 1, of the ith component (i.e., $\vec{x}$ is an element of $[0, 1]^{|\mathcal{E}|}$). With this notation, $\Phi(\vec{x}) = 1$ if the system is up when the configuration is $\vec{x}$, and 0 otherwise.

We may have different structure functions associated with the same system, each addressing a specific aspect of interest that must be evaluated (see below). Frequently (but not always) structure functions are *coherent*, corresponding to systems satisfying the following properties: (i) when all the components are down (up), the system is down (up); (ii) if the system is up (down) and we change the state of a component from 0 to 1 (from 1 to 0), the system remains up (down); (iii) all the components are relevant (a component i is irrelevant if the state of the system does not depend on the state of i). Formally, let us denote by $\vec{0}$ (by $\vec{1}$) a state vector having all its entries equal to 0 (equal to 1). We also denote by $\vec{x} \leq \vec{y}$ the relation $x_i \leq y_i$ for all i, by $\vec{x} < \vec{y}$ the fact that $\vec{x} \leq \vec{y}$ with, for some j, $x_j < y_j$, and by $(\vec{x}, 0_i)$ (by $(\vec{x}, 1_i)$) the state vector constructed from $\vec{x}$ by setting x_i to 0 (to 1). Then, Φ is coherent if and only if (i) $\Phi(\vec{0}) = 0$, $\Phi(\vec{1}) = 1$; (ii) if $\vec{x} < \vec{y}$ then $\Phi(\vec{x}) \leq \Phi(\vec{y})$; and (iii) for each component i there exists some state vector $\vec{x}$ such that $\Phi(\vec{x}, 0_i) \neq \Phi(\vec{x}, 1_i)$ (and thus, due to (ii), $\Phi(\vec{x}, 0_i) = 0$ and $\Phi(\vec{x}, 1_i) = 1$).

After specifying the function Φ, which defines how the system provides the service for which it was designed, a probabilistic structure must be added to take the failure processes into account. The usual framework is to assume that the state of the ith component is a random binary (Bernoulli) variable X_i with expectation $\mathbb{E}(X_i) = r_i$, and that the $|\mathcal{E}|$ random variables $X_1, \ldots, X_{|\mathcal{E}|}$ are independent. The numbers $r_i = \mathbb{P}(X_i = 1)$ (called the *elementary reliabilities*) are input data. Sometimes we will also use the notation q_i for the *unreliability* of link i, that is, $q_i = 1 - r_i$. The output parameter is the reliability R of the system, defined by

$$R = \mathbb{P}(\Phi(\vec{X}) = 1) = \mathbb{E}(\Phi(\vec{X})) \tag{7.1}$$

where $\vec{X} = (X_1, \ldots, X_{|\mathcal{E}|})$, or its *unreliability* $Q = 1 - R$. Observe that this is a static problem, that is, time is not explicitly used in the analysis. When time relations are considered, the context changes and the general framework in which the analysis is usually done is the theory of stochastic processes and, in particular, of Markov processes (see Chapter 2). For an exposition concerning the general theory (including dynamic models), see [6], [7] or [32].

The structure function can be specified by providing a table describing the mapping from $[0, 1]^{|\mathcal{E}|}$ into $[0, 1]$, a sort of exhaustive description, or, on the other side of the spectrum, by a program (or algorithm), which usually is a compact way of giving the function. An intermediate option is to define it by giving a *stochastic graph*, sometimes called a *network* in this context. These models are very useful, in particular, for communication network analysis. We will adopt them here as referencesystems. The lines of the communication network are modeled by the edges (or by the arcs in the directed case) of the graph, and the vertices represent the nodes. The basic model in this class (and in this chapter) is an undirected graph (lines are assumed to be bidirectional) with perfect nodes (corresponding to the situation where the reliability of a node is much higher than the reliability of a line), assumed to be connected and without loops. The state of line i at some instant of interest is a binary random variable X_i. The structure function Φ is then specified by means of some property of the graph. To be more specific, let us denote by $\mathcal{G} = (\mathcal{V}, \mathcal{E})$ the graph where $\mathcal{V}$ is the set of vertices and $\mathcal{E}$ is the set of edges. The set $\mathcal{E}'$ of operational lines at the fixed instant considered defines a random subgraph $G = (\mathcal{V}, \mathcal{E}')$ of $\mathcal{G}$. The reliability of the system is then the probability that G has some graph property. For instance, if we are interested in the fact that all the nodes can communicate with each other and we want to quantify the ability of the network to support this, the corresponding metric, called *all-terminal* reliability, is the probability that G is connected. Another important case is when the user is interested only in the communications between two particular nodes, usually called *source* and *terminal*. Denoting these nodes by s and t, the associated metric is the so-called *two-terminal* or *source-to-terminal* reliability, defined as the probability that there exists in G at least one path between s and t (that is, a path in $\mathcal{G}$ having all its lines operational in G). This last case of graphs having an 'entry' point s and an 'exit' point t (the terminology is used even if the graph is undirected), has a broad field of applications since it is a general tool describing the structure of a system, its *block diagram*, not only in the communications area. For instance, it is widely used in circuit analysis or more generally in the description of electrical systems. The previous considered metrics are particular cases of the $\mathcal{K}$*-terminal* reliability in which a subset $\mathcal{K}$ of nodes is defined and the associated measure is the probability that all the nodes in $\mathcal{K}$ can communicate, that is, the probability that the nodes of $\mathcal{K}$ belong to the same connected component of G. A large proportion of the research effort in the network reliability area has been done on the evaluation (exact or approximate) of this measure and the two particular cases described before (the all-terminal and two-terminal ones).

These problems (and several other related reliability problems) have received considerable attention from the research community (see [46, 18, 4, 55] for references) mainly because of the general applicability of these models, in particular in the communication network area, and because of the fact that in the general case the computation of these metrics is in the #P-complete class [59, 3], a family of NP-hard problems not known to be in NP. A #P-complete problem is equivalent to counting the number of solutions to an NP-complete one (see Chapter 8 for connections between counting problems and rare event simulation). This implies that a #P-complete problem is at least as hard as an NP-complete one. This last fact justifies the continued effort to find faster solution methods. Concerning network reliability, even if we limit the models to very particular classes, the problems remain #P-complete. For instance, this is the case if we consider the two-terminal reliability evaluation on a planar graph with vertex degree at most equal to 3.

It must be stated that all known *exact* techniques available to evaluate R are unable to deal with a network having, say, 100 elements (except, of course, in the case of particular types of topologies). For instance, in [51], the effective threshold is placed around 50 components. Our own experience confirms this figure. In [55] the different approaches that can be followed to evaluate these metrics numerically (exact combinatorial methods and bounding procedures, together with reduction techniques that allow the size of the models to be reduced) are discussed, and some examples illustrate the limits of the different possible techniques (including simulation). Let us observe that in the communication networks area, usual model sizes are often very large. For instance, in [33] the authors report on computational results analyzing (in a deterministic context) the topology of real fiber optic telephone networks. They give the sizes of seven networks provided by Bell Communications Research, ranging from 36 nodes and 65 edges to 116 nodes and 173 edges. They also say that in this type of communication system, the number of nodes in practical implementations is not larger than, say, 200. In [34], the same authors report on a realistic model of the link connections in the global communication system of a ship, having 494 nodes and 1096 edges. Monte Carlo algorithms appear, then, to be the only way to obtain (probabilistic) answers to reliability questions for networks having, for instance, more than 100 components. But, of course, specific techniques for dealing with the rare event case must be applied. This is the topic of this chapter.

The crude Monte Carlo technique in this context involves sampling the system configuration N times, that is, generating independent samples $\vec{X}^{(1)}, \ldots, \vec{X}^{(N)}$ of $\vec{X}$ and estimating the unknown parameter R by the unbiased estimator

$$\widehat{R} = \frac{1}{N}\sum_{n=1}^{N} \Phi(\vec{X}^{(n)}).$$

The evaluation of $\Phi(\vec{x})$ for a given configuration $\vec{x}$ takes the form of a graph exploration. For instance, in the source-to-terminal case, a depth first search procedure is typically implemented to check if source and terminal are connected in the graph resulting from the initial model when all lines corresponding to the zeros in $\vec{x}$ have been deleted.

The case of interest here is that of $R \approx 1$, and so, $Q = 1 - R \approx 0$. The rare event is '$\Phi(\vec{X}) = 0$', and the methods used to deal with it are the subject of the rest of the chapter. After a discussion in Section 7.2 of the many applications in this area through a literature review, Section 7.3 describes the main ideas used so far in order to analyze network reliability, focusing on the rare event situation. In Section 7.4 a specific approach is presented in more detail. Section 7.5 presents some numerical examples of the behavior of these techniques. Section 7.6 concludes the chapter.

7.2 Network reliability applications

There is a wide field of applications of network reliability techniques. We find these problems in evaluations of electrical power networks, transportation systems (especially urban transportation systems; see [56]), interconnection networks (i.e., networks connecting processors, memory and other devices inside a multiprocessor computer), fault-tolerant computer architectures, etc. As already stated, a central area of application is in the evaluation of communication systems. The usefulness of 'connectivity' measures such as those ones presented above is clear, for instance, in packet switching communication networks using dynamic routing which allows rerouting of data in the event of the failure of a link. It must be said that many modern packet switching networks are rather dense and that the reliability measures considered tend to be close to unity. The computation of the unreliability of the system systematically corresponds then to the evaluation of the probability of a rare event. In this section, we give some examples of applications in different contexts, such as the design of telecommunication networks and other systems [5, 20–25, 43, 44, 47, 50, 57], the design and evaluation of mobile ad hoc networks and of tactical radio networks (specially in military contexts) [19], the evaluation of transport networks, and the assessment of the reliability of road networks with respect to seismic hazards and other disasters [35, 45, 49, 61]. The aim of this section is to underline the wide range of application of these problems, and thus, of the methods proposed to solve them. Once again, let us recall that in most of the cases, the events of interest are rare.

The design of the topology of telecommunication and computer networks is one of the settings where the application of reliability models is more direct. As such, there are a number of papers which tackle different variants of this problem, which in general involves deciding which components (links, and sometimes nodes) to include in the network so that the communication among terminals is reliable and the cost is as low as possible. Such network design problems are in general NP-hard, so that most literature includes the use of combinatorial optimization heuristics (most often genetic algorithms) to find approximate solutions.

One of the first papers on applying genetic algorithms to solve reliable network design problems was published by Kumar *et al.* [43]. These authors tackled three different network design problems: maximization of reliability under a diameter constraint, maximization of diameter under a degree constraint, and maximization of average distance under a degree constraint. The solution method applied was based on a genetic algorithm, which solved very small instances of these problems (graphs up to 9 nodes), attaining optimal solutions. Even if the network size considered was very small, this work showed that genetic algorithms could be designed to tackle reliable network design problems.

Dengiz *et al.* [22, 23] study two variants of reliable network design: maximization of the all-terminal network reliability metric given a cost constraint, and minimization of the cost, given a reliability constraint. The node set is fixed, and the problem involves choosing which links to install. The problem is solved using an evolutionary approach, based on genetic algorithms plus a local search heuristic. Reliability is estimated using a specific heuristic, upper bounds, and Monte Carlo simulation. The authors evaluated their algorithm and an exact, branch-and-bound based alternative, using 79 randomly generated small test problems (with 6–20 nodes), and the results showed that both algorithms found the optimal solutions, and that the genetic algorithm was computationally the most efficient. Deeter and Smith [20, 21] also discussed the design of networks considering all-terminal reliability. These authors consider minimizing the network cost given a reliability constraint. In their setting, the nodes are given, and it is possible to choose which links to employ, and different 'link options', each having different reliability and cost values. A genetic algorithm is used to select the links and the level of link connection; Monte Carlo simulation is used to compute estimates of the network reliability. Experiments with different topologies showed the effectiveness of the approach in identifying low-cost solutions meeting the reliability requirements. Other more recent work by the same authors includes [1] and the genetic algorithm by Altiparmak *et al.* [2]. Other authors, such as Lin and Gen [44], have also studied the same all-terminal reliability network design problem and proposed alternative optimization methods resulting in improved performance.

Barán and Laufer [5] proposed a parallel asynchronous team algorithm applied to the reliable network design problem, where the nodes and links are fixed, but it is possible to choose (at a cost) a given reliability value for each link. This is a hybrid technique that combines different algorithms interacting to solve the same global problem. Two approaches were used to estimate network reliability in this paper: an upper bound of all the candidates included in the population is efficiently calculated, and after that, a Monte Carlo simulation is used to get good approximations of the all-terminal reliability. The empirical results show good values for medium-size networks. Duarte and Barán [24] addressed a multiobjective version of the previous problem, using a parallel asynchronous version of a genetic algorithm to search for optimal topologies for a network. The parallel version outperforms the sequential one, considering standard metrics in the multiobjective domain (where the solution is not just a topology, but a set of Pareto efficient ones). Later, Duarte *et al.* [25] published a comparison of several parallel multiobjective evolutionary algorithms for solving the same reliable network design problem.

Taboada *et al.* [57, 58] also look at multiobjective system reliability design problems, where it is necessary to decide the level of redundancy to allocate at each stage, and reliability, cost and weight are objective functions. In these papers different methodologies are explored: in [57], to help the decision maker make a selection, a pseudo-ranking scheme and clustering techniques to reduce the size of the Pareto optimal set are presented; in [58] a multiple objective genetic algorithm for solving the problem is given.

Marseguerra *et al.* [47] used a stochastic model for network reliability, considering a function of imperfectly known reliability parameters of network components. The problem to solve is again a multiobjective one, the aim being to find the network topologies that maximize the network reliability and minimize the variance of this estimation (taking into account the imperfectly known reliability parameters). The decision variable is the type and the redundancy level of components to be allocated within a fixed network topology, where each component has an associated reliability probability distribution. The optimization method is based on genetic algorithms, and a Monte Carlo evaluation algorithm is used to incorporate the uncertainty in the reliability values; the repeated evaluations of the good individuals are accumulated, to enhance the significance of the estimations. The numerical examples consider only very small networks (with 7 and 8 links), and allow the Pareto optimal solutions obtained to be examined and the differences in the configurations to be easily identified.

Premprayoon and Wardkein [50] tackle another variant of the reliable network design problem, where it is possible to define for each pair of nodes whether they will be connected by a link, whose characteristics (cost and reliability) can also be chosen from a given set. The objective is to minimize network cost subject to a requirement of attaining at least a given reliability level. The authors compare an ant colony optimization method, a tabu search method and a local search method; the network reliability evaluation is done by backtracking (as only very small network topologies are studied). The best computational results are obtained by the ant colony optimization.

Cook and Ramirez-Marquez [19] study mobile ad hoc wireless networks, in particular in a military context. These networks have their own characteristics, which this work describes, and a proposal is presented on how to adapt the classical analysis of network reliability to this new context. The methods proposed rely on considering the effect of node mobility and the continuous changes in the network's connectivity. Wakabayashi [61] also studies highway network reliability, taking normal and abnormal periods into account. The motivation of this work is to detect the critical link in the network, whose improvement will lead to reliability improvement for the larger network. This paper presents a comparative study between using a probability importance index (Birnbaum's structural importance) and a criticality importance based on network reliability measures, which address some problems in Birnbaum's index.

Li [45] proposes to employ an accessible node rate index based on two-terminal reliability to evaluate the anti-disaster level of a city road network. In particular, this author evaluates the connectedness to the start points of emergency vehicles, and provides a method of accessing such accessibility to individual residences in a real city by using GIS. Nojima [49] employs network reliability models to represent the risks on road networks caused by seismic activity. The performance measure of interest is defined as the system flow capacity of road networks subject to failure. In this paper, a variance reduction technique for Monte Carlo simulation is presented to perform efficient reliability analysis in terms of the system flow capacity. This method is used to define performance-based prioritization order; this results in a road prioritization strategy according to various levels of vulnerability and system requirement. Günnec and Salman [35] propose to assess the post-disaster performance of a road network under most likely disaster scenarios for the purpose of both strengthening the components of the network and planning post-disaster logistical activities. In this paper, the authors seek to measure the reliability and the expected post-disaster performance of a network under disaster risk. In particular, they evaluate the reliability of connection between different pairs of origin–destination nodes in the network, in terms of expected weighted sum of shortest travel time/distance between the origin–destination pairs. The estimation of this measure is done by Monte Carlo sampling.

7.3 Variance reduction techniques

Many generic variance reduction techniques have been proposed in order to improve the performance of Monte Carlo simulations, especially for the rare event case. Importance sampling is probably the most common, but we also find techniques based on antithetic variates, control variables, stratified sampling, etc. These techniques can be also applied to network reliability evaluation, with varying degrees of success. Nevertheless, the special characteristics of this problem provide the opportunity to develop more specialized variance reduction methods, sometimes inspired by the classical ones and sometimes completely original, which provide improved performance. As many methods have been proposed, it is not possible to describe each of them in detail. In this section, we will briefly present the main ideas which have appeared in the literature, and we will reference the publications which fall in the same broad categories. We also give an assessment of the most promising approaches.

7.3.1 Sampling techniques based on bounds

This family of methods can be interpreted as a hybrid of classical importance sampling with control variates. Its first application to network reliability problems was presented by Van Slyke and Frank [60], and later by Kumamoto *et al.* [41] and Fishman [31]. It can be applied to any reliability evaluation problem where there are two functions Φ^L and Φ^U which provide lower and upper bounds on the system structure function Φ. These functions must possess the following properties:

- $\Phi^L(\vec{x}) \leq \Phi(\vec{x}) \leq \Phi^U(\vec{x})$ for any state vector $\vec{x}$.

- For $k = 1, \ldots, |\mathcal{E}|$ and for any value assignment $\widetilde{x}^{(k)} = (\widetilde{x}_1, \ldots, \widetilde{x}_k)$ of the first k components of the state vector $\vec{x}$, the values

$$R_k^L(\widetilde{x}^{(k)}) \equiv \mathbb{P}(\Phi^L(\vec{X}) = 1 \mid X_1 = \widetilde{x}_1, \ldots, X_k = \widetilde{x}_k)$$

and

$$R_k^U(\tilde{x}^{(k)}) \equiv \mathbb{P}(\Phi^U(\vec{X}) = 1 \mid X_1 = \tilde{x}_1, \ldots, X_k = \tilde{x}_k)$$

can be computed in polynomial time.

The numbers R_0^L and R_0^U are defined by $R_0^L = \mathbb{P}(\Phi^L(\vec{X}) = 1)$ and $R_0^U = \mathbb{P}(\Phi^U(\vec{X}) = 1)$. For the bound-based sampling, we define the remaining state space

$$W = \{\vec{x} : \Phi^L(\vec{x}) = 0, \ \Phi^U(\vec{x}) = 1\}$$

from where the samples will be chosen proportionally to their probability in the original state space. From the estimator obtained there and the previous information, we construct an estimator of the system reliability. The variance reduction attained is directly proportional to the fraction of the total probability that is included in the subspace W.

We now give a more detailed description of the sampling routine $\mathcal{M}$ for the bound based sampling:

Input: network $\mathcal{G}$, terminal set $\mathcal{K}$, Φ^L and Φ^U
Output: an estimator of R ($\mathcal{K}$-terminal reliability)

Procedure $\mathcal{M}$:
Sample $\tilde{X}$; result: $\tilde{x} = (\tilde{x}_1, \ldots, \tilde{x}_{|\mathcal{E}|})$
Compute $R = R_0^L + \Phi(\tilde{x})\left(R_0^U - R_0^L\right)$
Return R

The sample $\tilde{X} = (\tilde{X}_1, \ldots, \tilde{X}_{|\mathcal{E}|})$ is chosen by sampling succesively, for $l = 1, \ldots, |\mathcal{E}|$, the state $\tilde{X}_l$ of link l following the Bernoulli distribution with parameter

$$\begin{aligned}\tilde{r}_l &= \mathbb{P}(X_l = 1 \mid X_1 = \tilde{x}_1, \ldots, X_{l-1} = \tilde{x}_{l-1} \text{ and } \Phi^U(\vec{X}) = 1, \Phi^L(\vec{X}) = 0) \\ &= \left[\frac{R_l^U(\tilde{x}^{(l-1)}) - R_l^L(\tilde{x}^{(l-1)})}{R_{l-1}^U(\tilde{x}^{(l-1)}) - R_{l-1}^L(\tilde{x}^{(l-1)})}\right] r_l.\end{aligned}$$

The variance of each sample of the methods of this family is

$$\begin{aligned}\mathrm{Var} &= R(R_0^U - R) - R_0^L(R_0^U - R) \\ &= R(1 - R) - (1 - R_0^U)R - R_0^L(R_0^U - R),\end{aligned}$$

which is lower than the crude Monte Carlo one ($R(1 - R)$). The difference will depend on the tightness of bounds Φ^L and Φ^U. The execution time performance depends on the computational complexity of the evaluation of these bounds.

In [28] a variant of this family is proposed for the case where all links have the same elementary reliability. The method is based on the efficient computation of a lower bound on the network reliability, in turn based on the evaluation of a subset of states with number of failed links less than the cardinality of the smallest minimal cutset of the network, and it does not employ any upper bound. The variance reduction obtained with this method is of order $1/\lfloor 1 - R_0^L \rfloor$, for a computation cost per sample similar to the crude Monte Carlo method. The authors classify their method within the antithetic variates based family.

7.3.2 Dagger sampling and other related techniques

Dagger sampling was proposed by Kumamoto *et al.* [42] and can be seen as an extension of the antithetic variates technique. The main idea behind this sampling method is the generation of sample blocks of size L such that within each block the random variables are chosen in order to induce negative correlations between the individual samples. The size of the blocks, L, is fixed in such a way that for each edge l the sequence of L replications can be partitioned into exactly N_l sub-blocks of size L/N_l, where $N_l = \lfloor 1/q_l \rfloor$. For each of these sub-blocks, a single position is randomly chosen; this position corresponds to a sample where link l will fail. In this way, the failure pattern is such that the sampled failure frequency for each link of the network is proportional to the link unreliability value. After all random variables have been sampled, the method checks every replication within the block, checking in each case whether the resulting network is connected or not, in order to obtain an estimate of R.

In this method, a single invocation of $\mathcal{M}$ corresponds to L crude Monte Carlo samples; this must be taken into account when comparing the computational complexity of the algorithms.

Input: network $\mathcal{G}$, terminal set $\mathcal{K}$
Output: an estimator of R ($\mathcal{K}$-terminal reliability)

Initialization $\mathcal{I}$:
Compute the integer vector $(N_l : l \in \mathcal{E})$: $N_l = \lfloor 1/q_l \rfloor$.
Choose the sample number: $L = \text{lcm}\{N_l : l \in \mathcal{E}\}$.

Procedure $\mathcal{M}$:
For each link l
For each sub-block of replications of size L/N_l,
Sample random state vectors $\vec{X}^{(j)}$ in this sub-block:
Choose randomly a replication from the sub-block:
Sample U uniformly on [0,1]; result: u; set
$k = \lceil u/q_l \rceil$

Link l is failed in that replication:
$X_l^{(j)} = 1, \forall j \neq k;\ X_l^{(k)} = 0$, if $k \leq L/N_l$.
endFor
endFor
Initialize $T = 0$.
For i from 1 to L
Count replications corresponding to an operational state of the network:
$T = T + \Phi(\vec{X}^{(i)})$.
endFor
Return $R = T/L$.

The complexity analysis of this algorithm, given in [30], shows that the execution time per sample has worst-case complexity $O(|\mathcal{E}|)$, as in the case of crude Monte Carlo. Nevertheless, as the number of random variables that are needed in the dagger method is much smaller than in the crude Monte Carlo, there may be a large gain in execution time, which is an important advantage of the method compared to the crude procedure.

The variance reduction obtained by the dagger method is based on the induced negative correlation among the samples which belong to the same sub-block for some link of the network, and is relatively small, approaching 0 when the link reliability is near 1. El Khadiri and Rubino [27] discuss some problems of the dagger method, which happen in particular when value L is too large. As it is necessary to generate and save into memory L states of the network, memory requirements grow linearly with L. The authors show that the block size L may be chosen arbitrarily, and they propose an alternative method, inspired by the dagger and applying a generalization of the standard antithetic approach. This new method obtains better results, both because L can be chosen arbitrarily, and because the sampling algorithm used incorporates some features in order to improve its efficiency. The algorithm employs mechanisms similar to those used in discrete event simulation, employing a list with the incumbent failure instants for each link, and finding which is the first replication including at least one failure. This implementation yields important gains in execution time and also results in minimal memory complexity (which depends linearly on the network size and is independent of L).

When all links have the same reliability p, the generalized antithetic algorithm has sample complexity $O(\alpha|\mathcal{E}|)$, where

$$\alpha = \max(1/L, 1 - p).$$

This represents an improvement over the crude method which grows with the block size L, up to a bound given by the inverse of the link reliabilities (in the worst case, $p = 0.5$, we have that the maximum gain that can be attained is a factor of 2, corresponding to setting $L = 2$).

7.3.3 Graph evolution models

The techniques described in this subsection, instead of using the static network reliability model, employ alternative timed models where the states of the links are assumed to change over time. These models correspond to Markov processes, whose properties can then be exploited to obtain efficient estimators of the classical reliability.

Easton and Wong [26] proposed the sequential construction and sequential destruction methods, which complement the previous idea with the use of an ordering of the network links. In the sequential construction method, all links are considered as being in a state of failure at an initial instant, and then they are successively repaired, one by one following the ordering chosen, until the network reaches an operative state. The reliability estimator can be interpreted as a function of the expectation of how much time is needed for the network to reach an operative state. The sequential destruction method is similar, but all links are considered to be operating at the initial time and are successively put into a failure state. These techniques can be classified as hybrids between stratified sampling and importance sampling procedures.

The sample space for the sequential construction method consists of pairs $(\tilde{x}, \tilde{\pi})$ where $\tilde{x}$ is a state vector or configuration and $\tilde{\pi} = (\tilde{\pi}_1, \ldots, \tilde{\pi}_{|\mathcal{E}|})$ is a permutation of the link indexes in $\mathcal{E}$. There exists an index k such that

$$\tilde{x}_{\tilde{\pi}_1} = \ldots = \tilde{x}_{\tilde{\pi}_k} = 1, \quad \tilde{x}_{\tilde{\pi}_{k+1}} = \ldots = \tilde{x}_{\tilde{\pi}_{|\mathcal{E}|}} = 0.$$

If we choose a vector $\tilde{x}$ following the system state probabilities (i.e, with the same distribution as $\vec{X}$) and we choose a permutation $\tilde{\pi}$ independently and uniformly over all the compatible permutations, then the probability of observing a given pair $(\tilde{x}, \tilde{\pi})$ is

$$\rho(\tilde{x}, \tilde{\pi}) = \frac{\mathbb{P}(\vec{X} = \tilde{x})}{k!(|\mathcal{E}| - k)!} = \frac{1}{|\mathcal{E}|!} C_k^{|\mathcal{E}|} \mathbb{P}(\vec{X} = \tilde{x}),$$

where k is the number of working links in $\tilde{x}$. The sequential construction method samples $\tilde{\pi}$, and considers simultaneously the set $\mathcal{P}_{\tilde{\pi}}$ of all possible pairs $(\tilde{x}, \tilde{\pi})$ such that $\tilde{x}$ is consistent with $\tilde{\pi}$ following the previous criterion. The reliability estimator R is then the conditional probability of operation of the system given $\mathcal{P}_{\tilde{\pi}}$, corresponding to the quotient of the sum of the probabilities of the pairs $(\tilde{x}, \tilde{\pi}) \in \mathcal{P}_{\tilde{\pi}}$ such that $\Phi(\tilde{x}) = 1$ divided by the probability of $\mathcal{P}_{\tilde{\pi}}$. We give below a more detailed description of the algorithm $\mathcal{M}$ for generating a sequential construction sample (no initialization is needed):

Input: network $\mathcal{G}$, terminal set $\mathcal{K}$
Output: an estimator of R ($\mathcal{K}$-terminal reliability)

Procedure $\mathcal{M}$:

Sample $\tilde{\pi} = (\tilde{\pi}_1, \ldots, \tilde{\pi}_{|\mathcal{E}|})$

For $k = 1, \ldots, |\mathcal{E}|$ (Define $\tilde{x}^{(k)}$)
$\tilde{x}^{(k)}_{\tilde{\pi}_1} = \ldots = \tilde{x}^{(k)}_{\tilde{\pi}_k} = 1,\ \tilde{x}^{(k)}_{\tilde{\pi}_{k+1}} = \ldots = \tilde{x}^{(k)}_{\tilde{\pi}_{|\mathcal{E}|}} = 0$
endFor
Determine first $r \in 0, \ldots, |\mathcal{E}|$
such that $\Phi(\tilde{x}^{(r)}) = 1$
Compute
$$R = \frac{\sum_{k=0}^{|\mathcal{E}|} \Phi(\tilde{x}^{(k)})\rho(\tilde{x}^{(k)}, \tilde{\pi})}{\sum_{k=0}^{|\mathcal{E}|} \rho(\tilde{x}^{(k)}, \tilde{\pi})} = \frac{\sum_{k=r}^{|\mathcal{E}|} C_k^{|\mathcal{E}|}\mathbb{P}(\vec{X}=\tilde{x}^{(k)})}{\sum_{k=0}^{|\mathcal{E}|} C_k^{|\mathcal{E}|}\mathbb{P}(\vec{X}=\tilde{x}^{(k)})}$$
Return R.

It is possible to show that the estimator obtained by this method has smaller variance than the one corresponding to a crude Monte Carlo sample. Much effort is needed to compute the first index r, this depending on the effort needed to compute $\Phi(\tilde{x}^{(k)})$, as it is necessary to determine when the network becomes operational as the links are repaired one by one. In the worst case, and using a depth first search for computing Φ, the complexity is $O(|\mathcal{E}| \max(|\mathcal{V}|, |\mathcal{E}|))$. Nevertheless, for the case of the two-terminal metric, it is possible to determine the value of r with a computation cost similar to a single computation of Φ, so that this method obtains a single sample at a cost similar to that of the crude method.

As already mentioned, the sequential destruction method is very similar; in the case of very low-reliability systems, it may exhibit better performance in computation terms, as the value of r would be determined in fewer iterations than needed by the sequential construction method.

Other methods based in graph evolution models have been published in [29]. In particular, three methods are discussed in that work: *destruction processes*, *construction processes*, and *merge processes*. All the methods rely on constructing a Markov chain $(\vec{Y}(t))$ such that at time $t = 1$ we have that $\mathbb{P}(\Phi(\vec{Y}(1)) = 1) = \mathbb{P}(\Phi(\vec{X}) = 1)$, so that computing the expectation of $\Phi(\vec{Y}(1))$ also gives the reliability R. Then the algorithm samples a permutation of the order at which links go up (in a creation or merge process) or down (in a destruction process), it identifies the critical link (that which causes the system to change from a down to an up state, or vice versa), and it computes (using a convolution of exponential random variables) the exact conditional probability that at time 1 the critical link will have changed its state. In the case of merge processes, they improve on previous ideas identifying irrelevant links and partitions of the subjacent network. In these methods, the sample complexity is higher, of order $O(|\mathcal{E}|^2)$, but the variance is much smaller than the one obtained by the sequential construction method. An important result is that, for fixed $\mathcal{E}$, the merge processes method has coefficient of variation uniformly bounded for all values of links' reliabilities. Another related work is [36], which gives a hybrid variant of crude Monte Carlo and graph evolution models, completed with the use of importance sampling to further speed up the simulations.

More recently, Hui *et al.* [37] have applied cross-entropy techniques to improve the performance of crude Monte Carlo and graph evolution methods, in particular creation processes and merge processes. The main idea is to apply an importance sampling scheme, changing the underlying network reliability parameters, and to use cross-entropy to search for an optimal change of measure. Their results show that cross-entropy does indeed give better accuracy; the improvement over crude Monte Carlo is quite large. In the case of construction and merge process based Monte Carlo, the application of cross-entropy results in much more modest improvements. Similar results have been obtained by Murray and Cancela [48], who compared the behavior of these methods (and of a generalized antithetic method) when evaluating the diameter constrained reliability of a network, a variant of the classical model taking into account a bound on path length.

A quite different approach to exploit the Markov process modeling a creation process of the network has been employed by Cancela *et al.* [15]. This work applies the well-know splitting technique (see Chapter 3), much employed for rare event simulation in the context of stochastic processes, to the stochastic process starting from an empty network and creating (or putting into operational states) the links one by one, taking independent exponential distributions for these times. As already mentioned, the state of this system at time $t = 1$ has the same distribution as the state of the static network model; in a highly reliable network, the network almost always becomes operational before time 1, and the rare event is to observe $\Phi(\vec{Y}(1)) = 0$. The splitting strategy developed in [15] involves taking a number of intermediate time thresholds, and splitting such trajectories of process $(\vec{Y}(t))$ that at these thresholds still satisfy $\Phi(\vec{Y}(t)) = 0$. The results show that this method is very robust and can achieve better performance than that of Huì *et al.* [37].

Finally, let us also mention [38], where the author proposes to directly estimate the reliability ranking of some edge relocated networks without estimating their reliabilities and compare the proposed approach to the traditional approach using the merge process estimation algorithm. Another recent related paper is [40], which is concerned with network planning. Here, the objective is to maximize network's reliability, subject to a fixed budget. The authors show how the cross-entropy method can easily be modified to tackle the noise introduced by the use of network reliability estimators in the objective function instead of exact evaluations.

7.3.4 Coverage method

The coverage method was proposed by Karp and Luby [39]. It can be seen as a hybrid variant of importance sampling and stratified sampling, and employs the list of the minimal cuts of the system to improve the crude sampling procedure. The main idea is to embed the set F of network failure events within a *universal weighted space* $(\mathcal{U}, w)$, where w is a non-negative weighting function in $\mathcal{U}$, satisfying the following criteria:

- $w(F) = \mathbb{P}(F) = Q$.

- $w(\mathcal{U})$ can be efficiently computed (in polynomial time). Moreover, it is possible to efficiently sample values in $\mathcal{U}$ with probability proportional to the weights of the elements of the set.
- It can be efficiently decided whether an element of $\mathcal{U}$ belongs to F.
- $w(\mathcal{U})/w(F)$ is bounded above by a value M for all the instances in the problem class considered.
- $w(C)$ is the total weight of the elements of $\mathcal{U}$ with second component equal to C.

If we take a sample from $\mathcal{U}$, and we obtain the estimator $\hat{Q}$ multiplying the proportion of elements of this sample that are included in F by $w(\mathcal{U})$, then $\hat{Q}$ is an unbiased estimator of Q.

Let $\mathcal{C}$ be the set of the $\mathcal{K}$-mincuts of network $\mathcal{G}$. We define the universal weighted sample space $\mathcal{U}$ composed of the pairs $(\vec{x}, C)$ where $\vec{x}$ is a state vector of the network, $C \in \mathcal{C}$ is a cut, and $x_l = 0$ for all links l belonging to C. This way every system failure state $\vec{x}$ will appear in $\mathcal{U}$ as many times as the number of failed mincuts in $\vec{x}$; in order to embed F in $\mathcal{U}$ it is necessary to assign to each $\vec{x}$ a single cut $C \in \mathcal{C}$. To do this we choose a node $s \in K$, then we find the set N of all nodes reachable from s following paths formed by operational links, and we select $C \equiv C(\vec{x})$ the set of links from N to $\mathcal{V} - N$. The elements of F appear then in $\mathcal{U}$ as pairs $(\vec{x}, C)$ such that $C = C(\vec{x})$, and it can be decided in linear time if an element from $\mathcal{U}$ belongs to F just by verifying the condition $C = C(\vec{x})$. The weighting function w is given by $w(\vec{x}, C) = \mathbb{P}(\vec{X} = \vec{x})$.

We give now pseudo-code for the initialization and sampling routines for the coverage method:

Input: network $\mathcal{G}$, terminal set $\mathcal{K}$, list of mincuts $\mathcal{C}$
Output: an estimator of R ($\mathcal{K}$-terminal reliability)

Initialization $\mathcal{I}$:
  For each $C \in \mathcal{C}$
    Compute $w(C) = \prod_{l \in C} q_l$
  endFor
  Compute $w(\mathcal{U}) = \sum_{C \in \mathcal{C}} w(C)$

Procedure $\mathcal{M}$:
    Sample $(\vec{X}, C)$ from $\mathcal{U}$ with distribution w:
    Sample a cut C with probability $w(C)/w(\mathcal{U})$
    Construct $\vec{X}$:
      $\forall l \in C$, $X_l = 0$
      $\forall l \notin C$, $X_l = 1$ with probability p_l,
        $X_l = 0$ otherwise.

If $C = C(\vec{X})$
  $R = 1 - w(\mathcal{U})$
Else
  $R = 1$
endIf
Return R.

This method can obtain good variance reduction levels, but has the drawback of depending on the previous calculation and storage of the list of all mincuts of the network considered. As the size of this list grows exponentially with the size of the network, the requirements of time and space can quickly make its application impractical. Also, there are some exact methods that compute the reliability in time polynomial in the number of mincuts of the network [52], further reducing the attractiveness of the approach.

7.3.5 State space partitioning and conditioning methods

A number of methods are based on sampling within the space of the state vectors of the network, using techniques related to partitioning this space and/or to conditionally sampling within it.

One of these is the total hazard method. Random hazard variables, and in particular the total hazard ones, have been employed in different contexts to simulate stochastic models [53]. Ross [54] developed a total hazard estimation to compute the reliability R.

Let C_1 be a $\mathcal{K}$-mincut. The first hazard, h_1, is the probability that all the components in C_1 are failed (implying that the network is not $\mathcal{K}$-connected); so,

$$h_1 = \prod_{i \in C_1} q_i.$$

The total hazard method involves simulating the state of all the links belonging to C_1. If all the links are failed, the procedure ends. If at least one link is operational, we fix the states of the simulated links, and we look for a new mincut C_2 in the modified network. From this new cut we compute the second hazard, h_2, given by

$$h_2 = \prod_{i \in C_2} q_i.$$

Then we simulate the state of the links belonging to C_2 and the previous process is repeated, generating new networks until all the components of a mincut are failed or until a trivial network is reached, with no mincuts (all the links' states have been fixed). For r hazards, the total hazard is given by

$$H(\mathcal{G}) = \sum_{i=1}^{r} h_i,$$

and it is an unbiased estimator of Q. The implementation suggested in [53] employs the list of all mincuts of the system under consideration, and it updates it as the states of the links are fixed; the mincut is chosen at each step in order to result in the maximum risk.

Inputs: network $\mathcal{G}$, set $\mathcal{K}$, list of $\mathcal{K}$-mincuts of $\mathcal{G}$
Output: an estimate for R

Initialization $\mathcal{I}$:
$H = 0$.

Procedure $\mathcal{M}$:
Select a $\mathcal{K}$-mincut C.
Simulate the state of the links in C
Repeat until all the links in the selected mincut are failed
Update the list of mincuts of network $\mathcal{G}$.
Compute the hazard: $h = \prod_{i \in C} q_i$.
Accumulate in H: $H = H + h$.
Select a $\mathcal{K}$-mincut C.
Simulate the state of the links in C.
endRepeat
Return $R = 1 - H$.

The variance reduction that can be obtained with this method depends strongly on how the $\mathcal{K}$-mincut C is chosen. A heuristic with good behavior is to select at each step the mincut with the highest associated hazard. This implies a significant computational overhead, as it must be implemented as a search in the list of mincuts (whose size is exponential in the size of the graph), or employing a maximal flow algorithm at each iteration of the method. The analysis of the computational complexity of generating a sample strongly depends on this step. With the computationally less costly choice for the mincut selection, the computational complexity per sample is of order $O(|\mathcal{E}|)$.

In [9], Cancela and El Khadiri highlighted that there are some cases where the total hazard estimator is less efficient than crude Monte Carlo. They proposed a modification leading to a more precise estimator whose variance is always lower than that of the crude Monte Carlo method.

7.4 The RVR sampling principle

Another family of related methods are the recursive variance reduction (RVR) ones. These methods, first proposed in [8], have been extensively discussed and adapted to different contexts [10–14, 16, 17]. RVR methods combine different ideas to obtain good performance estimators. On one hand, they employ either one or more cutsets, pathsets, or both, of the network of interest, in order to partition the state vector space into subsets depending on the operational/failed status of the links belonging to the chosen sets. Some of the elements of this partition correspond to network configurations known in advance (corresponding to either a failed or an operational network). Then the rest of the state vector space is explored by recursively sampling one of the subsets in the partition, which corresponds to a subnetwork of the original one, including some particular links failed and others operational; once this subnetwork has been randomly chosen, an RVR method recursively searches for new cutsets and/or pathsets, and restarts the whole process.

In this section we provide some details about the RVR approach. For the presentation of the RVR principle, we consider the two-terminal problem where we look at the unreliability between two given nodes s and t in $\mathcal{G}$, the version using series–parallel simplification for reducing the size of the network and a selected cut for transforming a network reliability problem into a smaller one and then recursively until the network has unreliability equal to 0 or to 1. This estimator was proposed in [11].

If s and t are not connected in $\mathcal{G}$, we define $Z(\mathcal{G}) = 1$. Otherwise, let us denote by sp-red($\mathcal{G}$) the result of making all possible series–parallel reductions in $\mathcal{G}$. As these reductions preserve the unreliability, we set $Z(\mathcal{G}) = Z(\text{sp-red}(\mathcal{G}))$. Let γ be an st-cut in sp-red($\mathcal{G}$), $\gamma = \{l_1, l_2, \ldots, l_H\}$ where $l_1, l_2, \ldots$ are the links in the cut. Let L_h be the event 'link l_h is down'. If Ω is the set of all possible configurations in the model, consider the partition $\Omega = (E_0, E_1, \ldots, E_H)$ where

$$
\begin{aligned}
E_0 &= L_1 L_2 \cdots L_H = \text{all links in } \gamma \text{ are down},\\
E_1 &= L_1^c = \text{at least one link in } \gamma \text{ is up, and the first such link is } l_1,\\
E_2 &= L_1 L_2^c = \text{at least one link in } \gamma \text{ is up, and the first such link is } l_2,\\
E_3 &= L_1 L_2 L_3^c = \text{at least one link in } \gamma \text{ is up, and the first such link is } l_3,\\
&\cdots\\
E_H &= L_1 L_2 \cdots L_H^c = \text{at least one link in } \gamma \text{ is up, and the first such link is } l_H.
\end{aligned}
$$

We have $\mathbb{P}(E_0) = q_1 \cdots q_H$, where $q_h = \mathbb{P}(L_h) = 1 - r_h$, and $\mathbb{P}(E_h) = q_1 q_2 \cdots q_{h-1} r_h$ for $h = 1, 2, \ldots, H$. To simplify the notation, call π_h the product $\pi_h = q_1 q_2 \cdots q_h$ for $h = 1, 2, \ldots, H$, $\pi_0 = 1$. We have $\mathbb{P}(E_0) = \pi_H$ and $\mathbb{P}(E_h) = \pi_{h-1} r_h$ for $h = 1, 2, \ldots, H$.

Let I be the random variable 'index in γ of the first link up', with $I = 0$ if all links in π are down. We have $\mathbb{P}(I = h) = \mathbb{P}(E_h)$. Now define the random variable V on $\{1, 2, \ldots, H\}$ by

$$
\mathbb{P}(V = h) = \mathbb{P}(I = h \mid I \neq 0) = \frac{\pi_{h-1} r_h}{1 - \pi_H}.
$$

Last, for $h = 1, 2, \ldots, H$, denote by $\mathcal{G}_h = \text{sp-red}(\mathcal{G}) \mid E_h$ the network obtained from sp-red($\mathcal{G}$) by deleting links $l_1, l_2, \ldots, l_{h-1}$ and contracting link l_h. We are now ready to give the estimator proposed in [11]:

$$Z(\mathcal{G}) = \pi_H + (1 - \pi_H) \sum_{h=1}^{H} \mathbb{1}(V = h) Z(\mathcal{G}_h).$$

Let us denote by v a sample from the distribution of V. Then a sample $Z^{(k)}(\mathcal{G})$ of $Z(\mathcal{G})$ can be deduced from a sample $Z^{(k)}(\mathcal{G}_v)$ of $Z(\mathcal{G}_v)$ by

$$Z^{(k)}(\mathcal{G}) = \pi_H + (1 - p_H) Z^{(k)}(\mathcal{G}_v).$$

If s and t are merged into a final single node, the unreliability of $\mathcal{G}_v$ is equal to 0 and then $Z^{(k)}(\mathcal{G}) = \pi_H + (1 - p_H) \times 0 = \pi_H$. If s and t are not connected, the unreliability of $\mathcal{G}_v$ is equal to 1 and then $Z^{(k)}(\mathcal{G}) = \pi_H + (1 - p_H) \times 1 = 1$. Otherwise, we have found an st-cut in $\mathcal{G}_v$ and we proceed again as before. The main interest of this procedure is that $\mathcal{G}_v$ is smaller than $\mathcal{G}$, and sometimes much smaller, because of the series–parallel simplifications, deletions and contractions performed.

A function which returns a trial of $Z(\mathcal{G})$ can be summarized as follows:

TRIAL-RVR($\mathcal{G}$, K)

1. Check end recursion condition:
   If $|K| = 1$ return(0)
   If $\mathcal{G}$ is not $\mathcal{K}$-connected return(1)
2. Construct sp-red($\mathcal{G}$) by applying series–parallel reductions to $\mathcal{G}$
3. Find a $\mathcal{K}$-cut π in sp-red($\mathcal{G}$): $\pi = \{l_1, \ldots, l_H\}$
4. Compute the probability π_H that all links in π are down
5. Compute the probability mass function distribution of the random variable V
6. Generate a trial v of V
7. Construct the network $\mathcal{G}_v$ = sp-red($(\mathcal{G} - l_1 - l_2 - \ldots - l_{v-1}) * l_v$)
8. Recursive step: return($\pi_H + (1 - \pi_H) \times$ Trial-RVR($\mathcal{G}_v$, K_v)).

The memory space complexity of the function TRIAL-RVR($\mathcal{G}, K$) is of order $O(|\mathcal{E}|(|\mathcal{E}| + |\mathcal{V}|))$ and time complexity is, in the worst case, of order $O(|\mathcal{E}|(|\mathcal{K}|^2|\mathcal{E}||\mathcal{V}|^2))$. The worst case corresponds to a version using a maximal flow procedure in order to select a $\mathcal{K}$-cut π at step 3 of the above algorithm.

By calling the function TRIAL-RVR($\mathcal{G}, K$) N times, we obtain N independent trials $Z^{(k)}(\mathcal{G})$ of $Z(\mathcal{G})$, $1 \leq k \leq N$. The sample mean $\widehat{Z}(\mathcal{G})$ of these trials leads to an estimate of $Q = 1 - R$ and the variance is estimated by

$$\widehat{V}_{RVR} = \frac{1}{N(N-1)} \sum_{k=1}^{N} \left(\widehat{Z}(\mathcal{G}) - Z^{(k)}(\mathcal{G})\right)^2.$$

In [12] it is shown how the computational complexity of the RVR method can be improved by generating the N samples simultaneously, thus avoiding replicating a large part of the computations, and [13] discusses the sensitivity of the RVR's accuracy to the strategy of choosing cuts. Instead of using cuts to recursively change the original problem into a smaller one, the method in [10] exploits paths and that in [14] exploits both paths and cuts leading to a more interesting behavior than for versions based on only paths or cuts.

7.5 Numerical results and conclusions

To the best of our knowledge, the CE-MP method [37] which uses the cross-entropy technique to further improve the performance of the merge process method [29] and the RVR technique which exploits series–parallel reductions and a minimum cost st-cut strategy [13], where each link l has value $-\ln(q_l)$, are the most suitable procedures published in the literature to compute network reliability in a rare event context. In this section we present some numerical illustrations of these methods.

For the examples, we consider highly reliable grid topologies G_3 and G_6 (see Figure 7.1), where links are assigned equal unreliability $q = 10^{-3}$ or $q = 10^{-6}$ as in [37] and $\mathcal{K}$ is the set of the four corner nodes. For those networks exact values of $Q = 1 - R$ are tabulated in column 3 of Table 7.1. Each exact unreliability Q is used in the computation of the relative error parameter which helps to appreciate the quality of the estimates produced by the two estimators considered. Tables 7.2 and 7.3 show that both methods lead to small relative errors and the RVR method offers the most accurate estimates.

In the general case, we do not know the exact values. Then, the best estimator in terms of accuracy is the one with smallest variance for a fixed sample size N, leading to smallest lengths of confidence intervals. Column 6 of Table 7.3 shows that the RVR method significantly reduces the variance with respect to the CE-MP method, and the best gains are obtained for highly reliable cases.

To illustrate the behavior of the RVR method on dense networks, let us now consider the evaluation of complete topologies for which we calculate the exact

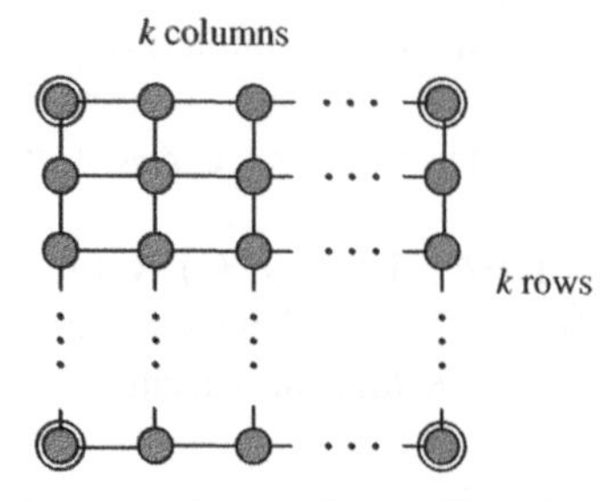

Figure 7.1 G_k: the grid network topology. The four corner nodes are the terminals.

Table 7.1 Exact unreliabilities of grid networks (see Figure 7.1) used for numerical illustrations [37]

Network	Common link unreliability q	Q
G_3	10^{-3}	4.01199×10^{-6}
G_3	10^{-6}	4.00001×10^{-12}
G_6	10^{-3}	4.00800×10^{-6}
G_6	10^{-6}	4.00001×10^{-12}

Table 7.2 Performance of the MP-CE method for the evaluation of G_3 and G_6. Terminals are the four corner nodes and $N = 10^6$

Network	q	Estimate of Q [37]	RE (%)	Variance[37]
G_3	10^{-3}	4.01172×10^{-6}	6.73×10^{-3}	1.84515×10^{-17}
G_3	10^{-6}	3.99876×10^{-12}	3.12×10^{-2}	1.85116×10^{-29}
G_6	10^{-3}	4.00239×10^{-6}	1.40×10^{-1}	3.74067×10^{-17}
G_6	10^{-6}	3.99869×10^{-12}	3.30×10^{-1}	3.75850×10^{-29}

Table 7.3 Performance of the RVR method for the evaluation of G_3 and G_6. Terminals are the four corner nodes and $N = 10^6$

Network	q	Estimate of Q	RE (%)	Variance	V_{CE-MP}/V_{RVR}
G_3	10^{-3}	4.01208×10^{-6}	2.16×10^{-3}	3.01610×10^{-18}	6.12×10^{0}
G_3	10^{-6}	3.99992×10^{-12}	2.31×10^{-3}	1.00018×10^{-35}	1.85×10^{6}
G_6	10^{-3}	4.00803×10^{-6}	7.49×10^{-4}	4.02467×10^{-21}	9.29×10^{3}
G_6	10^{-6}	4.00001×10^{-12}	5.00×10^{-5}	3.99998×10^{-36}	9.40×10^{6}

values of Q by a Maple program; see column 2 of Table 7.4. We consider a sample size $N = 10$ for each network. The estimates obtained by the RVR method are given in column 3, and associated relative errors and variances are in column 4 and 5 respectively. In column 6 we give the variance gains when the RVR method is compared to the crude Monte Carlo. The latter's variance is equal to $Q(1 - Q)/N$. We can see that relative errors are acceptable for all cases considered even if the sample size is small ($N = 10$) and substantial gains in variance are obtained in all cases. In particular, the improvement of RVR over crude Monte Carlo increases with the rarity of the event considered.

7.6 Conclusions

As the reader can appreciate, the literature on this topic is considerable, and the number of ideas that have been explored so far to deal with rare events in the

Table 7.4 Performance of the RVR method for the evaluation of complete networks. $\mathcal{K} = \mathcal{V}$, the common link unreliability $q = 0.55$ and the sample size N is equal to 10

Network	Q	Estimate	RE (%)	Variance	V_{CMC}/V_{RVR}
C_{10}	4.58481×10^{-2}	4.67262×10^{-2}	1.92	6.57617×10^{-6}	6.65×10^{2}
C_{20}	2.33295×10^{-4}	2.32376×10^{-4}	0.394	2.07036×10^{-11}	1.13×10^{6}
C_{30}	8.86419×10^{-7}	8.74893×10^{-7}	1.30	4.59076×10^{-16}	1.93×10^{8}
C_{40}	2.99368×10^{-9}	3.00302×10^{-9}	0.312	1.07423×10^{-2}	2.79×10^{11}
C_{50}	9.47855×10^{-12}	9.58856×10^{-12}	1.16	1.84357×10^{-26}	5.14×10^{13}

network reliability family of metrics is large. One reason for this is probably the fact that the cost of the exact computation of these metrics is extremely high.

In the chapter, we underlined the quality of some of the methods that have been presented, and we can say that the development of algorithms in the area is such that good perfomance can now be achieved. Most of the methods combine, on the one hand, the application of some general probabilistic properties, and, on the other, the exploitation of the particular structure of the network reliability evaluation problem, in order to reach an efficient solution. The ideas based on putting the problem in terms of a dynamic auxiliary model, and the methods that operate recursively on the network while using polynomial reduction techniques appear to be the most promising ones. In both cases, even though results are available on the complexity of the procedures as a function of some graph properties, and on their theoretical efficiencies, considerable research effort is still needed to better understand their behavior.

References

[1] F. Altiparmak, B. Dengiz, and A. E. Smith. Optimal design of reliable computer networks: A comparison of metaheuristics. *Journal of Heuristics*, **9**(6): 471–487, 2003.

[2] F. Altiparmak, M. Gen, B. Dengiz, and A. Smith. A network-based genetic algorithm for design of communication networks. *Journal of Society of Plant Engineers Japan*, **15**(4): 184–190, 2004.

[3] M. O. Ball. Computational complexity of network reliability analysis: An overview. *IEEE Transactions on Reliability*, **R-35**(3), 1986.

[4] M. O. Ball, C. J. Colbourn, and J. S. Provan. Network reliability. In M. O. Ball, T. L. Magnanti and C. L. Monma, eds, *Network Models*, Handbooks in Operations Research and Management Science, Vol. **7**, pp. 673–762. Elsevier North-Holland, Amsetrdam, 1995.

[5] B. Barán and F. Laufer. Topological optimization of reliable networks using A-teams. *Third World Multiconference on Systemics, Cybernetics and Informatics (SCI'99) and Fifth International Conference on Information Systems Analysis and Synthesis (ISAS'99)*, Orlando, USA, Aug 1999. IIIS.

[6] E. E. Barlow and F. Proschan. *Statistical Theory of Reliability and Life Testing*. Holt, Rinehart & Winston, New York, 1975.

[7] Y. K. Belyayev, B. V. Gnedenko, and A. D. Solovyev. *Mathematical Methods of Reliability Theory*. Academic Press, New York, 1969.

[8] H. Cancela and M. El Khadiri. A recursive variance-reduction algorithm for estimating communication-network reliability. *IEEE Transactions on Reliability*, **44**(4): 595–602, 1995.

[9] H. Cancela and M. El Khadiri. An improvement to the total hazard method for system reliability simulation. *Probability in the Engineering and Informational Sciences*, **10**(2): 187–196, 1996.

[10] H. Cancela and M. El Khadiri. A simulation algorithm for source-terminal communication network reliability. In *Proceedings of the 29th Annual Simulation Symposium*, pp. 155–161, New Orleans, LA, April 1996. IEEE Computer Society Press.

[11] H. Cancela and M. El Khadiri. Series-parallel reductions in Monte Carlo network reliability evaluation. *IEEE Transactions on Reliability*, **47**(2): 159–164, 1998.

[12] H. Cancela and M. El Khadiri. On the RVR simulation algorithm for network reliability evaluation. *IEEE Transactions on Reliability*, **52**(2): 207–212, 2003.

[13] H. Cancela and M. El Khadiri. On the accuracy of the RVR estimator of K-terminal unreliability parameter. In *7th International Workshop on Rare Event Simulation (RESIM 2008)*, University of Rennes 1, France, September 2008.

[14] H. Cancela, M. El Khadiri, and G. Rubino. An efficient simulation method for K-network reliability problem. In *6th International Workshop on Rare Event Simulation (RESIM 2006)*, University of Bamberg, Germany, October 2006.

[15] H. Cancela, L. Murray, and G. Rubino. Splitting in source-terminal network reliability estimation. In *7th International Workshop on Rare Event Simulation (RESIM 2008)*, University of Rennes 1, France, September 2008.

[16] H. Cancela, G. Rubino, and M. E. Urquhart. Path set based conditioning for transient simulation of highly dependable networks. In M. Zaremba, J. Sasiadek and H. Erbe, eds, *7th IFAC Symposium on Cost Oriented Automation*, Elsevier Science, Gatineau/Ottawa, Canada, June 2004.

[17] H. Cancela and M. E. Urquhart. RVR simulation techniques for residual connectedness network reliability evaluation. *IEEE Transactions on Computers*, **51**(4): 439–443, 2002.

[18] C. J. Colbourn. *The Combinatorics of Network Reliability*. Oxford University Press, New York, 1987.

[19] J. L. Cook and J. E. Ramirez-Marquez. Two-terminal reliability analyses for a mobile ad hoc wireless network. *Reliability Engineering and Systems Safety*, **92**(6): 821–829, 2007.

[20] D. Deeter and A. Smith. Heuristic optimization of network design considering all-terminal reliability. In *Proceedings of the Annual Reliability and Maintainability Symposium*, pp. 194–199, Philadelphia, PA, USA, 1997.

[21] D. Deeter and A. Smith. Economic design of reliable networks. *IIE Transactions*, **30**(12): 1161–1174, 1998.

[22] B. Dengiz, F. Altiparmak, and A. E. Smith. Efficient optimization of all-terminal reliable networks, using an evolutionary approach. *IEEE Transactions on Reliability*, **46**(1): 18–26, 1997.

[23] B. Dengiz, F. Altiparmak, and A. E. Smith. Local search genetic algorithm for optimal design of reliable networks. *IEEE Transactions on Evolutionary Computation*, **1**(3): 179–188, 1997.

[24] S. Duarte and B. Barán. Multiobjective network design optimisation using parallel evolutionary algorithms [in Spanish]. In *XXVII Conferencia Latinoamericana de Informática*, Merída, Venezuela, 2001.

[25] S. Duarte, B. Barán, and D. Benítez. Telecommunication network design with parallel multi-objective evolutionary algorithms. In *Proceedings of the 2003 IFIP/ACM Latin America Conference Network (LANC'03)*, pp. 1–11. ACM, La Paz, Bolivia, 2003.

[26] M. C. Easton and C. K. Wong. Sequential destruction method for Monte Carlo evaluation of system reliability. *IEEE Transactions on Reliability*, **R-29**(1): 27–32, 1980.

[27] M. El Khadiri and G. Rubino. A Monte-Carlo method based on antithetic variates for network reliability computations. Technical Report 626, IRISA, Rennes, France, 1992. http://www.irisa.fr/doccenter/overview

[28] M. El Khadiri and G. Rubino. Reliability evaluation of communication networks. In *SAFECOMP'92, International Conference on Safety, Security and Reliability of Computers*, Zurich, October 1992.

[29] T. Elperin, I. Gertsbakh, and M. Lomonosov. Estimation of network reliability using graph evolution models. *IEEE Transactions on Reliability*, **40**(5), 1991.

[30] G. S. Fishman. A comparison of four Monte Carlo methods for estimating the probability of s-t conectedness. *IEEE Transactions on Reliability*, **R-35**(2), 1986.

[31] G. S. Fishman. A Monte Carlo sampling plan for estimating network reliability. *Operational Research*, **34**(4): 581–594, 1986.

[32] I. B. Gertsbakh. *Statistical Reliability Theory*. Marcel Dekker, New York, 1989.

[33] M. Grötschel, C. L. Monma, and M. Stoer. Computational results with a cutting plane algorithm for designing communication networks with low-connectivity constraints. Technical Report 188, Institüt fur Mathematik, Universität Augsburg, Germany, 1989.

[34] M. Grötschel, C. Monma, and M. Stoer. *Reliability of Computer and Communication Networks*. In Fred Roberts, Frank Hwang and Clyde Monma, eds, Polyhedral approaches for network survivability. 121–142, AMS, DIMACS and ACM, 1991.

[35] D. Günnec and F. S. Salman. Assessing the reliability and the expected performance of a network under disaster risk. In *Proceedings of the International Network Optimization Conference, INOC 2007*, Spa, Belgium, 2007.

[36] K.-P. Hui, N. Bean, M. Kraetzl, and D. Kroese. The tree cut and merge algorithm for estimation of network reliability. *Probability in the Engineering and Informational Sciences*, **17**: 23–45, 2003.

[37] K.-P. Hui, N. Bean, M. Kraetzl, and D. Kroese. The cross-entropy method for network reliability estimation. *Annals of Operations Research*, **134**: 101–118, 2005.

[38] K.-P. Hui. Monte Carlo network reliability ranking estimation. *IEEE Transactions on Reliability*, **56**(1): 50–57, 2007.

[39] R. Karp and M. G. Luby. A new Monte Carlo method for estimating the failure probability of an n-component system. *Computer Science Division*, University of California (Berkeley), 1983.

[40] D. P. Kroese, K.-P. Hui, and S. Nariai. Network reliability optimization via the cross-entropy method. IEEE *Transactions on Reliability*, **56**(2): 275–287, 2007.

[41] H. Kumamoto, K. Tanaka, and K. Inoue. Efficient evaluation of system reliability by Monte Carlo method. *IEEE Transactions on Reliability*, **R-26**(5), 1977.

[42] H. Kumamoto, K. Tanaka, K. Inoue, and E. J. Henley. Dagger-sampling Monte Carlo for system unavailability evaluation. *IEEE Transactions on Reliability*, **R-29**(2), 1980.

[43] A. Kumar, R Pathak, and M Gupta. Genetic algorithm based approach for designing computer network topology. In *Proceedings of the 1993 ACM Conference on Computer Science*, pp. 358–365, 1993.

[44] L. Lin and M. Gen. A self-controlled genetic algorithm for reliable communication network design. In *IEEE Congress on Evolutionary Computation*, pp. 640–647, Vancouver, BC, Canada, July 16–21, 2006.

[45] Y. Li. Measuring individual residence's accessible probability by using geographical information systems. In *Proceedings of the 2nd Symposium on Transportation Network Reliability (INSTR 2004)*, Vol. 3, pp. 239–244, Christchurch, New Zealand, 2004.

[46] M. O. Locks and A. Satyanarayana, eds. *Network Reliability–The State of the Art*, volume R-35, No. 3. IEEE Transactions on Reliability, 1986.

[47] M. Marseguerra, E. Zio, L. Podofillini, and D. Coit. Optimal design of reliable network systems in presence of uncertainty. *IEEE Transactions on Reliability*, **54**(2): 243–253, 2005.

[48] L. Murray and H. Cancela. Comparison of five Monte Carlo methods to estimate the network diameter constrained reliability. In *Proceedings of CLEI 2007 (Latin American Conference on Informatics)*, San José, Costa Rica, October 2007.

[49] N. Nojima. Prioritization in upgrading seismic performance of road network based on system reliability analysis. In Hu, Yuxian, Shiro Takada and Leon R. L. Wang, eds, *Third China–Japan–US Trilateral Symposium on Lifeline Earthquake Engineering*, pp. 323–330, Kunming, China, 1998.

[50] P. Premprayoon and P. Wardkein. Topological communication network design using ant colony optimization. In *The 7th International Conference on Advanced Communication Technology*, pp. 1147–1151, Phoenix Park, Republic of Korea, February 21–23, 2005.

[51] J. S. Provan. Bounds on the reliability of networks. *IEEE Transactions on Reliability*, **R-35**(3), 1986.

[52] J. S. Provan and M. O. Ball. Computing network reliability in time polynomial in the number of cuts. *Operational Research*, 32, 1984.

[53] S. Ross. System reliability evaluation by simulation: random hazards versus importance sampling. *Probability in the Engineering and Informational Sciences*, **6**: 119–126, 1992.

[54] S. M. Ross. Variance reduction in simulation via random hazard. *Probability in the Engineering and Informational Sciences*, 4, 1990.

[55] G. Rubino. Network reliability evaluation. In K. Bagchi, J. Walrand and G. W. Zobrist, eds, *State-of-the-Art in Performance Modeling and Simulation*. Gordon and Breach Books, Amsterdam, 1998.

[56] B. Sanso and F. Soumis. Communication and transportation networks reliability using routing models. *IEEE Transactions on Reliability*, **R-30**(5), 1981.

[57] H. A. Taboada, F. Baheranwala, and D. W. Coit. Practical solutions for multi-objective optimization: An application to system reliability design problems. *Reliability Engineering and System Safety*, **92**: 314–322, 2007.

[58] H. A. Taboada, J. Espiritu, and D. W. Coit. Moms-ga: A multi-objective multi-state genetic algorithm for system reliability optimization design problems. *IEEE Transactions on Reliability*, **57**(1), 2007.

[59] L. G. Valiant. The complexity of enumeration and reliability problems. *SIAM Journal of Computing*, **8**: 410–421, 1979.

[60] R. M. Van Slyke and H. Frank. Network reliability analysis: part I. *Networks*, 1, 1972.

[61] H. Wakabayashi. Network reliability improvement: Probability importance and criticality importance. In *Proceedings of the 2nd Symposium on Transportation Network Reliability (INSTR 2004)*, Vol. 3, pp. 204–210, Christchurch, New Zealand, 2004.

8

Rare event simulation and counting problems

José Blanchet and Daniel Rudoy

8.1 Introduction

Randomized approximation algorithms for counting problems have been the subject of many papers and monographs in theoretical computer science (see [13, 15, 25, 26]). At the same time, rare event simulation methodology has a long history of development within the applied probability and operations research communities. In this chapter, we offer a primer on the subject of approximate counting using rare event simulation techniques, thereby connecting two distinct points of view. The use of rare event simulation techniques for counting is very recent (see [5, 22]) and we hope that this chapter will motivate researchers in the rare event simulation community to consider the types of problems that we will discuss.

Our focus, consequently, is on the theoretical properties of randomized approximation algorithms for counting and their connections to rare event simulation. Even though powerful heuristic algorithms based on rare event simulation ideas already exist and enjoy empirical success, their efficiency is not yet rigorously understood [22, 23]. The key to our development is that not only has the machinery for measuring efficiency in counting problems been developed in theoretical computer science, but also these concepts have a natural correspondence to well-known notions of efficiency in rare event simulation. Therefore, it should be relatively easy for researchers in both communities

Rare Event Simulation using Monte Carlo Methods Edited by G. Rubino and B. Tuffin

to quickly get familiar with this correspondence which we hope to elucidate throughout this chapter.

Counting problems are important from both the applied and theoretical points of view. As an illustration, consider a canonical example of counting the number of bipartite graphs with a given degree sequence. Since the adjacency matrix of each graph has prescribed row and column sums, the counting problem is equivalent to that of counting the number of binary contingency tables with fixed row and column sums. In some cases, as illustrated in the detailed data analysis performed in [8] in the context of biological data for competitive species, statisticians are concerned with the problem of testing the null hypothesis that a given table, generated from some collected data, is consistent with a typical sample obtained uniformly from the space of tables that satisfy the row and column sums. In other words, we wish to see if the column and row sums are sufficient statistics for the distribution of zeros and ones inside the table.

One way to test such a hypothesis is to simulate many independent realizations of tables that are sampled uniformly over the space of all tables with the specified column and row sums and then to see if a statistic computed from the data lies in the tails of the histogram obtained from the simulated samples. If this happens then one would reject the null hypothesis. The statistical motivation is basically to estimate the expectation of an object whose law is determined by the uniform distribution of the space of tables with given column and row sums, not necessarily to count the number of tables. However, these problems are intimately related because, as we will show, a fast counting procedure often provides an efficient way to estimate expectations via importance sampling (IS). Many other statistically motivated counting problems present a real challenge when designing efficient algorithms for hypothesis testing including, for instance, general n-ary contingency tables and multidimensional contingency tables (e.g., the so-called two- or three-way tables).

Counting is an instance of the more general problem of computing normalizing constants of a discrete probability distribution. This problem arises in science (e.g., computing the free energy of a Gibbs distribution) and engineering (e.g., computing the normalizing constant in the steady-state distribution of large so-called Erlang loss networks); other engineering applications are described in [1]. We believe that the rare event simulation techniques that we describe here can be applied to more general problems involving computing normalizing constants.

From a theoretical perspective, counting problems are very interesting from the point of view of complexity theory and as such have attracted much attention from the theoretical computer science community. It is now widely believed that no exact polynomial-time algorithm can be developed for most counting problems and, therefore, it is of interest to develop polynomial-time approximation algorithms that provide accurate estimates with high confidence. The most common technique used in the design of approximate counting methods is the Markov chain Monte Carlo (MCMC) method based on a splitting-type representation, writing the quantity of interest as a telescoping product of ratios, each of which can be estimated by running a fast-mixing Markov chain. IS, on the other hand, has not yet been fully explored as a viable method for many counting problems.

In contrast to the MCMC method, which in some cases has been shown to apply to rather general input sequences (e.g., [4, 15]), IS may require regularity conditions (e.g., when counting the number of graphs with given row and column sums, requiring the maximum degree to grow slowly relative to the sum of the degrees) to deliver a provably efficient estimator (see [4] for counterexamples). Nevertheless, under appropriate constraints, IS has been rigorously shown to deliver the fastest algorithms for counting bipartite and simple graphs (see [1, 5]). These two papers, written independently, appear to be the first to rigorously justify the excellent empirical performance of two particular IS estimators [8]. In this setting, there are a number of theoretical properties of IS-based counting algorithms that remain unexplored. It is not known, for example, if counting problems (such as counting bipartite graphs) belong to some interesting complexity class under the regularity constraints necessary for IS to work (which are themselves still not fully characterized).

The rest of this is organized as follows. In Section 8.2 we provide a complexity-theoretic treatment of counting problems, thereby motivating the need for approximation algorithms. Subsequently, we present two notions of measuring efficiency of such approximation schemes and show their correspondence. In Section 8.3 we elucidate the relationship between counting and sampling problems and point out the roles played by MCMC and IS. In Section 8.4 we provide a rigorous analysis of an IS-based approximation algorithm for counting the number of binary contingency tables with prescribed row and column sums. We conclude the chapter, in Section 8.5, by discussing how to arrive at novel approximate counting schemes by combining IS together with the splitting-type decomposition common in MCMC approaches.

8.2 Background material

We are interested in studying approximation algorithms for counting problems and gaining a rigorous understanding of their efficiency. At the most basic level, we are interested in characterizing the space complexity (how much memory is utilized) and the time complexity (number of arithmetic operations) of each algorithm. In this section we show that the evaluation of efficiency in these terms has been a common ingredient in analyses of algorithms appearing both in the theoretical computer science and the rare event simulation literatures. We begin by discussing coarse notions of the complexity of counting together with examples of a number of counting problems.

8.2.1 Complexity theory and counting

A cornerstone of the analysis of algorithms is the quantification of their complexity as a function of the input problem size. Thus, once a suitable encoding for a problem is found, the space- and time-complexity of an associated algorithm is then reported as a function of the size of this representation. This approach necessitates a generic procedure to quantify the size of inputs uniformly across a great variety of problems. This can be done by encoding every problem as a binary string. In particular, let $\Sigma^* = \{0, 1\}^*$ denote the set of all binary strings of arbitrary length over which problem instances together with their solutions may be encoded and define a binary relation $R \subseteq \Sigma^* \times \Sigma^*$. This relation maps problems encoded by some $x \in \Sigma^*$ to the set of possible solutions:

$$R(x) = \{y \in \Sigma^* | (x, y) \in R\}.$$

To illustrate the above with an example, suppose we wish to find a Hamiltonian path (i.e., a path that visits each vertex exactly once) in some graph G. Then G, or equivalently its adjacency matrix A, is encoded into a binary string x whose length $|x|$ is a function of the number of edges in the graph or the number of ones in A. The string y would be the encoding of some Hamiltonian path (if it exists) and the relation R represents the algorithm used to find y given x. In this setting, quantifying efficiency boils down to analyzing the complexity of R in terms of $|x|$ and $|y|$. One important set of relations correspond to problems whose solutions are 'easy' to check.

Definition 1. *p-relation.*
Let $R \subseteq \Sigma^* \times \Sigma^*$ be a relation. We say that R is a p-relation if the following two conditions hold:

- *There exists a polynomial p such that, for all $x, y \in \Sigma^*$,*

$$(x, y) \in R \rightarrow |y| \leq p(x).$$

- *The predicate $(x, y) \in R$ can be tested in polynomial time $p(|x| + |y|)$.*

Now let $x \in \Sigma^*$ and take some p-relation R. The question of whether or not $R(x)$ is the empty set corresponds to the complexity class NP, which is the class of problems for which solutions y can be verified in time polynomial in $|x| + |y|$. On the other hand, problems for which a solution y may be found *and* verified in polynomial time correspond the class P. Our interest, however, is not in the decision problem, but in associated counting problems which are members of the so-called #P (pronounced sharp-P) complexity class.

Definition 2. *The class #P.*
The class #P is the set of counting problems associated with a p-relation. Specifically, $\beta : \Sigma^* \rightarrow \{0, 1, 2, \ldots\}$ belongs to #P if and only if there exists a p-relation R such that

$$\beta(x) = |\{y | (x, y) \in R\}|.$$

Here the function β counts the number of solutions to a problem instance. Continuing with our example, one might be interested in counting all Hamiltonian paths in some graph, rather than merely checking if one exists. Surely, problems in #P must be at least as difficult as their decision counterparts because counting the total number of solutions will determine if the solution set is non-empty. As a result, many of the counting problems encountered in practice can be shown to be NP-hard (problems that are at least as hard as any NP problem although possibly even harder!), NP-complete (problems that are in NP and are also NP-hard, so these are the hardest NP problems) or even #P-complete. The latter problems are the 'hardest' counting problems because their solution would imply that a solution for all other #P problems can be found with no more than a polynomial factor difference in complexity. We now proceed to make these definitions concrete through a number of examples.

- The first counting problem that was shown to be #P-complete is that of counting the number of perfect matchings in a bipartite graph [26]. A matching on $G = (V, E)$ is a set $E' \subset E$ of pairwise non-adjacent edges and is called a *perfect matching* when every vertex is incident to an edge in E'. This is equivalent to computing the permanent of the associated 0–1 incidence matrix A where $A(i, j) = 1$ if $(v_i, v_j) \in E$. This counting problem is known to be NP-hard, but it is interesting to note that the decision problem of checking whether or not G has at least one perfect matching is solvable in polynomial time.

- The decision counterpart of the #3-SAT problem was the first problem shown to be NP-complete. Its counting counterpart is one of the few problems that has been considered in the simulation community through the application of adaptive IS as described in [22]. At present, however, no rigorous analysis has been provided for this approach.

- A well-known counting problem is counting the number of directed graphs that are compatible with a given degree sequence. If we were to think of the underlying graphs in terms of their incidence matrices, then the prescribed degree sequences are nothing more than constraints on the row and column sums. In other words, this problem is equivalent to that of counting the number of binary contingency tables with fixed row and column sums.

- Another famous #P-hard problem is approximating the volume of a convex body. It turns out that methods used for solving counting problems (i.e., MCMC) have also been successfully applied to this problem [17]. Moreover, we believe that rare event simulation techniques can also be used to address this problem–forming a basis for one line of current research.

- Many other examples can be found in the theoretical computer science literature, for example, counting the number of self-avoiding random walks, which is a problem believed to be #P-complete and is particularly interesting because it is not 'self-reducible' like the approximate counting problems that are addressed using the MCMC approach. Other problems such as counting the number of k-colorings and others are described in the survey [14], and the monograph [13].

8.2.2 Approximation algorithms and efficiency

In light of the aforementioned complexity results, it is unlikely that exact solutions can be found for many counting problems of practical interest. This has motivated the development of efficient approximation algorithms for counting problems together with the theoretical tools required not just to quantify the approximation error, but to do so as a function of computational complexity. In other words, we seek approximation schemes which use the smallest amount of computation both in space and time in order to achieve a desired relative precision ε with $1-\delta$ confidence. This notion can be and has been formalized both in the theoretical computer science and rare event simulation communities. It is the purpose of this section to explain these formalisms and show their correspondence precisely.

Most of the successful approximation algorithms known for #P problems, with the notable exception of correlation-decay-based methods [11, 28], are randomized schemes. Suppose we are interested in a counting problem $\beta : \Sigma^* \to N$. In other words, we get a problem instance x (e.g., a particular bipartite graph) and are interested in estimating the output of $\beta(x)$ using some estimator $\widehat{\beta}(x)$. Analysis of the behavior of such an estimator rests on the following definition (cf. [21, p. 254]).

Definition 3. *Fully-Polynomial Randomized Approximation Scheme (FPRAS). A randomized approximation scheme is a randomized algorithm such that when given a problem instance x together with error and confidence parameters $0 < \varepsilon, \delta < 1$, the output $\hat{\beta}(x)$ has the following property for all x:*

$$\Pr[(1-\varepsilon)\beta(x) \leq \widehat{\beta}(x) \leq (1+\varepsilon)\beta(x)] \geq 1-\delta$$

If the algorithm runs in time polynomial in $|x|, \varepsilon^{-1}$, and $\log(1/\delta)$, then it is referred to as a fully polynomial randomized approximation scheme (FPRAS).

Here the quantity $|x|$ is a measure of the size of the problem instance whose particular form is informed by the precise asymptotic regime that we may wish to study. For example, when working with binary contingency tables it is convenient to think of $|x|$ as a function of d, the number of ones in the table (i.e., $|x| = O(d \log d)$). This allows us to study the efficiency of the approximation scheme asymptotically in d (i.e., as the size of the problem grows large). In particular, we adopt the notation $\hat{\beta}_d$ to underscore the fact that we are interested in studying the behavior of the estimators as a function of the input size. Given a specific counting problem, we suppress the explicit dependence on the encoding x and consider the family of estimators $\{\hat{\beta}_d : d \geq 1\}$ indexed implicitly as a function of $|x|$ through the parameter d. Note that some authors do not stress the dependence of order $\log(1/\delta)$ and instead appear to be satisfied with a dependence of order $O(1/\delta)$ in their definition (cf. [13, pp. 25–27]).

We now turn our attention to concepts of efficiency that often appear in the rare event simulation literature. In this setting, it is common to work with a family of estimators $\left\{\widehat{\beta}_{d,k} : d \geq 1\right\}$. Here, the explicit dependence on the problem encoding is also suppressed and k denotes the number of independently and identically distributed (i.i.d.) replications of $\hat{\beta}_d$ required to produce $\widehat{\beta}_{d,k}$ by means of a sample average. Of interest are estimators $\widehat{\beta}_{d,k}$, with the property that for a given $0 < \varepsilon, \delta < 1$, k is the number of replications k so that $\left|\widehat{\beta}_{d,k} - \beta_d\right| \leq \beta_d \varepsilon$ with probability $(1 - \delta)$. If $\widehat{\beta}_{d,k}$ has this property, we say that $\widehat{\beta}_{d,k}$ has *ε-relative precision with $1 - \delta$ confidence*.

There a number of notions of efficiency that are often used to quantify the performance of these estimators as it relates to their variance. Here we consider two such notions, namely strong and exponential efficiency, and show that the latter (which also implies the former) is intimately connected to the notion of an FPRAS. We begin with a formal definition of strong efficiency.

Definition 4. *Strong efficiency.*
Let $\{\hat{\beta}_d : d \geq 1\}$ be a family of estimators with ε-relative precision and $1 - \delta$ confidence and let $\sigma_d^2 = Var(\hat{\beta}_d)$ denote the underlying variance. This family of estimators is said to be strongly efficient if the corresponding coefficient of variation, $cv_d \triangleq \sigma_d / \beta_d$, is uniformly bounded for $d \geq 0$.

One can interpret strong efficiency as a measure of computational complexity in terms of the number of i.i.d. replications required. This number can be easily obtained via Chebyshev's inequality as follows:

$$P\left(\left|\widehat{\beta}_{d,k} - \beta_d\right| \geq \varepsilon \beta_d\right) \leq \frac{\sigma_d^2}{k \varepsilon^2 \beta_d^2}.$$

Therefore $k \geq \varepsilon^{-2} \delta^{-1} (\sigma_d / \beta_d)^2$ replications are required to produce an estimator that achieves ε-relative precision with $1 - \delta$ confidence. The computational complexity of the estimator depends not only on the number of required replications k, but also on the cost associated with generating each one, which we denote by $\kappa(d)$ since it depends on size of the input. We aim to have $\kappa(d) = O(d^p)$ for some $p \in (0, \infty)$. Hence, in the presence of strong efficiency, setting $k = O(\varepsilon^{-1} \delta^{-1})$ implies that computing $\widehat{\beta}_{d,k}$ requires $O(\kappa(d) \varepsilon^{-2} \delta^{-1})$ operations in order to achieve ε-relative precision with $1 - \delta$ confidence. According to Definition 3, an FPRAS corresponds to a stronger notion of efficiency that we shall call *exponential efficiency* since the computational complexity of an FPRAS is polynomial in d, $1/\varepsilon$, and $\log(1/\delta)$, rather than in d, $1/\varepsilon$, and $1/\delta$.

Definition 5. *Exponential efficiency.*
We say that the family of estimators $\left(\hat{\beta}_d : d \geq 1\right)$ is exponentially efficient for estimating β_d if there exists $\theta > 0$ such that

$$\psi(\theta) \triangleq \sup_{d \geq 1} \log E \exp\left(\theta \hat{\beta}_d / \beta_d\right) < \infty.$$

We will relate exponential efficiency to an FPRAS using the following uniform version of Chernoff's bound, proved in [5], instead of the Chebyshev inequality.

Lemma 1. *Suppose that the family of estimators* $\left(\hat{\beta}_d : d \geq 1\right)$ *is exponentially efficient for estimating* β_d. *Then for* $\varepsilon > 0$ *we have*

$$P\left(\left|\widehat{\beta}_{d,k} - \beta_d\right| \geq \varepsilon\beta_d\right) \leq 2\exp\left(-k\min(I(\varepsilon), I(-\varepsilon)\right), \quad (8.1)$$

where $I(\varepsilon) = \sup_\theta(\theta(1+\varepsilon) - \psi(\theta))$ *and* $\psi(\theta)$ *is the cumulant generating function. Moreover,* $I(\varepsilon), I(-\varepsilon) > 0$ *and* $I(\varepsilon) \geq \rho\varepsilon^2$ *for some* $\rho > 0$.

An immediate consequence of the previous results is that if the family $(\hat{\beta}_d : d \geq 1)$ is exponentially efficient and $\kappa(d)$ operations are required to generate a single replication, then $\widehat{\beta}_{d,k}$ requires $O(\kappa(d)\varepsilon^{-2}\log(\delta^{-1}))$ operations to achieve ε-relative precision with $1-\delta$ confidence. If $\kappa(d)$ grows polynomially in the size of the problem d, then the estimator $\widehat{\beta}_{d,k}$ is also an FPRAS. We have therefore established the correspondence among the different notions of efficiency that appear in the literature.

8.3 Approximate counting, sampling and Markov chain Monte Carlo

In this section, we describe the relationship between approximate sampling and approximate counting. It has been rigorously established in [16] that the ability to exactly or approximately sample from the space of problem solutions we wish to count enables the construction of an FPRAS for the counting problem. Here, we first discuss how an approximate sampler may be used to construct an FPRAS for counting in the context of the well-studied example of matchings. In Section 8.4, we will show another reduction for the binary contingency tables problem. Then we discuss how the MCMC method has been employed to design good samplers and underscore that IS methods can also be used for this purpose.

8.3.1 From approximate sampling to approximate counting

To show how to use an approximate sampling algorithm to construct an FPRAS, we begin with a rigorous complexity-theoretic definition of an approximate sampler.

Definition 6. *Fully Polynomial Approximate Uniform Sampler (FPAUS).*
Given an instance $x \in \Sigma^*$, *let* Ω *denote the set of all solutions to* x *and let* π *denote the uniform distribution over* Ω. *An almost uniform sampler takes as input an instance* x *and an error tolerance* ε *and outputs a sample* $Z \sim \mu$ *from a distribution* μ *such that*

$$\|\mu - \pi\|_{TV} \triangleq \max_{A \subset \Omega} |\mu(A) - \pi(A)| < \varepsilon.$$

If the number of steps taken by the sampler is polynomial in $|x|$ *and in* $\log(\varepsilon^{-1})$ *then the sampler is called a fully polynomial almost uniform sampler (FPAUS).*

It is interesting to note that the complexity of an FPRAS must be polynomial in $1/\varepsilon$ whereas an FPAUS must be polynomial only in $\log(1/\varepsilon)$. This is because the error in an FPAUS is propagated, as we shall see, through the product of ratios.

We now show a canonical example of how to convert an FPAUS to an FPRAS for the particular problem of counting matchings; our exposition closely follows the presentation in [27]. We begin by assuming that an exact, not approximate, sampler is available. We then generalize the argument to the approximate sampling case via the incorporation of the appropriate error probabilities (an alternative approach via Chebyshev's inequality can be found in [13]). Let $G = (V, E)$ be a graph and denote by $\mathcal{M}(G)$ the set of matchings of G. Furthermore, suppose that we have an algorithm $\mathcal{A}$ that can generate matchings uniformly at random from $\mathcal{M}(G)$ in time polynomial in $|V| = m$. We show how to use $\mathcal{A}$ to construct an FPRAS for estimating $|\mathcal{M}(G)|$.

We begin by constructing a sequence of graphs $G_0, \ldots, G_m$ recursively by setting $G_0 = G$ and iteratively removing one edge at a time, so that $G_i = (V, E_{i-1} \backslash e_i)$, until we reach the empty graph G_m. The edge selected for removal is randomly selected. This allows us to restate the problem as a telescoping product:

$$|\mathcal{M}(G)| = \frac{|\mathcal{M}(G_0)|}{|\mathcal{M}(G_1)|} \frac{|\mathcal{M}(G_1)|}{|\mathcal{M}(G_2)|} \cdots \frac{|\mathcal{M}(G_{m-1})|}{|\mathcal{M}(G_m)|} |\mathcal{M}(G_m)| = \prod_{i=0}^{m-1} \frac{1}{p_i}, \tag{8.2}$$

where $p_i = \frac{|\mathcal{M}(G_{i+1})|}{|\mathcal{M}(G_i)|}$ and $|\mathcal{M}(G_m)| = 1$ since G_m is the empty graph. Notice that by construction G_{i+1} is a subgraph of G_i and, therefore, $\mathcal{M}(G_{i+1}) \subseteq \mathcal{M}(G_i)$, which implies that $0 < p_i \leq 1$. Equation (8.2) is closely related to the splitting technique (see Chapter 3).

We design an FPRAS by using the exact sampler $\mathcal{A}$ to estimate each p_i to within a relative precision of ε/m with confidence $1 - \delta/m$ in turn as follows. We generate N matchings uniformly at random from $\mathcal{M}(G_i)$ and count how many of these are also matchings in $\mathcal{M}(G_{i+1})$. Now, let Z_{ij} be a 0–1 random variable that denotes whether or not the jth matching produced by $\mathcal{A}$ is in $\mathcal{M}(G_{i+1})$. Then we have that $\widehat{p_i} = N^{-1} \sum_{j=1}^{N} Z_{ij}$ is an estimator of p_i and quantifying the error in this estimate allows us to evaluate the efficiency of the resultant approximate scheme for estimating $|\mathcal{M}(G)|$. Observe that

$$|\mathcal{M}(G_i) \backslash \mathcal{M}(G_{i+1})| \leq |\mathcal{M}(G_{i+1})|,$$

since we can associate each matching $\mathcal{M} \in \mathcal{M}(G_i)$ with $\mathcal{M} \backslash e_{i+1} \in \mathcal{M}(G_{i+1})$. Coupled with the fact that $\mathcal{M}(G_{i+1}) \subset \mathcal{M}(G_i)$, this implies that $p_i > 1/2$. By Chernoff's inequality, setting $N = O\left((3m/\varepsilon)^2 \log(2m/\delta)\right)$ yields

$$\Pr\left(|p_i - \widehat{p_i}| > \frac{\varepsilon}{m}\right) \leq \frac{\delta}{m}.$$

Since the output of the overall approximation scheme is

$$|\widehat{\mathcal{M}(G)}| = \prod_{i=0}^{m-1} \frac{1}{\widehat{p}_i},$$

the associated error, using the bound just developed, is given by

$$\Pr(|\widehat{\mathcal{M}(G)}| \notin (1 \pm \varepsilon)\, |\mathcal{M}(G)|) \leq \delta.$$

Suppose that the exact sampler $\mathcal{A}$ runs in time $p(|G|)$ polynomial in the size of the graph $|G|$. Clearly, we have obtained an FPRAS since the complexity of the overall procedure is $O(m^3 p(m) \log(m\delta^{-1})\varepsilon^{-2})$. This illustrates how to reduce exact sampling to approximate counting in the case of matchings. Now, suppose that instead of an exact sampler $\mathcal{A}$, we have access to an FPAUS that takes time $p(m, 1/\eta)$ to output a sample that is η-close in total variation to the target distribution. Then the previous estimates are easily adapted. Let $\widetilde{Z}_{ij}$ be a 0-1 random variable with mean $\widetilde{p}_i$, that denotes whether or not the ith matching produced by $\mathcal{A}$ is in $\mathcal{M}(G_{i+1})$ so that $|p_i - \tilde{p}_i| \leq \eta$. If we set $\eta = \varepsilon/12m$, then by applying, once more, Chernoff's inequality we obtain, for $N = O((m/\varepsilon)^2 \log(2m/\delta))$,

$$\Pr\left(|q_i - \tilde{p}_i| > \frac{\varepsilon}{12m}\right) \leq \frac{\delta}{m}.$$

Then for all $1 \leq i \leq m$, we have that, with confidence $1 - \delta$, $p_i(1 - \varepsilon/6m) \leq q_i \leq p_i(1 + \varepsilon/6m)$. Since $(1 - \varepsilon/6m)^m \geq (1 - \varepsilon)$ and $(1 + \varepsilon/6m)^m \leq (1 + \varepsilon)$, we again see that

$$\Pr(|\widehat{\mathcal{M}(G)}| \notin (1 \pm \varepsilon)\, |\mathcal{M}(G)|) \leq \delta.$$

Thus, we have shown that in the case of counting matchings, one can reduce approximate sampling to approximate counting so that the complexity of the resultant FPRAS is $O(m^3 p(m, 1/\eta) \log(m\delta^{-1})\varepsilon^{-2})$. The essence of this reduction is to use the FPAUS to approximate each of the m factors in (8.2) to within ε/m with confidence $1 - \delta/m$. The proof relies on the fact that the number of samples to achieve this is not too great. Generalizing, it is essential that the ratios are polynomially bounded in the size of the input. The way to achieve this in a general setting is to relate the problem at hand to solutions of smaller instances. This substructure is a crucial ingredient that, while not always easy to find, is required for the reduction of counting to sampling and has been formalized as *self-reducibility* in [16] where the authors prove that for any self-reducible relations an FPAUS may be converted to an FPRAS for the associated counting problem.

The remaining ingredient in designing an FPRAS that has not yet been discussed is how to obtain an FPAUS for the problem at hand. The key observation is that in some cases both MCMC and IS can form the basis of a viable solution to this problem. The MCMC method is widely used in theoretical computer science for the design of approximately uniform samplers. The basic approach is to define a Markov chain on the space of all solutions (which is discrete, but often exponentially large) with a uniform stationary distribution. If it can then be proven that the chain has a geometric rate of convergence, with a rate that degrades slowly with the input size (i.e., it is rapidly mixing), then approximately uniform samples from the set of solutions can be obtained using a polynomial number of steps. Consequently, much effort has been made to develop bounds on convergence rates of chains in this setting; methods based on both coupling [10, 19], and spectral analysis [9, 24, 25] have been explored. On the other hand, there has been little work on the design of FPRASs using IS methods. We believe that the techniques that are well known in the rare event simulation community can be brought to bear on this issue, as we discuss next.

8.4 Illustrating counting strategies

In this section we discuss how IS may be used to construct exponentially efficient estimators for counting problems. In particular, we consider the problem of counting the number of bipartite graphs with a given degree sequence using state-dependent importance samplers. Recall that the adjacency matrix A of any bipartite graph can be thought of as binary contingency table with specified row and column sums. Let $\{r_1, \ldots, r_m\}$ and $\{c_1, \ldots, c_n\}$ be non-negative integers representing the row and column sums, also referred to as margins, of A respectively. Furthermore, let $d = \sum c_j = \sum r_i$ be the total number of ones inside of A. We are interested in computing $M(\mathbf{r}, \mathbf{c})$, the number of binary contingency tables with given column and row sums, where $\mathbf{c} = (c_1, \ldots, c_n)$ and $\mathbf{r} = (r_1, \ldots, r_m)$. Equivalently, we are interested in counting the number of solutions to the following system of equations, where each $x_{i,j} \in \{0, 1\}$:

$$\sum_{j=1}^{n} x_{i,j} = r_i \text{ for } 1 \leq i \leq m, \qquad \sum_{i=1}^{m} x_{i,j} = c_j \text{ for } 1 \leq j \leq n.$$

We study the efficiency of the IS algorithm first presented in [8], though under a more restrictive set of assumptions than in [5] for clarity. Specifically, we wish to understand the efficiency of the family of underlying estimators asymptotically in m and n (and therefore in d). Moreover, we shall assume that there exists $\kappa \in (0, \infty)$ such that $\max_{j \leq n, i \leq m}\{c_j, r_i\} \leq \kappa$ for all $n, m \geq 1$. After describing the algorithm in depth, we show that in this setting, the family of estimators is exponentially efficient (thereby yielding an FPRAS). We conclude by discussing how to blend IS together with MCMC for counting problems.

8.4.1 Importance sampling for counting binary contingency tables

The core idea behind the IS strategy for counting binary contingency tables presented in [8] is to construct a sample $\mathbf{X} = (X_1, \ldots, X_n)$ from the uniform distribution on the set of all tables satisfying the row and columns sums, by sequentially assigning the values of each column X_k. The columns are assigned starting from the one with the largest margin X_{k_0} (i.e., $k_0 = \arg\max_j c_j$) down to the one with the smallest margin X_{k_n} (i.e., $k_n = \arg\min_j c_j$) in a 'convenient' way.

In order to be more precise, we reformulate the algorithm of [8] in a way that will allow us to relate counting binary contingency tables to a certain rare event estimation problem. Let $\mathbf{X}_1, \mathbf{X}_2, \ldots, \mathbf{X}_n$ be a sequence of independent binary vectors living in $\{0, 1\}^m$, so that the distribution of $\mathbf{X}_k$ is uniform over the space

$$C_k = \{(x_1, x_2, \ldots, x_m) : x_i \in (0, 1), x_1 + \cdots + x_m = c_k\}.$$

Next, we define $\mathbf{S}_0 = \mathbf{r}$ where $(\mathbf{r} = (r_1, \ldots, r_m))$ and set $\mathbf{S}_{k+1} = \mathbf{S}_k - \mathbf{X}_{k+1}$ for $0 \leq k \leq n-1$. Given $\boldsymbol{\sigma} = (\sigma_1, \ldots, \sigma_m)$ and $\boldsymbol{\rho} = (\rho_1, \ldots, \rho_{n-k})$, we write $P_{\boldsymbol{\sigma},\boldsymbol{\rho}}(\cdot)$ to denote the probability measure corresponding to $(\mathbf{S}_k, \mathbf{S}_{k+1}, \ldots, \mathbf{S}_n)$ given that $\mathbf{S}_k = \boldsymbol{\sigma}$ and $\rho_j = c_{j+k}$ for $1 \leq j \leq n-k$. The Markov transition kernel associated with the $\mathbf{S}_k$ is given by

$$K_{k-1}(\boldsymbol{\sigma}, \boldsymbol{\sigma} + \mathbf{z}) = I(z_1 + \cdots + z_m = c_k)\binom{m}{c_k}^{-1},$$

where $\mathbf{z} = (z_1, \ldots, z_m) \in \{0, 1\}^m$. Also observe that

$$M(\mathbf{r}, \mathbf{c}) = P_{\mathbf{r},\mathbf{c}}(\mathbf{S}_n = 0) \times \binom{m}{c_1} \times \cdots \times \binom{m}{c_n}.$$

Thus, our problem is equivalent to efficiently estimating the probability that $P_{\mathbf{r},\mathbf{c}}(\mathbf{S}_n = 0)$ as $m, n \nearrow \infty$ (and so as $d \nearrow \infty$), which is equivalent to obtaining a sample $\mathbf{X}$ that satisfies all the row and column sums.

The strategy of [8] is based on the application of state-dependent IS by assigning the components of the $\mathbf{X}_k$ in a way that helps to induce the occurrence of the event $\{\mathbf{S}_n = 0\}$. Their biasing strategy is based on the so-called *conditional Poisson distribution* [7] defined as follows.

Definition 7. *Conditional Poisson distribution.*

Let $(Z_1, \ldots, Z_m)$ be independent Bernoulli random variables with parameters $(p_1, \ldots, p_m)$. The joint distribution of $(Z_1, \ldots, Z_m)$ given that $S_Z = Z_1 + \cdots + Z_m = a$ is called the conditional Poisson distribution and is given by

$$P(Z_1 = z_1, \ldots, Z_m = z_m) = \frac{1}{w}\prod_{j=1}^{m} \gamma_j^{z_j},$$

where $\gamma_j = p_j/(1 - p_j)$ and the normalizing constant w can be computed as

$$w = \sum_{(z_1,\ldots,z_m)} \prod_{j=1}^{m} \gamma_j^{z_j} \mathbb{I}(z_1 + \cdots + z_m = a).$$

It is important to note that in [7] the conditional Poisson distribution is motivated as the zero-variance IS distribution of a related problem. It turns out that if we require only the first column sum to be c_1, but do not place any restrictions on the rest of the columns and require the row sums to be $(r_1, r_2, \ldots, r_m)$, then X_1 has the conditional Poisson distribution. Several methods for efficiently sampling according to the conditional Poisson distribution are described in [7]. The recursive procedure adopted in [8] is known as the drafting method which is a sequential procedure that allows to sample c units without replacement from the set $A_m = \{1, 2, \ldots, m\}$; the i-th unit has a probability proportional to w_i. Let A_k, $0 \le k \le c$, be the set of selected units after k draws, so that $A_0 = \oslash$ and A_c is the final sample to be obtained. At the kth step (with $1 \le k \le c$), a unit $j \in A_{k-1}^c$ is selected into the sample with probability:

$$p\left(j, A_{k-1}^c\right) = \frac{\widetilde{w}\left(c-k, A_{k-1}^c \backslash \{j\}\right) w_j}{(c-k+1)\,\widetilde{w}\left(c-k+1, A_{k-1}^c\right)},$$

where

$$\widetilde{w}(i, A) = \sum_{C \subseteq A, card(C)=i} \left(\prod_{i \in C} w_i \right),$$

$\widetilde{w}(0, A) = 1$ for all $A \subseteq A_m$ and $\widetilde{w}(i, A) = 0$ for $i > card(A)$. The computation of the $\widetilde{w}(i, A)$'s is performed using the recursion:

$$\widetilde{w}(i, A) = \widetilde{w}(i, A \backslash \{j\}) + \widetilde{w}(i-1, A \backslash \{j\})\, w_j.$$

For instance, to compute $\widetilde{w}(c, A_m)$ we apply the recursion

$$\widetilde{w}\left(i, A_j\right) = \widetilde{w}\left(i, A_j \backslash \{j\}\right) + \widetilde{w}\left(i-1, A_j \backslash \{j\}\right) w_j$$

for $1 \le i \le c$ and $i \le j \le m$. It follows that computing $\widetilde{w}(c, A_m)$ takes $O(cm)$ operations. Evaluating $p(j, A_m) = p\left(j, A_0^c\right)$ then takes $O\left(cm^2\right)$ operations. Each of the $p\left(j, A_k^c\right)$'s can be evaluated similarly, however, it is more convenient to use Lemma 1 of [7], which states that:

$$p\left(j, A_k^c\right) = \frac{w_{i_k} p\left(j, A_{k-1}^c\right) - w_j p\left(j, A_{k-1}^c\right)}{(c-k)(w_{i_k} - w_j) p\left(i_k, A_{k-1}^c\right)},$$

for$1 \le k \le c-1$ and $j \in A_k^c$, where i_k is the element selected in the kth iteration of the drafting procedure. Therefore, we conclude that using the drafting method it takes $O\left(cm^2\right)$ operations to generate a sample from (1.3).

Using the drafting method, an IS strategy is introduced in [8] that biases $\mathbf{X}_k$ according to an appropriate conditional Poisson distribution, thereby inducing the assignment of ones in rows with high margins. In particular, we propose choosing $\gamma_j(k) = \sigma_j + \theta_{k-1}\sigma_j^2/d_{k-1}$ where $d_{k-1} = \sigma_1 + \cdots + \sigma_m$ and $a = c_k$.

The values of the θ_k will be selected in such a way as to improve the complexity of the overall estimator. The transition kernel corresponding to this sampling strategy is given by

$$\widetilde{K}_{k-1}(\boldsymbol{\sigma}, \boldsymbol{\sigma}+\mathbf{z}) = \frac{I(z_1 + \cdots + z_m = c_k)}{w_{k-1}(\boldsymbol{\sigma})} \prod_{j=1}^{m} \left(\sigma_j + \frac{\theta_{k-1}}{d_{k-1}}\sigma_j^2\right)^{z_j}.$$

Therefore, the corresponding IS estimator is

$$\beta_{m,n} = \prod_{j=1}^{n} \frac{K_{j-1}(\mathbf{S}_{j-1}, \mathbf{S}_j)}{\widetilde{K}_{j-1}(\mathbf{S}_{j-1}, \mathbf{S}_j)} I(\mathbf{S}_n = 0).$$

8.4.2 Analysis of efficiency

In order to analyze the efficiency of the estimator $\beta_{m,n}$ as $m, n \nearrow \infty$, we first provide an asymptotic estimate for $M(\mathbf{r}, \mathbf{c})$. Such an approximation is not difficult to develop, especially under the assumption of bounded row and column sums as in our context. To begin, consider a set of m cells of type A and n cells of type B where the jth cell of type A contains r_j tokens and the ith cell of type B contains c_i tokens for a total of $2d = \sum c_i + \sum r_j$ tokens. Now, suppose that both types of tokens are labeled from 1 to d and consider a set of pairings of the form $\{(a_1, b_1), \ldots, (a_d, b_d)\}$ where the a_j and b_i correspond to tokens of type A and B, respectively. Since the order of the pairs is not important there are a total of $d!$ such pairings. Two pairs, say (a_{j_1}, b_{j_1}) and (a_{j_2}, b_{j_2}) are said to be parallel if a_{j_1} and a_{j_2} belong to the same cell and the same occurs for b_{j_1} and b_{j_2}. Consider a pairing sampled uniformly at random (among the $d!$ possibilities) and let N be the number of parallel pairs in the sample; then it follows (since the tokens are labeled within each cell) that

$$d!P(N = 0) = r_1! \cdots r_m! c_1! \cdots c_n! M(\mathbf{r}, \mathbf{c}).$$

Therefore,

$$M(\mathbf{r}, \mathbf{c}) = \frac{d!P(N = 0)}{r_1! \cdots r_m! c_1! \cdots c_n!},$$

A formal Poisson approximation $P(N = 0) \approx \exp(-E(N))$ allows us to estimate $M(\mathbf{r}, \mathbf{c})$. In particular, it is not difficult to see that

$$E(N) = \frac{2}{d(d-1)} \sum_{j=1}^{n} \binom{c_j}{2} \sum_{i=1}^{m} \binom{r_i}{2}.$$

Therefore we obtain

$$M(\mathbf{r}, \mathbf{c}) \approx \frac{d! \exp(-\alpha(\mathbf{r}, \mathbf{c}))}{\mathbf{r}!\mathbf{c}!}, \tag{8.3}$$

where $\mathbf{r}! = r_1! \cdots r_m!$, $\mathbf{c}! = c_1! \cdots c_n!$ and

$$\alpha(\mathbf{r}, \mathbf{c}) = \frac{2}{d^2} \sum_{j=1}^{n} \binom{c_j}{2} \sum_{i=1}^{m} \binom{r_i}{2}.$$

Approximation (8.3) can be rigorously justified under our assumptions of bounded row and column sums [2]. We record these observations in the following result. See [5, 12, 20] for additional extensions.

Theorem 1. *Assume that* $\sup_{m,n\geq 1} \max_{j\leq n, i\leq m}\{c_j, r_i\} \leq \kappa < \infty$. *Then as* $d \nearrow \infty$,

$$M(\mathbf{r}, \mathbf{c}) = \frac{d! \exp(-\alpha(\mathbf{r}, \mathbf{c}))}{\mathbf{r}!\mathbf{c}!}(1 + o(1)).$$

Now we return to showing the exponential efficiency of the estimator $\beta_{m,n}$ as $d \nearrow \infty$. The idea is to show that there exists a constant $\kappa_1 \in (0, \infty)$ such that

$$\beta_{m,n} \leq \kappa_1 P_{\mathbf{r},\mathbf{c}} (S_n = 0) .$$

For $0 \leq k \leq n-1$, given $\mathbf{S}_k = \boldsymbol{\sigma}$ and $\boldsymbol{\rho} = (c_{k+1}, \ldots, c_n)$ with $\rho_j = c_{j+k}$, let us define

$$\beta (\boldsymbol{\sigma}, \boldsymbol{\rho}) = \prod_{j=k+1}^{n} \frac{K_{j-1} (\mathbf{S}_{j-1}, \mathbf{S}_j)}{\widetilde{K}_{j-1} (\mathbf{S}_{j-1}, \mathbf{S}_j)}$$

and set

$$v (\boldsymbol{\sigma}, \boldsymbol{\rho}) = \frac{d_k! \exp (-\alpha (\boldsymbol{\sigma}, \boldsymbol{\rho}))}{\boldsymbol{\sigma}!\boldsymbol{\rho}!} \binom{m}{\rho_1}^{-1} \cdots \binom{m}{\rho_{n-k}}^{-1},$$

where $d_k = \sigma_1 + \cdots + \sigma_m$. Now let $\widetilde{\boldsymbol{\sigma}} = \mathbf{S}_{k+1}$, $\widetilde{\boldsymbol{\rho}} = (c_{k+2}, \ldots, c_n)$ and note that

$$\frac{\beta (\boldsymbol{\sigma}, \boldsymbol{\rho})}{v (\boldsymbol{\sigma}, \boldsymbol{\rho})} = \frac{\beta (\widetilde{\boldsymbol{\sigma}}, \widetilde{\boldsymbol{\rho}})}{v (\widetilde{\boldsymbol{\sigma}}, \widetilde{\boldsymbol{\rho}})} \times \frac{K_k (\boldsymbol{\sigma}, \widetilde{\boldsymbol{\sigma}})}{\widetilde{K}_k (\boldsymbol{\sigma}, \widetilde{\boldsymbol{\sigma}})} \times \frac{v (\widetilde{\boldsymbol{\sigma}}, \widetilde{\boldsymbol{\rho}})}{v (\boldsymbol{\sigma}, \boldsymbol{\rho})} = \frac{\beta (\widetilde{\boldsymbol{\sigma}}, \widetilde{\boldsymbol{\rho}})}{v (\widetilde{\boldsymbol{\sigma}}, \widetilde{\boldsymbol{\rho}})} \times \frac{K_k (\boldsymbol{\sigma}, \widetilde{\boldsymbol{\sigma}})}{\widetilde{K}_k (\boldsymbol{\sigma}, \widetilde{\boldsymbol{\sigma}})}$$
$$\times \frac{(d_k - c_{k+1})!\boldsymbol{\sigma}!c_{k+1}!}{d_k!\widetilde{\boldsymbol{\sigma}}!} \binom{m}{c_{k+1}} \times \exp (\alpha (\boldsymbol{\sigma}, \boldsymbol{\rho}) - \alpha (\widetilde{\boldsymbol{\sigma}}, \widetilde{\boldsymbol{\rho}})) .$$

We write $\boldsymbol{\sigma} - \widetilde{\boldsymbol{\sigma}} = \mathbf{z} = (z_1, \ldots, z_m)$ (with $z_1 + \cdots + z_m = c_{k+1}$) and note that

$$\frac{K_k (\boldsymbol{\sigma}, \widetilde{\boldsymbol{\sigma}})}{\widetilde{K}_k (\boldsymbol{\sigma}, \widetilde{\boldsymbol{\sigma}})} \frac{(d_k - c_k)!\boldsymbol{\sigma}!c_k!}{d_k!\widetilde{\boldsymbol{\sigma}}!} \binom{m}{c_k} \tag{8.4}$$
$$= w_k (\boldsymbol{\sigma}) \prod_{j=1}^{m} \left(1 + \sigma_j \theta_k / d_k\right)^{-z_j} \times \frac{d_k^{-c_{k+1}} c_{k+1}!}{\left(1 - d_k^{-1}\right) \cdot \ldots \cdot \left(1 - (c_{k+1} - 1) d_k^{-1}\right)}.$$

Observe that

$$\prod_{j=1}^{c_{k+1}-1}\left(1-\frac{j}{d_k}\right)^{-1}=\exp\left(\frac{c_{k+1}\left(c_{k+1}-1\right)}{2d_k}+O\left(\frac{1}{d_k^2}\right)\right),$$

and therefore, the expression in (8.4) is bounded by

$$w_k\left(\boldsymbol{\sigma}\right)\frac{c_{k+1}!}{d_k^{c_{k+1}}}\exp\left(-\frac{\theta_k}{d_k}c_{k+1}+\frac{c_{k+1}\left(c_{k+1}-1\right)}{2d_k}+O\left(\frac{1}{d_k^2}\right)\right). \tag{8.5}$$

The next proposition provides an estimate for $w_k\left(\boldsymbol{\sigma}\right)$.

Proposition 1. *There exists a constant $\kappa_2 \in (0,\infty)$ such that:*

$$w_k\left(\boldsymbol{\sigma}\right)=\frac{d_k^{c_{k+1}}}{c_{k+1}!}\left(1+\frac{\theta_k}{d_k^2}\sum_{j=1}^{m}\sigma_j^2\right)^{c_{k+1}}\exp\left(-\binom{c_{k+1}}{2}\sum_{j=1}^{m}\frac{\sigma_j^2}{d_k^2}\right)\times\exp\left(\kappa_2\frac{\theta_k}{d_k^2}\right).$$

Proof. We give a sketch of the proof of this result. Let $J_1,\ldots,J_{c_{k+1}}$ be i.i.d. random variables such that

$$P\left(J=j\right)=\sigma_j\left(1+\frac{\sigma_j\theta_k}{d_k}\right)\frac{1}{\eta}$$

where $\eta=\sum_{j=1}^{m}\sigma_j\left(1+\sigma_j\theta_k/d_k\right)=d_k(1+\theta_k\sum_{j=1}^{m}\sigma_j^2/d_k^2)$. Let $I_{i,j}=1$ when $J_i=J_j$ and set $N=\sum_{i<j}I_{i,j}$. Then, we have that

$$w_k\left(\boldsymbol{\sigma}\right)=\frac{1}{c_{k+1}!}\eta^{c_{k+1}}P\left(N=0\right)=\frac{1}{c_{k+1}!}\eta^{c_{k+1}}\exp\left(-E(N)+O\left(\frac{1}{d_k^2}\right)\right).$$

This Poisson-based approximation, which can be rigorously justified using the inclusion–exclusion principle, is the only missing piece in the proof of this proposition. Note that

$$E(N)=\binom{c_{k+1}}{2}\sum_{j=1}^{m}\frac{\sigma_j^2\left(1+\sigma_j\theta_k/d_k\right)^2}{d_k^2\left(1+\theta_k\sum_{j=1}^{m}\sigma_j^2/d_k^2\right)^2}=\binom{c_{k+1}}{2}\sum_{j=1}^{m}\frac{\sigma_j^2}{d_k^2}+O\left(\frac{\theta_k}{d_k^2}\right).$$

On the other hand, we have that

$$\eta^{c_{k+1}}=d_k^{c_{k+1}}\left(1+\frac{\theta_k}{d_k^2}\sum_{j=1}^{m}\sigma_j^2\right)^{c_{k+1}}.$$

Therefore,

$$w_k\left(\boldsymbol{\sigma}\right)=\frac{d_k^{c_{k+1}}}{c_{k+1}!}\left(1+\frac{\theta_k}{d_k^2}\sum_{j=1}^{m}\sigma_j^2\right)^{c_{k+1}}\exp\left(-\binom{c_{k+1}}{2}\sum_{j=1}^{m}\frac{\sigma_j^2}{d_k^2}\right)\exp\left(O\left(\frac{\theta_k}{d_k^2}\right)\right).$$

Using the proposition above together with the upper bound in (8.5), we obtain the following estimate of the right-hand side of (8.4):

$$\frac{\beta(\boldsymbol{\sigma},\boldsymbol{\rho})}{\upsilon(\boldsymbol{\sigma},\boldsymbol{\rho})} \le \frac{\beta(\widetilde{\boldsymbol{\sigma}},\widetilde{\boldsymbol{\rho}})}{\upsilon(\widetilde{\boldsymbol{\sigma}},\widetilde{\boldsymbol{\rho}})} \prod_{j=1}^{m} \left(1+\frac{\sigma_j\theta_k}{d_k}\right)^{-z_j} \exp\left(\frac{c_{k+1}(c_{k+1}-1)}{2d_k}\right) \tag{8.6}$$

$$\times \exp\left(\frac{\theta_k}{d_k^2}\sum_{j=1}^{m}\sigma_j^2 - \binom{c_{k+1}}{2}\sum_{j=1}^{m}\frac{\sigma_j^2}{d_k^2} + O\left(\frac{\theta_k}{d_k^2}\right)\right)$$

$$\times \exp\left(\alpha(\boldsymbol{\sigma},\boldsymbol{\rho}) - \alpha(\widetilde{\boldsymbol{\sigma}},\widetilde{\boldsymbol{\rho}})\right).$$

Finally, to bound the difference $\alpha(\boldsymbol{\sigma},\boldsymbol{\rho}) - \alpha(\widetilde{\boldsymbol{\sigma}},\widetilde{\boldsymbol{\rho}})$, note that it equals

$$\alpha(\boldsymbol{\sigma},\boldsymbol{\rho}) - \alpha(\widetilde{\boldsymbol{\sigma}},\widetilde{\boldsymbol{\rho}}) = 2\chi_k\sum\sigma_j z_j - 2\chi_k c_{k+1} - 2\chi_k\frac{c_{k+1}}{d_k}\sum\sigma_j(\sigma_j-1)$$

$$+\chi_k\frac{c_{k+1}^2\sum\sigma_j(\sigma_j-1)}{d_k^2} + \frac{c_{k+1}(c_{k+1}-1)\sum\sigma_j(\sigma_j-1)}{2d_k^2},$$

where $\chi_k = \sum_{j=k+2}^{n} c_j(c_j-1)/(2(d_k-c_{k+1})^2) = O(1/d_k)$. We can now provide a convenient expression for θ_k by observing that if we select $\theta_k = 2d_k\chi_k$ we obtain that

$$\alpha(\boldsymbol{\sigma},\boldsymbol{\rho}) - \alpha(\widetilde{\boldsymbol{\sigma}},\widetilde{\boldsymbol{\rho}}) = \frac{\theta_k}{d_k}\sum\sigma_j z_j - \frac{\theta_k}{d_k}c_{k+1} - \frac{\theta_k c_{k+1}}{d_k^2}\sum\sigma_j(\sigma_j-1)$$

$$+\frac{c_{k+1}(c_{k+1}-1)\sum\sigma_j(\sigma_j-1)}{2d_k^2} + O\left(\frac{\theta_k}{d_k^2}\right).$$

Combining the terms in the previous expression into (8.6) and grouping the terms that do not depend on θ_k directly yields

$$\frac{c_{k+1}(c_{k+1}-1)}{2d_k} - \binom{c_{k+1}}{2}\sum\frac{\sigma_j^2}{d_k^2} + \frac{c_{k+1}(c_{k+1}-1)\sum\sigma_j(\sigma_j-1)}{2d_k^2}$$

$$= \binom{c_{k+1}}{2}\frac{1}{d_k} - \binom{c_{k+1}}{2}\frac{1}{d_k^2}\sum\sigma_j = \binom{c_{k+1}}{2}\frac{1}{d_k} - \binom{c_{k+1}}{2}\frac{1}{d_k} = 0.$$

On the other hand,

$$-\frac{\theta_k}{d_k}c_{k+1} - \frac{\theta_k c_{k+1}}{d_k^2}\sum\sigma_j(\sigma_j-1) + \frac{\theta_k}{d_k^2}\sum\sigma_j^2$$

$$= -\frac{\theta_k c_{k+1}}{d_k^2}\sum\sigma_j^2 + \frac{\theta_k}{d_k^2}\sum\sigma_j^2 \le 0.$$

Therefore, we conclude that

$$\begin{aligned}\frac{\beta(\boldsymbol{\sigma},\boldsymbol{\rho})}{v(\boldsymbol{\sigma},\boldsymbol{\rho})} &\leq \frac{\beta(\widetilde{\boldsymbol{\sigma}},\widetilde{\boldsymbol{\rho}})}{v(\widetilde{\boldsymbol{\sigma}},\widetilde{\boldsymbol{\rho}})}\prod_{j=1}^{m}\left(1+\frac{\sigma_j\theta_k}{d_k}\right)^{-z_j}\exp\left(\frac{\theta_k}{d_k}\sum\sigma_j z_j + O\left(\frac{\theta_k}{d_k^2}\right)\right)\\ &= \frac{\beta(\widetilde{\boldsymbol{\sigma}},\widetilde{\boldsymbol{\rho}})}{v(\widetilde{\boldsymbol{\sigma}},\widetilde{\boldsymbol{\rho}})}\prod_{j=1}^{m}\left(1+\frac{\sigma_j\theta_k}{d_k}\right)^{-z_j}\prod_{j=1}^{m}\left(1+\frac{\sigma_j\theta_k}{d_k}\right)^{z_j}\exp\left(O\left(\frac{\theta_k}{d_k^2}\right)\right)\\ &= \frac{\beta(\widetilde{\boldsymbol{\sigma}},\widetilde{\boldsymbol{\rho}})}{v(\widetilde{\boldsymbol{\sigma}},\widetilde{\boldsymbol{\rho}})}\exp\left(O\left(\frac{\theta_k}{d_k^2}\right)\right).\end{aligned}$$

This analysis yields the following result.

Theorem 2. *Suppose that* $\max_{j\leq n, i\leq m} c_j, r_i \leq \kappa$, *then there exists a deterministic constant* $\kappa^* \in (0,\infty)$ *such that:*

$$\frac{\beta_{m,n}}{P_{\mathbf{r},\mathbf{c}}(S_n = 0)} \leq \kappa^*.$$

As a consequence, the estimator $\beta_{m,n}$ *is exponentially efficient and since* $O\left(d^2\right)$ *operations are required to generate a copy of* $\beta_{m,n}$ *and the corresponding IS algorithm is an FPRAS.*

Proof. The result follows by noting that the θ_k remain uniformly bounded as $n, m \nearrow \infty$ and since $d_k \geq 1/k$ there must exist a constant $\widetilde{\kappa} \in (0,\infty)$ such that

$$\frac{\beta_{n,m}}{v(\mathbf{r},\mathbf{c})} = \frac{\beta(\mathbf{r},\mathbf{c})}{v(\mathbf{r},\mathbf{c})} \leq \exp\left(\sum_{j=1}^{n}\frac{\widetilde{\kappa}}{j^2}\right).$$

The conclusion of the theorem then follows by noting that the kth increment requires $O(c_k m + n)$ operations to be generated (the term $c_k m$ comes from the conditional Poisson sampling and n arises in the computation of θ_k). □

The previous result can be extended to degree sequences that satisfy certain growth conditions [5]. It is important to note that the selection of $\theta_k = 2\chi_k d_k$ seems crucial in order to guarantee that the resultant importance estimator is exponentially efficient. On the other hand, the excellent numerical performance reported in [8] corresponds to the selection $\theta_k = 1$. The selection of $\theta_k = 2\chi_k d_k$ can be motivated from another perspective. Indeed, as has been discussed in previous chapters, note that the zero-variance change of measure, which corresponds to sampling the $\mathbf{S}_k$ conditional on the event $\mathbf{S}_n = 0$, is Markovian and can be described by the transition kernel

$$K_{k-1}^*(\boldsymbol{\sigma},\boldsymbol{\sigma}+\mathbf{z}) = \binom{m}{c_k}^{-1} I(z_1+\cdots+z_m = c_k)\,\frac{u(\boldsymbol{\sigma}+\mathbf{z},\widetilde{\boldsymbol{\rho}})}{u(\boldsymbol{\sigma},\boldsymbol{\rho})},$$

where $\mathbf{S}_k = \boldsymbol{\sigma}$, $\boldsymbol{\rho} = (c_{k+1}, \ldots, c_n)$, $\widetilde{\boldsymbol{\rho}} = (c_{k+2}, \ldots, c_n)$ and $u(\boldsymbol{\sigma}, \boldsymbol{\rho}) = P_{\sigma,\rho}(\mathbf{S}_{n-k} = 0)$. As a consequence, given that $u(\boldsymbol{\sigma}, \boldsymbol{\rho}) \approx v(\boldsymbol{\sigma}, \boldsymbol{\rho})$ it seems natural to mimic the zero-variance change of measure directly using $v(\cdot)$ by means of the Markov transition kernel

$$\widetilde{K}_{k-1}(\boldsymbol{\sigma}, \boldsymbol{\sigma} + \mathbf{z}) = \binom{m}{c_k}^{-1} I(z_1 + \cdots + z_m = c_k) \frac{v(\boldsymbol{\sigma} + \mathbf{z}, \widetilde{\boldsymbol{\rho}})}{w(\boldsymbol{\sigma}, \boldsymbol{\rho})},$$

where $w(\boldsymbol{\sigma}, \boldsymbol{\rho}) = E_{\sigma,\rho}(v(\boldsymbol{\sigma} + \mathbf{X}_{k+1}, \widetilde{\boldsymbol{\rho}}))$ is the normalizing constant that makes $\widetilde{K}_{k-1}$ a well-defined Markov transition kernel. It turns out that generating increments according to $\widetilde{K}_{k-1}$ corresponds to a strategy based on sampling the $\mathbf{X}_k$ by means of a conditional Poisson distribution with $\gamma_j(k) = \sigma_j \exp(2\chi_k \sigma_j) \approx \sigma_j (1 + \theta_k \sigma_j / d_k)$ (where $\mathbf{S}_k = \boldsymbol{\sigma}$), therefore our selection of θ_k, thereby motivating our choice of θ_k directly. The analysis of an IS strategy based on $\widetilde{K}_{k-1}$ is studied in [5]. A similar approximation procedure is also discussed briefly in [8]. They mention that numerical experiments were also performed with the selection $\gamma_j(k) = \sigma_j \exp(2\chi_k \sigma_j)$ and report very similar empirical performance of algorithms corresponding to this selection and that based on $\theta_k = 1$.

Another example of a successful IS strategy for counting is given in [1] in the context of counting simple graphs. This problem is equivalent to counting the number of symmetric binary tables with given margins and with zeros on the main diagonal. It turns out that one can also pose this counting problem as a rare event estimation problem involving a suitably defined random walk as in our previous example. In [1] a change of measure is proposed that also can be shown to be exponentially efficient. An alternative approach, based on approximating the zero-variance change of measure (as previously discussed) is studied in [6].

8.5 Blending importance sampling and Markov chain Monte Carlo

It turns out that the machinery developed to analyze the efficiency of the importance sampler in Section 8.4.1 can be used together with the splitting decomposition (analogous to equation (8.2)) to arrive at an FPRAS based on an MCMC approach. Consider any set of m-dimensional binary columns $\{\mathbf{z}_1, \ldots, \mathbf{z}_n\}$ which satisfy their associated column and row sums (i.e., $\sum_{j=1}^{m} z_{k,j} = c_k$ and $\sum_{k=1}^{n} z_{k,j} = r_j$). Let $\mathbf{s}_0 = \mathbf{r}$, $\mathbf{s}_k = \mathbf{s}_{k-1} - \mathbf{z}_k$ for $1 \le k \le n$, and set $\boldsymbol{\rho}_k = (c_{k+1}, \ldots, c_n)$ (where $\boldsymbol{\rho}_n$ is the empty vector). Then by letting $M(\mathbf{s}_n, \boldsymbol{\rho}_n) = 1$, we obtain the decomposition

$$\frac{1}{M(\mathbf{s}_0, \boldsymbol{\rho}_0)} = \frac{M(\mathbf{s}_1, \boldsymbol{\rho}_1)}{M(\mathbf{s}_0, \boldsymbol{\rho}_0)} \times \frac{M(\mathbf{s}_2, \boldsymbol{\rho}_2)}{M(\mathbf{s}_1, \boldsymbol{\rho}_1)} \times \cdots \times \frac{M(\mathbf{s}_n, \boldsymbol{\rho}_n)}{M(\mathbf{s}_{n-1}, \boldsymbol{\rho}_{n-1})}. \tag{8.7}$$

The ratio $M(\mathbf{s}_k, \boldsymbol{\rho}_k) / M(\mathbf{s}_{k-1}, \boldsymbol{\rho}_{k-1})$ is the probability that a table drawn uniformly from the set of tables satisfying the column and row sums $\boldsymbol{\rho}_{k-1}$ and

$\mathbf{s}_{k-1}$, respectively, also satisfies the columns and row sums $\boldsymbol{\rho}_k$ and $\mathbf{s}_k$ induced by the removal of its first column. Assuming bounded row and column sums (as $m, n \nearrow \infty$), it is easy to verify using Theorem 1 that there exist constants $c^* \in (0, \infty)$ and $\theta > 0$ such that

$$p_k = \frac{M\left(\mathbf{s}_k, \boldsymbol{\rho}_k\right)}{M\left(\mathbf{s}_{k-1}, \boldsymbol{\rho}_{k-1}\right)} \geq \theta d^{-c^*}. \tag{8.8}$$

Thus, an FPRAS can be developed along similar lines to the matchings example in Section 8.3. However, note that the lower bound on each ratio is not uniform in k and θd^{-c^*} decreases to zero as m, n, and therefore k, grow. Nevertheless, since the convergence to zero is polynomial, we still obtain a polynomial running time. In particular, if we have access to an exact sampler for generating binary contingency tables then, using Chernoff's inequality as in Section 8.3, we obtain that $N = O\left(n \log(2n/\delta)\, d^{c^*} \varepsilon^{-2}\right)$ i.i.d. Bernoulli replicates, $Z_1, \ldots, Z_N$, with parameter p_k, are required in order to obtain an estimator $\widehat{p}_k := N^{-1} \sum_{j=1}^{N} Z_j$ satisfying $P\left(|\widehat{p}_k - p_k| > \varepsilon/n\right) \leq \delta/n$. Then if the exact sampler runs in polynomial time, $\kappa(d)$, the overall complexity is $O(n^2 \kappa(d) \log(2n/\delta)\, d^{c^*} \varepsilon^{-2})$.

It is important to note that MCMC method has already been applied to this problem based upon direct analysis of the mixing rate of the underlying chain. Using conductance estimates, for example, it is shown in [18] how to obtain an almost uniform sampler (FPAUS) in $O\left(n^{12} \log(1/\varepsilon)\right)$ operations assuming bounded degree sequences (see their Corollary 4.2 combined with Theorem 2.1), which yields an $O\left(n^{14} d^{c^*} \varepsilon^{-2} \log(n/\delta)\right)$ FPRAS. An even faster FPAUS, applicable to general degree sequences, whose complexity is (roughly) $O\left(d^3 (nm)^2 \max c_i, r_j\right)$ operations is provided in [3]. As a final remark, we note that the reduction based on the splitting formula (8.7) and, consequently, the analysis for the IS-based MCMC method may not be the best possible. It is conceivable that a better selection of the sequence $\mathbf{s}_0, \ldots, \mathbf{s}_{n-1}$ might yield improved estimates for the lower bounds (8.8) together with a tighter analysis of the running time of the associated counting algorithm.

Acknowledgement

This research was partially supported by NSF grant DMS 0595595.

References

[1] M. Bayati, J. Kim, and A. Saberi. A sequential algorithm for generating random graphs. *RANDOM 2007*, 2007.

[2] A. Bekessy, P. Bekessy, and J. Komlos. Asymptotic enumeration of regular matrices. *Studia Scientiarum Mathematicarum Hungarica*, pp. 343–353, 1972.

[3] I. Bezakova, N. Bhatnagar, and E. Vigoda. Sampling binary contingency tables with a greedy start. *Random Structures and Algorithms*, **30**(1–2): 682–687, 2006.

[4] I. Bezakova, A. Sinclair, D. Stefankovic, and E. Vigoda. Negative examples for sequential importance sampling of binary contingency tables. *Submitted*, 2007.

[5] J. Blanchet. Efficient importance sampling for binary contingency tables. *Submitted*, 2007.

[6] J. Blanchet and J. Blitzstein. Fast counting of simple graphs via importance sampling. Preprint, 2008.

[7] X.-H. Chen, A. P. Dempster, and J. S. Liu. Weighted finite population sampling to maximize entropy. *Biometrika*, **81**: 457–469, 1994.

[8] Y. Chen, P. Diaconis, S. P. Holmes, and J. S. Liu. Sequential Monte Carlo method for statistical analysis of tables. *Journal of the American Statistical Association*, **100**: 109–120, 2005.

[9] P. Diaconis and D. Stroock. Geometric bounds for eigenvalues of Markov chains. *Annals of Applied Probability*, **1**: 36–61, 1991.

[10] M. Dyer, L. A. Goldberg, C. Greenhill, M. Jerrum, and M. Mitzenmacher. An extension of path coupling and its application to the Glauber dynamics for graph colorings. *SIAM Journal of Computing*, **30**(6): 1962–1975, 2001.

[11] D. Gamarnik and D. Katz. Correlation decay and deterministic FPTAS for counting list-colorings of a graph. *Proceedings of 18th ACM-SIAM Symposium on Discrete Algorithms (SODA)*, 2007.

[12] C. Greenhill, B. D. Mckay, and X. Wang. Asymptotic enumeration of sparse 0–1 matrices with irregular row and column sums. *Journal of Combinatorial Theory, Series A*, **113**: 291–324, 2006.

[13] M. Jerrum. *Counting, Sampling and Integrating: Algorithms and Complexity*. Birkhäuser, Basel, 2001.

[14] M. Jerrum and A. Sinclair. The Markov chain Monte Carlo method: an approach to approximate counting and integration. In D. S. Hochbaum, ed., *Approximation Algorithms for NP-hard Problems*. PWS, Boston, 1996.

[15] M. Jerrum, A. Sinclair, and E. Vigoda. A polynomial-time approximation algorithm for the permanent of a matrix with non-negative entries. *Journal of the Association for Computing Machinery*, **51**(4): 671–697, 2004.

[16] M. Jerrum, L. Valiant, and V. Vazirani. Random generation of combinatorial structures from a uniform distribution. *Theoretical Computer Science*, **43**: 169–188, 1986.

[17] R. Kannan, L. Lovasz, and M. Simonovits. Random walks and an $O^*(n^5)$ volume algorithm for convex bodies. *Random Structures and Algorithms*, **11**(1): 1–50, 1997.

[18] R. Kannan, P. Tetali, and S. Vempala. Simple Markov-chain algorithms for generating bipartite graphs and tournaments. In *Proceedings of the 8th Annual ACM-SIAM Symposium on Discrete Algorithms (SODA)*, pp. 193–200, 1997.

[19] M. Luby and E. Vigoda. Approximately counting to four (extended abstract). *Proceedings of the 29th Annual ACM Symposium on Theory of Computing (STOC)*, pp. 682–687, 1997.

[20] B. D. McKay. Asymptotics for 0–1 matrices with prescribed line sums, enumeration and design. In *Enumeration and Design* Academic Press, Canada, pp. 225–238, 1984.

[21] M. Mitzenmacher and E. Upfal. *Probability and Computing: Randomized Algorithms and Probabilistic Analysis*. Cambridge University Press, Cambridge, 2005.

[22] R. Y. Rubinstein. How many needles are in a haystack, or how to solve fast #P-complete counting problems. *Methodology and Computing in Applied Probability*, **11**: 5–49, 2007.

[23] R. Y. Rubinstein and D. P. Kroese. *The Cross-Entropy Method*. Springer, New York, 2004.

[24] A. Sinclair. Improved bounds for mixing rates of Markov chains and multicommodity flow. *Combinatorics, Probability and Computing*, **1**: 351–370, 1992.

[25] A. Sinclair. *Algorithms for Random Generation and Counting*. Birkhäuser, Boston, 1993.

[26] L. G. Valiant. The complexity of computing the permanent. *Theoretical Computer Science*, **8**: 189–201, 1979.

[27] E. Vigoda. Relationship between counting and sampling. Online course notes. http://www.cc.gatech.edu/~vigoda/MCMC_Course/Sampling-Counting.pdf, 2006.

[28] D. Weitz. Counting down the tree. *Proceedings of the 38th Annual Symposium on the Theory of Computing (STOC)*, 2006.

9

Rare event estimation for a large-scale stochastic hybrid system with air traffic application

Henk A. P. Blom, G. J. (Bert) Bakker and Jaroslav Krystul

9.1 Introduction

This study is motivated by the problem of safety verification for a future air traffic operations concept through the analysis of reach probabilities. From a control-theoretic perspective, such an advanced operations concept is a blueprint for a controlled stochastic hybrid system (SHS) which satisfies the strong Markov property [13]. Recently, Sastry and co-workers [2, 1] studied the optimization of the control policy of a discrete-time SHS, such that the probability of staying within some prescribed safe set remains above some prescribed minimum level. Specifically, Amin *et al.* [2] developed a theoretical framework which expressed the reach probability as a multiplicative function, and this was used to develop a dynamic programming-based approach to computing probabilistic maximal safe sets, that is, initial states of a system for which control policies exist that assure that the reach probability stays below some given value. Subsequently,

Rare Event Simulation using Monte Carlo Methods Edited by G. Rubino and B. Tuffin

Abate *et al.* [1] showed this problem to be complementary to the problem of optimizing the control policy of an SHS such that the reach probability of some prescribed unsafe set remains below some given maximum level, and that the same dynamic programming-based computation of maximal safe sets can be used. The dynamic programming approach becomes computationally intractable when the SHS considered is of large-scale type. Prandini and Hu [39] developed a Markov chain approximation based method for the computation of reach probabilities for a continuous-time SHS. This way the dynamic programming challenge is avoided, but the computational load of their method prohibits its application to a large-scale SHS. Prajna *et al.* [38] developed an approach which obtains an upper bound of the reach probability, but this cannot handle large-scale SHS either.

In theory, reach probability estimation can be done by simulating many trajectories of the process considered, and counting the fraction of cases where the simulated trajectory reaches the unsafe set within some given period T. When the reach probability value is very small then the number of straightforward Monte Carlo (MC) simulations needed is impractically large. The rare event estimation literature forms a potentially rich source of information for speeding up MC simulation, for example by combining methods from large-deviation and importance sampling theories [11, 29, 31]. An early successful development in this area is sequential MC simulation for the estimation of the intensity of radiation that penetrates a shield of absorbing material in nuclear physics (see [10]). More recently this approach has also found application in non-nominal delay time and loss estimation in telecommunication networks [3]. L'Ecuyer *et al.* [36] provide a very good recent overview of these sequential MC simulation developments.

In order to exploit rare event estimation theory within probabilistic reachability analysis of controlled SHS, we need to establish a theoretically unambiguous connection between the two concepts. Implicitly, this connection has recently been elaborated by Del Moral and co-workers [16–18, 20, 21]. They embedded theoretical physics equations, which supported the development of advanced MC simulations, within the stochastic analysis setting that is typically used for probabilistic reachability analysis. They subsequently showed that this embedding provides a powerful background for the development and analysis of sequential MC simulation for rare event simulation. In Chapter 3 of the present volume this novel development is well explained in the broader context of splitting techniques in rare event simulation.

The aim of this chapter is to present a part of the framework developed by Del Moral *et al.* [16–18, 20, 21] in a probabilistic reachability setting, to further develop this for a large-scale SHS, and to demonstrate its practical use for safety verification of an advanced air traffic operation. In [8, 9], the practical use of the approach of Del Moral [16–18, 20, 21] for safety verification of an advanced air traffic operation has already been demonstrated for some specific scenarios. In these scenarios, the main contributions to the reach probability value came from diffusion behavior. It also became clear that the same sequential MC simulation approach failed to work for scenarios of the same air traffic operation where the reach probability is determined by rare switching between modes. This chapter aims to tackle such more demanding rare event estimation problems for large-scale controlled SHSs. Essentially the approach is to introduce an aggregation of the discrete mode process, and to develop importance switching and Rao–Blackwellization relative to these aggregated modes.

The chapter is organized as follows. Section 9.2 develops a factorization of the reach probability. Section 9.3 explains the approach of [16, 17, 20, 21]. Section 9.4 presents an extension of this approach to hybrid systems. Section 9.5 develops the aggregation mode process and characterizes key relations with the controlled SHS. Section 9.6 develops a novel sequential MC simulation approach for estimating reach probabilities. Section 9.7 briefly describes the free flight air traffic example considered. Section 9.8 applies the novel approach to estimate reach probabilities for this air traffic example. Section 9.9 presents concluding remarks. An early version of this chapter is [5].

9.2 Factorization of reach probability

Throughout this and the following sections, all stochastic processes are defined on a complete stochastic basis $(\Omega, \mathcal{F}, \mathbb{F}, P, \mathrm{T})$ with $(\Omega, \mathcal{F}, P)$ a complete probability space, and $\mathbb{F}$ an increasing sequence of sub-σ-algebras on the positive time line $\mathrm{T} = \mathbb{R}_+$, i.e. $\mathbb{F} \triangleq \{\mathcal{J}, (\mathcal{F}_t, t \in \mathrm{T}), \mathcal{F}\}$, $\mathcal{J}$ containing all P-null sets of $\mathcal{F}$ and $\mathcal{J} \subset \mathcal{F}_s \subset \mathcal{F}_t \subset \mathcal{F}$ for every $s < t$.

Let us denote $E' = \mathbb{R}^n \times \mathbb{M}$, with $\mathbb{M}$ a discrete set. Let $\mathcal{E}'$ be the Borel σ-algebra of E'. We consider a time-homogeneous strong Markov process which is also a generalized stochastic hybrid process $\{x_t, \theta_t\}$ [32, 12, 14, 35], with $\{x_t\}$ assuming values in $\mathbb{R}^n$ and $\{\theta_t\}$ assuming values in $\mathbb{M}$. The first component of $\{x_t\}$ equals t and the other components of $\{x_t\}$ form an $\mathbb{R}^{n-1}$-valued càdlàg process $\{s_t\}$. The $\mathbb{M}$-valued process $\{\theta_t\}$ is a càdlàg switching process. Incorporating t as a state component allows any time-inhomogeneous strong Markov process $\{s_t, \theta_t\}$ to be represented as a time-homogeneous strong Markov process $\{t, s_t, \theta_t\}$ [19]. The problem considered is to estimate the probability that $\{s_t\}$ hits a given 'small' closed subset $D \subset \mathbb{R}^{n-1}$ within a given time period $[0, T)$, i.e. $P(\exists t \in [0, T);\ s_t \in D)$.

Following Del Moral and co-workers [16, 17, 20, 21], this probability can be characterized in the form of a multiplicative function the terms of which are defined through an arbitrarily assumed nested sequence of closed (time-invariant) subsets $D = D_m \subset D_{m-1} \subset \ldots \subset D_1$, with the constraint that $P(s_0 \in D_1) = 0$ and each component of $\{x_t\}$, which may hit any D_k, is a pathwise continuous process. In order to derive a multiplicative functional characterization of the hitting probability, we set $\tau_0 = 0$ and define τ_k, $k = 1, \ldots, m$, as the first moment that $\{s_t\}$ hits subset k, that is,

$$\tau_k = \inf\{t > 0;\ s_t \in D_k\}, \tag{9.1}$$

which implies $P(\exists t \in [0, T);\ s_t \in D_m) = P(\tau_m < T)$.

We also define {0, 1}-valued random variables $\{\chi_k, k = 0, \ldots, m\}$ as follows:

$$\chi_k = \begin{cases} 1, & \text{if } \tau_k < T \text{ or } k = 0, \\ 0, & \text{otherwise.} \end{cases}$$

By using these τ_k and χ_k definitions and the assumption that each component of $\{s_t\}$ that may hit any D_k, $k = 1, \ldots, m$, has continuous paths (i.e., $\{s_t\}$ cannot enter D_k by jumping over the boundary of D_k) we can write the probability of $\{s_t\}$ hitting D before T as a product of conditional probabilities of reaching D_k given D_{k-1} has been reached at some earlier moment in time, that is,

$$P(\tau_m < T) = \mathbb{E}[\chi_m] = \mathbb{E}\left[\prod_{k=1}^{m} \chi_k\right] = \prod_{k=1}^{m} \mathbb{E}[\chi_k | \chi_{k-1} = 1]$$

$$= \prod_{k=1}^{m} P(\tau_k < T | \tau_{k-1} < T) = \prod_{k=1}^{m} \gamma_k \tag{9.2}$$

with $\gamma_k \triangleq P(\tau_k < T | \tau_{k-1} < T)$.

With this, the problem can be seen as one of estimating the conditional probabilities γ_k in such a way that the product of the estimators $\tilde{\gamma}_k$ is unbiased. Because of the multiplication of the various individual $\tilde{\gamma}_k$ estimators, which depend on each other, in general such a product may be heavily biased. Garvels *et al.* [27, 28] showed for a discrete-time Markov process, that estimating the γ_k's in (9.2) by an appropriate sequential MC simulation approach, which is known as the splitting method, guarantees unbiased estimation of $P(\tau_m < T)$. The key innovation of [16–18, 20, 21] was to develop such a convergence type of proof for a sequential MC simulation approach towards the estimation of the γ_k's in (9.2) under the much weaker condition that $\{s_t\}$ is embedded in (or is) a strong Markov process.

9.3 Sequential Monte Carlo simulation

For the process $\{x_t, \theta_t\}$ we follow the approach of [16, 17, 20, 21] to characterize how the evolution proceeds from $\tau_{k-1} \wedge T$ to $\tau_k \wedge T$. For any $B \in \mathcal{E}'$, let $p_{\xi_k|\chi_k}(B|1)$ denote the conditional probability of $\xi_k = (x_{\tau_k \wedge T}, \theta_{\tau_k \wedge T}) \in B$ given $\chi_k = 1$. Under the assumption that $P(s_0 \in D_1) = 0$, we characterize the following recursive sequence of transformations:

$$p_{\xi_{k-1}|\chi_{k-1}}(\cdot|1) \xrightarrow{\text{prediction}} p_{\xi_k|\chi_{k-1}}(\cdot|1) \xrightarrow{\text{conditioning}} p_{\xi_k|\chi_k}(\cdot|1).$$

$$\downarrow$$

$$\gamma_k$$

Because $\{x_t, \theta_t\}$ is a strong Markov process, $\{\xi_k\}$ is a Markov sequence. Hence the prediction step satisfies a Chapman–Kolmogorov equation:

$$p_{\xi_k|\chi_{k-1}}(B|1) = \int_{E'} p_{\xi_k|\xi_{k-1}}(B|\xi) p_{\xi_{k-1}|\chi_{k-1}}(d\xi|1). \tag{9.3}$$

Next we characterize the conditional probability of reaching the next subset:

$$\begin{aligned} \gamma_k &= P(\tau_k < T|\tau_{k-1} < T) = P(\chi_k = 1|\chi_{k-1} = 1) \\ &= \mathbb{E}[\chi_k|\chi_{k-1} = 1] = \int_{E'} 1_{Q_k}(\xi) p_{\xi_k|\chi_{k-1}}(d\xi|1), \end{aligned} \tag{9.4}$$

where $Q_k \triangleq (0, T) \times D_k \times \mathbb{M}$. Similarly, the condition step satisfies, for any $B \in \mathcal{E}'$,

$$p_{\xi_k|\chi_k}(B|1) = \frac{\int_B 1_{Q_k}(\xi) p_{\xi_k|\chi_{k-1}}(d\xi|1)}{\int_{E'} 1_{Q_k}(\xi') p_{\xi_k|\chi_{k-1}}(d\xi'|1)}. \tag{9.5}$$

With this, the γ_k's in (9.2) are characterized as a solution of the set of recursive equations 9.3–9.5. Following [16, 17, 20, 21], this recursive characterization can numerically be approximated through a sequential MC simulation to estimate $P(\tau_m < T)$. This is referred to as the interacting particle system (IPS) algorithm, and works as follows.

Simulate N_p random trajectories of $\{x_t, \theta_t\}$ over $[0, T)$, each of which starts from a random initial condition $((0, s_0), \theta_0)$, with $s_0 \notin D_1$. Each simulated trajectory stops at $\tau_1 \wedge T$, that is, upon hitting Q_1 or when the first x-component reaches T. The full hybrid states of these trajectory end points form an empirical density $\tilde{\pi}_1$ as an approximation of $p_{\xi_1|\chi_1}(\cdot|1)$. This empirical density is used to generate (i.e., to resample) N_p initial conditions of trajectories which are subsequently simulated until hitting Q_2 or when the first x-component reaches T; the end points in Q_2 form an empirical density $\tilde{\pi}_2$ as an approximation of $p_{\xi_2|\chi_2}(\cdot|1)$. This cycle repeats from Q_2 to $Q_3, \ldots$, and finally from Q_{m-1} to $Q_m = Q$. During the kth cycle, a fraction $\widetilde{\gamma}_k$ of the N_p simulated trajectories arrives at Q_k. The product of these m fractions forms an estimator for $P(\tau_m < T)$.

Using the recursive characterization of the conditional density, it has also been shown[16, 20] that the product of these fractions $\tilde{\gamma}_k$ forms an unbiased estimate of the probability of $\{s_t\}$ hitting the set D within the time period $[0, T)$, that is,

$$\mathbb{E}\left[\prod_{k=1}^{m} \tilde{\gamma}_k\right] = \prod_{k=1}^{m} \gamma_k = P(\tau < T).$$

In addition, there is a bound on the L^1 estimation error [16, 20],

$$\mathbb{E}\left(\prod_{k=1}^{m} \tilde{\gamma}_k - \prod_{k=1}^{m} \gamma_k\right) \leq \frac{c_p}{\sqrt{N_p}},$$

with c_p a finite constant which depends on the simulated scenario and the sequence of nested subsets adopted. These convergence results assume that the resampling of the empirical density $\tilde{\pi}_k$ is done uniformly, hence there is a chance of resampling some particles more than once, and other particles not at all. Furthermore, Cérou *et al.* [18] developed some complementary error bounds, and showed convergence under an alternate resampling approach.

Application of this IPS algorithm to air traffic operation may work well for specific scenarios where rare discrete modes are not significantly contributing to the reach probability [8, 9]. However, there also are relevant scenarios which do not satisfy the latter condition. To tackle this problem, [32–34] proposed hybrid versions of the baseline IPS algorithm. These approaches work well only if the size of space $\mathbb{M}$ is not too big. However, in many realistic scenarios the state space $\mathbb{M}$ of the discrete valued component $\{\theta_t\}$ is usually very large. Therefore another extension was proposed, namely, the hierarchical hybrid IPS algorithm (HHIPS), which will be presented and applied in this chapter. For the reader's convenience, however, before addressing HHIPS, we first introduce the hybrid IPS (HIPS) algorithm of [33].

9.4 Importance switching based hybrid IPS algorithm

Although in theory the IPS approach is applicable virtually to any strong Markov process, in practice the straightforward application of this approach to stochastic hybrid processes may fail to produce reasonable estimates within a reasonable amount of simulation time. First, there may be few or no particles in modes with small probabilities (i.e., 'light' modes). This happens because each resampling step tends to sample more 'heavy' particles from modes with higher probabilities, thus, 'light' particles in the 'light' modes tend to be discarded. Second, if the switching rate is small then it is highly unlikely to observe even one switch during a simulation run. In such cases, the possible switching between modes is not properly taken into account. Together with the first problem, this badly affects IPS estimation performance. By increasing the number of particles the IPS estimates should improve but only at the cost of substantially increased simulation time which makes the performance of IPS approach similar to that of the standard Monte Carlo.

The HIPS algorithm of [33] incorporates sampling per mode (stratified sampling with modes defining the strata) to cope with large differences in mode weights, and importance switching (a form of importance sampling for the discrete-valued component $\{\theta_t\}$) to cope with rare mode switching. In what follows, we outline the HIPS algorithm.

If the initial probabilities of some particular modes are very small then it is highly unlikely that particles will be drawn in these modes. To avoid this, at the initial sampling step we start with a fixed number of particles in each mode however small the initial probability is and adjust the weights appropriately. Let

N_p denote the initial number of particles in each mode $\theta \in \mathbb{M}$. In total the system of particles will consist of $N = N_p \cdot |\mathbb{M}|$ particles. Let J^θ denote the ordered set of indices of particles which are in mode θ ($J^\theta \cap J^\eta = \varnothing$ for $\theta \neq \eta$). The whole set of indices is defined by

$$J \triangleq \bigcup_{\theta \in \mathbb{M}} J^\theta = \{1, 2, \ldots, N\}, \quad |J| = N.$$

At the initial sampling step we will have $|J^\theta| = N_p$ particles in each mode $\theta \in \mathbb{M}$. As particles evolve and switch from one mode to another, the numbers of particles in different modes will change, as will the index sets J^θ. But at each resampling step we will again sample N_p particles for each mode $\theta \in \mathbb{M}$ from a conditional empirical distribution.

Let $\tilde{\gamma}_k$ and $\tilde{\pi}_k$ denote numerical approximations of γ_k and $p_{\xi_k|\chi_k}(\cdot|1)$, respectively. We choose these numerical approximations in the form of the weighted empirical distributions associated with the particle system $\{\xi_k^i, \omega_k^i\}_{i=1}^N$, where $\xi_k \triangleq (x_{\tau_k \wedge T}, \theta_{\tau_k \wedge T})$ and $\omega \in [0, 1]$. When simulating from $\tau_{k-1} \wedge T$ to $\tau_k \wedge T$, only a fraction γ_k of the Monte Carlo simulated trajectories will reach Q_k. The HIPS algorithm estimates these fractions and their product in a recursive way using the following steps:

Step 0 generates, for each $\theta \in \mathbb{M}$ value, N_p initial particles at $k = 0$ and then starts the cycling through steps 1 through 3 for $k := 1, 2, \ldots, m$.

Step 1 extrapolates each particle from $\tau_{k-1} \wedge T$ to $\tau_k \wedge T$ in time steps of length h, using importance switching for the new value of the $\{\theta_t\}$ component.

Step 2 evaluates the particles that have arrived at Q_k. For this, use is made of equations 9.4–9.5.

Step 3 resamples with replacement, for each $\theta \in \mathbb{M}$, N_p particles that have arrived at Q_k; the weights must be adjusted accordingly.

Each of these steps is specified in detail below.

Hybrid interacting particle system (HIPS)

HIPS Step 0: Initial sampling for $k = 0$.

- For each $\theta \in \mathbb{M}$, sample N_p independent initial $\mathbb{R}^n$ values outside D_1:

$$x_0^j \sim p_{x_0|\theta_0}(\cdot|\theta). \text{ Set } \theta_0^j = \theta, \text{ then } \xi_0^j = (x_0^j, \theta_0^j), \ j \in J^\theta.$$

- Assign initial weights:

$$\omega_0^j = P_{\theta_0}(\theta) / N_p, \quad j \in J^\theta, \quad \theta \in \mathbb{M}.$$

- Then $\tilde{\gamma}_0 = 1$ and

$$p_{x_0,\theta_0}(dx,\theta) \approx \sum_{\theta\in\mathbb{M}}\sum_{j\in J^\theta} \omega_0^j \delta_{(x_0^j,\theta_0^j)}(dx,\theta) = \sum_{i=1}^{N} \omega_0^i \delta_{(x_0^i,\theta_0^i)}(dx,\theta).$$

Iteration cycle: For $k = 1, \ldots, m$, cycle over steps 1–3:

HIPS Step 1: Prediction.

- For $i = 1, \ldots, N$, using importance switching for the $\{\theta_t\}$ component,[1] generate path starting at $\xi_{k-1}^i = (x_{\tau_{k-1}\wedge T}^i, \theta_{\tau_{k-1}\wedge T}^i)$ until the kth set Q_k is reached.
- The weight of each particle must be adjusted recursively in time (i.e., at each time discretization step):

$$\omega_{t+h}^i = \omega_t^i \cdot L_{t+h|t}(\theta_{t+h}^i|\theta_t^i, x_t^i),$$

where

$$L_{t+h|t}(\theta_{t+h}^i|\theta_t^i, x_t^i) = \frac{p_{\theta_{t+h}|\theta_t,x_t}(\theta_{t+h}^i|\theta_t^i, x_t^i)}{\tilde{p}_{\theta_{t+h}|\theta_t,x_t}(\theta_{t+h}^i|\theta_t^i, x_t^i)}$$

is the likelihood ratio corresponding to the change of switching rates of the $\{\theta_t\}$ component.

- This yields a new set of particles $\{\xi_k^i, \omega_k^i\}_{i=1}^N$.

HIPS Step 2: Evaluation of the Q_k-arrived particles.

- Particles which do not reach the set Q_k are killed, i.e. we set $\hat{\omega}_k^i = 0$, else set $\hat{\omega}_k^i = \omega_k^i$ and $\hat{\xi}_k^i = \xi_k^i$.
- The new set of particles is $\{\hat{\xi}_k^i, \hat{\omega}_k^i\}_{i=1}^N$.
- Approximation of γ_k:

$$\gamma_k \approx \tilde{\gamma}_k = \sum_{i=1}^{N} \hat{\omega}_k^i.$$

If all particles are killed, i.e. $\tilde{\gamma}_k = 0$, then the algorithm stops and $P_{hit}(0, T) \approx 0$.

- If $k = m$, then stop HIPS with the estimate

$$P_{hit} \approx \prod_{k=1}^{m} \tilde{\gamma}_k.$$

[1] In order to increase the frequency of switchings at each time discretization step we replace the original transition probabilities $p_{\theta_{t+h}|\theta_t,x_t}(\cdot|\theta, x)$ by some known transition probabilities $\tilde{p}_{\theta_{t+h}|\theta_t,x_t}(\cdot|\theta, x)$, which guarantees higher switching rates (see [32, 33] for details).

- For each $i = 1, \ldots, N$, set $\tilde{\xi}_k^i = \hat{\xi}_k^i$ and normalize the weights: $\tilde{\omega}_k^i = \hat{\omega}_k^i / \tilde{\gamma}_k$.
- This yields a new set of particles $\{\tilde{\xi}_k^i, \tilde{\omega}_k^i\}_{i=1}^N$.
- The estimated $p_{\xi_k|\chi_k}(\cdot|1)$ satisfies

$$p_{\xi_k|\chi_k}(dx, \theta|1) \approx \tilde{\pi}_k(dx, \theta) = \sum_{i=1}^{N} \tilde{\omega}_k^i \delta_{(\tilde{x}_k^i, \tilde{\theta}_k^i)}(dx, \theta).$$

HIPS Step 3: Resampling step.

- For each mode $\theta \in \mathbb{M}$, resample with replacement N_p values of $\tilde{\xi}_k$ from the unnormalized conditional empirical measure

$$p_{\xi_k|\chi_k,\theta_k}(\cdot|1, \theta) \approx \tilde{\pi}_k(dx, \theta|\theta_k = \theta) = \sum_{j \in J^\theta} \tilde{\omega}_k^j \delta_{(\tilde{x}_k^j, \tilde{\theta}_k^j)}(dx, \theta)$$

and adjust the weights as follows:

$$\omega_k^j = \frac{\sum_{s \in J^\theta} \tilde{\omega}_k^s}{N_p}, \qquad j \in J^\theta, \quad \theta \in \mathbb{M}.$$

- This yields a new set of particles $\{\xi_k^i, \omega_k^i\}_{i=1}^N$.
- If $k < m$, then repeat steps 1–3 for $k := k + 1$.

9.5 Aggregation of modes

In [32–34], hybrid versions of the baseline IPS algorithm [16, 17, 20, 21] have been developed, which take into account that rare discrete modes may contribute significantly to the reach probability to be estimated. As explained in Section 9.4, the hybrid IPS version of [33] simulates a more frequent switching $\mathbb{M}$-valued process $\{\breve{\theta}_t\}$, and compensates importance weights for the difference between $\{\breve{\theta}_t\}$ and $\{\theta_t\}$. In [34] another hybrid IPS version has been developed, which makes use of Rao–Blackwellization, that is, using exact probabilistic equations for certain components and simulated particles for all other components [15]. For filtering of a stochastic hybrid process $\{x_t, \theta_t\}$ two Rao–Blackwellization versions have been developed [22, 6]: that of [22] uses exact probabilistic equations for $\{x_t\}$ and particle simulation for $\{\theta_t\}$, while that of [6] uses exact probabilistic equations for $\{\theta_t\}$ and particle simulation for $\{x_t\}$. In [34] the latter approach is combined with IPS. The resulting hybrid IPS version uses exact probabilistic equations for the evolution of $\{\theta_t\}$ and simulates particles for the Euclidean valued $\{x_t\}$. This Rao–Blackwellization based hybrid IPS version also resamples at the end of each IPS cycle N_p x-values from $\tilde{\pi}_k(\cdot, \theta)$ for each mode $\theta \in \mathbb{M}$, leading to a total of $N_p \times |\mathbb{M}|$ particles, where $|\mathbb{M}|$ is the number of elements in $\mathbb{M}$. Since the computational load increases linearly with $|\mathbb{M}|$, these hybrid IPS approaches are intractable when $|\mathbb{M}|$ is very large. This condition applies to the air traffic example (where $|\mathbb{M}| \approx 10^{25}$) considered later in this chapter.

The idea is to improve the situation for very large $|\mathbb{M}|$ by developing a hybrid IPS approach not for $\{\theta_t, x_t\}$, but for $\{\kappa_t, (\theta_t, x_t)\}$, where $\{\kappa_t\}$ is some complementary $\mathbb{K}$-valued process with $|\mathbb{K}| \ll |\mathbb{M}|$. In order to accomplish this, we group modes that have large differences in mode switching frequencies. This defines a partition $\{\mathbb{M}_\kappa,\ \kappa \in \mathbb{K}\}$, that is, $\bigcup_{\kappa\in\mathbb{K}} \mathbb{M}_\kappa = \mathbb{M}$ and $\mathbb{M}_\kappa \bigcap \mathbb{M}_{\kappa'} = \emptyset$ for $\kappa \neq \kappa'$, and a $\mathbb{K}$-valued aggregation mode process $\{\kappa_t\}$ as follows:

$$\kappa_t(\omega) = \kappa, \quad \text{if } \theta_t(\omega) \in \mathbb{M}_\kappa. \tag{9.6}$$

Because the evolution of the aggregation mode process $\{\kappa_t\}$ depends on the evolution of $\{\theta_t\}$, $\{\kappa_t\}$ may inherit rare mode switching from $\{\theta_t\}$. In order to avoid these rare effects in the evolution of particles, we also define a $\mathbb{K}$-valued Markov chain $\{\breve{\kappa}_t\}$ with known non-rare transition rates, and use the transition rates of $\{\breve{\kappa}_t\}$ to determine for each particle a new $\breve{\kappa}$-value at some time step h later. The particle weight is compensated with the corresponding importance switching ratio

$$p_{\kappa_{\tau+h}|\kappa_\tau,x_\tau,\theta_\tau}(\breve{\kappa}|\kappa, x, \theta)/p_{\breve{\kappa}_{\tau+h}|\breve{\kappa}_\tau}(\breve{\kappa}|\kappa),$$

where κ, x, θ denote the given $(\kappa_\tau, x_\tau, \theta_\tau)$ particle value, and $\breve{\kappa}$ denotes the value newly sampled for $\breve{\kappa}_{\tau+h}$.

Next, the prediction of the new $\theta_{\tau+h}$ particle from the (x_τ, θ_τ) particle values is done conditional on the newly sampled $\breve{\kappa}$-value. Theorem 1 provides a probabilistic characterization of such $\breve{\kappa}$-conditional θ-prediction.

Theorem 1. *($\breve{\kappa}$-conditional θ-prediction) For an arbitrary stopping time τ,*

$$p_{\theta_{\tau+h}|x_\tau,\theta_\tau,\kappa_{\tau+h}}(\eta|x, \theta, \breve{\kappa}) = \frac{1_{\mathbb{M}_{\breve{\kappa}}}(\eta) p_{\theta_{\tau+h}|x_\tau,\theta_\tau}(\eta|x, \theta)}{\sum_{\eta'\in\mathbb{M}} 1_{\mathbb{M}_{\breve{\kappa}}}(\eta') p_{\theta_{\tau+h}|x_\tau,\theta_\tau}(\eta'|x, \theta)}. \tag{9.7}$$

Proof. Using Bayes yields:

$$p_{\theta_{\tau+h}|x_\tau,\theta_\tau,\kappa_{\tau+h}}(\eta|x, \theta, \breve{\kappa}) = \frac{p_{\kappa_{\tau+h}|\theta_{\tau+h}}(\breve{\kappa}|\eta) p_{\theta_{\tau+h}|x_\tau,\theta_\tau}(\eta|x, \theta)}{\sum_{\eta'\in\mathbb{M}} p_{\kappa_{\tau+h}|\theta_{\tau+h}}(\breve{\kappa}|\eta') p_{\theta_{\tau+h}|x_\tau,\theta_\tau}(\eta'|x, \theta)}.$$

Substituting $p_{\kappa_{\tau+h}|\theta_{\tau+h}}(\breve{\kappa}|\eta) = 1_{\mathbb{M}_{\breve{\kappa}}}(\eta)$ yields (9.7). □

The prediction of the x-part of the particle over time step h is done by drawing a sample from $p_{x_{\tau+h}|x_\tau,\theta_\tau,\theta_{\tau+h}}(\cdot|x, \theta, \eta)$. In order to identify all particles that arrive at Q_k before time T, the prediction over time step h has to be done up to T/h times. After these prediction steps, there is no guarantee that for each $\breve{\kappa} \in \mathbb{K}$ some minimum number of particles have arrived at Q_k. Hence, we resample the Q_k-arrived particles such that we regain N_p particles for each $\kappa \in \mathbb{K}$. In order to make this possible, in Theorem 2 we provide a characterization of the (conditional) probabilities $p_{\kappa_{\tau+h}}$ and $p_{x_\tau,\theta_\tau|\kappa_{\tau+h}}$ as a function of p_{x_τ,θ_τ}, for arbitrary stopping time τ and time step h. This characterization allows us to sample a fixed number of particles per aggregation mode $\kappa \in \mathbb{K}$, and to sample for each particle a novel θ-value conditional on the aggregation mode value.

Theorem 2. *(Hierarchical interaction) If $p_{\kappa_{\tau+h}}(\kappa) > 0$ for an arbitrary stopping time τ, then*

$$p_{x_\tau,\theta_\tau|\kappa_{\tau+h}}(dx,\theta|\kappa) = \sum_{\eta\in\mathbb{M}_\kappa} p_{\theta_{\tau+h}|x_\tau,\theta_\tau}(\eta|x,\theta)p_{x_\tau,\theta_\tau}(dx,\theta)/p_{\kappa_{\tau+h}}(\kappa), \tag{9.8}$$

$$p_{\kappa_{\tau+h}}(\kappa) = \sum_{\theta\in\mathbb{M}}\int_{\mathbb{R}^n}\sum_{\eta\in\mathbb{M}_\kappa} p_{\theta_{\tau+h}|x_\tau,\theta_\tau}(\eta|x,\theta)p_{x_\tau,\theta_\tau}(dx,\theta). \tag{9.9}$$

Proof. By definition of the partitioning $\{\mathbb{M}_\kappa,\ \kappa \in \mathbb{K}\}$, we have

$$\begin{aligned} p_{\kappa_{\tau+h},x_\tau,\theta_\tau}(\kappa,dx,\theta) &= \sum_{\eta\in\mathbb{M}_\kappa} p_{\theta_{\tau+h},x_\tau,\theta_\tau}(\eta,dx,\theta) \\ &= \sum_{\eta\in\mathbb{M}_\kappa} p_{\theta_{\tau+h}|x_\tau,\theta_\tau}(\eta|x,\theta)p_{x_\tau,\theta_\tau}(dx,\theta). \end{aligned}$$

Dividing the left- and right-hand sides by $p_{\kappa_{\tau+h}}(\kappa)$ yields (9.8). From the law of total probability we have:

$$p_{\kappa_{\tau+h}}(\kappa) = \sum_{\theta\in\mathbb{M}}\int_{\mathbb{R}^n} p_{\kappa_{\tau+h},x_\tau,\theta_\tau}(\eta,dx,\theta).$$

Substitution of the latter into the former yields (9.9). □

In order to see what Theorem 2 means for the empirical kind of densities that will be used, we assume $p_{x_\tau,\theta_\tau}(\cdot)$ equals an empirical density:

$$p_{x_\tau,\theta_\tau}(dx,\theta) = \sum_{\kappa\in\mathbb{K}}\sum_{i=1}^{N^\kappa}\omega^{\kappa,i}\delta_{(x^{\kappa,i},\theta^{\kappa,i})}(dx,\theta) \tag{9.10}$$

with $\{x^{\kappa,i},\theta^{\kappa,i},\omega^{\kappa,i}\}_{i=1}^{N^\kappa}$, $\kappa \in \mathbb{K}$, a given set of particles. Substituting (9.10) into (9.8) and evaluating yields:

$$\begin{aligned} &p_{x_\tau,\theta_\tau|\kappa_{\tau+h}}(dx,\theta|\kappa) \\ &= \sum_{\eta\in\mathbb{M}_\kappa} p_{\theta_{\tau+h}|x_\tau,\theta_\tau}(\eta|x,\theta)\sum_{\kappa'\in\mathbb{K}}\sum_{i=1}^{N^{\kappa'}}\omega^{\kappa',i}\delta_{(x^{\kappa',i},\theta^{\kappa',i})}(dx,\theta)/p_{\kappa_{\tau+h}}(\kappa) \\ &= \sum_{\kappa'\in\mathbb{K}}\sum_{i=1}^{N^{\kappa'}}\sum_{\eta\in\mathbb{M}_\kappa} p_{\theta_{\tau+h}|x_\tau,\theta_\tau}(\eta|x^{\kappa',i},\theta^{\kappa',i})\omega^{\kappa',i}\delta_{(x^{\kappa',i},\theta^{\kappa',i})}(dx,\theta)/p_{\kappa_{\tau+h}}(\kappa). \end{aligned} \tag{9.11}$$

Similarly, substituting (9.10) into (9.9) yields

$$p_{\kappa_{\tau+h}}(\kappa) = \sum_{\kappa' \in \mathbb{K}} \sum_{i=1}^{N^{\kappa'}} \sum_{\eta \in \mathbb{M}_\kappa} p_{\theta_{\tau+h}|x_\tau,\theta_\tau}(\eta | x^{\kappa',i}, \theta^{\kappa',i}) \omega^{\kappa',i}. \tag{9.12}$$

The idea is to use equation (9.11) for resampling N_p particles from $p_{x_{\tau_k},\theta_{\tau_k}|\kappa_{\tau_k+h}}(\cdot|\kappa)$ for each κ-value once at the beginning of a prediction cycle from τ_k to τ_{k+1}. Equation (9.12) is used to compensate each particle weight for this resampling.

9.6 Hierarchical hybrid IPS algorithm

Similar as in the IPS algorithm for an SHS [9,8], a particle is defined as a triplet (x, θ, ω), $\omega \in [0, 1]$, $x \in \mathbb{R}^n$ and $\theta \in \mathbb{M}$. Numerical approximations $\widetilde{\gamma}_k$ and $\widetilde{\pi}_k$ are used for γ_k and $p_{\xi_k|\chi_k}(\cdot|1)$ respectively. When simulating from $\tau_{k-1} \wedge T$ to $\tau_k \wedge T$, a fraction $\widetilde{\gamma}_k$ of the Monte Carlo simulated trajectories only will reach Q_k. The Hierarchical Hybrid Interacting Particle System (HHIPS) algorithm estimates these fractions and their product in a recursive way, using the following steps:

Step 0 generates for each κ-value N_p initial particles at $k = 0$, and then starts the cycling over steps 1 through 3 for $k := 1, 2, \ldots, m$.

Step 1 extrapolates each particle from $\tau_{k-1} \wedge T$ to $\tau_k \wedge T$ in time steps of length h, using importance switching for the new κ-value and κ-conditional sampling of a new θ-value. For the latter use is made of the κ-conditional θ-prediction characterization in Theorem 1.

Step 2 evaluates the particles that have arrived at Q_k. For this, use is made of equations 9.4–9.5.

Step 3 resamples from the particles that have arrived at Q_k. In order to draw N_p samples per κ-value, use is made of the hierarchical interaction characterization in Theorem 2.

Each of these steps is specified in detail below.

Hierarchical hybrid interacting particle system

HHIPS Step 0: Initial sampling for $k = 0$.

- At time $t = 0$ we start with a set of $N^\kappa := N_p$ particles for each aggregation mode $\kappa \in \mathbb{K}$: $\{x^{\kappa,i}, \theta^{\kappa,i}, \omega^{\kappa,i}\}_{i=1}^{N_p}$, $\kappa \in \mathbb{K}$, where the particles are obtained as follows. First, the $\theta^{\kappa,i}$ are independently drawn from $p_{\theta_0|\kappa_0}(\cdot|\kappa)$. Then, the $x^{\kappa,i} \in \{0\} \times \mathbb{R}^{n-1}/D_1$ are independently drawn from $p_{x_0|\theta_0}(\cdot|\theta^{\kappa,i})$

with the first component of $x^{\kappa,i}$ equal to zero. The initial weights satisfy

$$\omega^{\kappa,i} = \frac{p_{\kappa_0}(\kappa)}{N_p}, \qquad i = 1, \ldots, N_p, \quad \kappa \in \mathbb{K}.$$

- With this we have $\tilde{\gamma}_0 = 1$ and

$$\tilde{p}_{x_0}, \theta_0(dx, \theta) = \sum_{\kappa \in \mathbb{K}} \sum_{i=1}^{N^\kappa} \omega^{\kappa,i} \delta_{(x^{\kappa,i}, \theta^{\kappa,i})}(dx, \theta).$$

- Identify a sufficiently large number J of equal discretization steps of time length $h = T/J$, which allows a numerical integration time step h to be used.

- Identify an appropriate positive value for $\alpha < 1/J$.
Iteration cycle: For $k = 1, \ldots, m$, cycle over steps 1–3:

HHIPS Step 1: Prediction.

- Start with empty sets S_k^κ, $\kappa \in \mathbb{K}$, to store all particles that arrive at $Q_k = (0, T) \times D_k \times \mathbb{M}$.

- For $j = 1, \ldots, J$, iterate over substeps 1(a)–1(c).

Substep 1(a). Sample $\kappa_{\tau+h}$ using importance switching.
If $k > 1$ and $j = 1$, then go to substep 1(b), else for each $\kappa \in \mathbb{K}$ and $i = 1, \ldots, N^\kappa$:

- If $\omega^{\kappa,i} = 0$ then $\breve{\omega}^{\kappa,i} := 0$ and $\breve{\kappa}^{\kappa,i} := \kappa$; else sample a $\breve{\kappa}^{\kappa,i} \in \mathbb{K}$ with probability α for each of the values in $\mathbb{K} \setminus \{\kappa\}$, and with probability $1 - \alpha(|\mathbb{K}| - 1)$ for the value κ, and correct the corresponding weight according to this importance switching, that is,

$$\breve{\omega}^{\kappa,i} = \begin{cases} \omega^{\kappa,i} \frac{p_{\kappa_{\tau+h}|x_\tau,\theta_\tau}(\breve{\kappa}^{\kappa,i}|x^{\kappa,i},\theta^{\kappa,i})}{1-\alpha(|\mathbb{K}|-1)}, & \text{if } \breve{\kappa}^{\kappa,i} = \kappa, \\ \omega^{\kappa,i} \frac{p_{\kappa_{\tau+h}|x_\tau,\theta_\tau}(\breve{\kappa}^{\kappa,i}|x^{\kappa,i},\theta^{\kappa,i})}{\alpha}, & \text{if } \breve{\kappa}^{\kappa,i} \neq \kappa. \end{cases}$$

- The resulting sets of particles are $\{\bar{x}^{\kappa',l}, \bar{\theta}^{\kappa',l}, \breve{\omega}^{\kappa',l}, \breve{\kappa}^{\kappa',l}\}_{l=1}^{N^{\kappa'}}$, $\kappa' \in \mathbb{K}$. For each $\kappa \in \mathbb{K}$, collect from these particles those N^κ particles for which $\breve{\kappa}^{\kappa',l} = \kappa$, that is,

$$N^\kappa := \sum_{\kappa' \in \mathbb{K}, N^{\kappa'} \neq 0} \sum_{l=1}^{N^{\kappa'}} 1_{\{\kappa\}}(\breve{\kappa}^{\kappa',l}).$$

- For each $\kappa \in \mathbb{K}$, renumber the indices of these N^κ particles such that the first index equals κ and the second index runs over $\{1, \ldots, N^\kappa\}$. This yields for each $\kappa \in \mathbb{K}$ the following new set of particles $\{x^{\kappa,i}, \theta^{\kappa,i}, \omega^{\kappa,i}\}_{i=1}^{N^\kappa}$ if $N^\kappa \neq 0$, and an empty set $\varnothing$ if $N^\kappa = 0$.

Substep 1(b). $\kappa_{\tau+h}$-conditional prediction of $(x_{\tau+h}, \theta_{\tau+h})$.
For each $\kappa \in \mathbb{K}$, determine the new set of particles $\{\bar{x}^{\kappa,i}, \bar{\theta}^{\kappa,i}, \bar{\omega}^{\kappa,i}\}_{i=1}^{N^\kappa}$ as follows:

- For each κ, i for which $\omega^{\kappa,i} = 0$, set $\bar{x}^{\kappa,i} := x^{\kappa,i}$ and $\bar{\theta}^{\kappa,i} := \theta^{\kappa,i}$. Else use Theorem 1 to sample a new value $\bar{\theta}^{\kappa,i}$ from

$$p_{\theta_{\tau+h}|x_{\tau},\theta_{\tau},\kappa_{\tau+h}}(\eta|x^{\kappa,i},\theta^{\kappa,i},\kappa)$$
$$= \frac{1_{\mathbb{M}_\kappa}(\eta)p_{\theta_{\tau+h}|x_\tau,\theta_\tau}(\eta|x^{\kappa,i},\theta^{\kappa,i})}{\sum_{\eta'\in\mathbb{M}_\kappa} 1_{\mathbb{M}_\kappa}(\eta')p_{\theta_{\tau+h}|x_\tau,\theta_\tau}(\eta'|x^{\kappa,i},\theta^{\kappa,i})}$$

and a new value $\bar{x}^{\kappa,i}$ from

$$p_{x_{\tau+h}|\theta_{\tau+h},x_\tau,\theta_\tau}(dx|\bar{\theta}^{\kappa,i}, x^{\kappa,i}, \theta^{\kappa,i}).$$

- The weights are not changed, that is, $\bar{\omega}^{\kappa,i} := \omega^{\kappa,i}$.

Substep 1(c). Memorizing particles that arrived at Q_k.

- If $(\bar{x}^{\kappa,i}, \bar{\theta}^{\kappa,i}) \in Q_k$ and $\omega^{\kappa,i} \neq 0$, then a copy of the particle $\{\bar{x}^{\kappa,i}, \bar{\theta}^{\kappa,i}, \bar{\omega}^{\kappa,i}\}$ is stored in the set S_k^κ.
- Subsequently, we set $\bar{\omega}^{\kappa,i} := 0$ in the original particle.
- If $j = J$, then step 1 is complete, hence go to step 2, else repeat substeps 1(a)–1(c) for $j := j + 1$.

HHIPS Step 2. Evaluate the Q_k-arrived particles.

- The particles which are memorized in S_k^κ, $\kappa \in \mathbb{K}$, provide an estimate of $p_{\xi_k|\chi_k}(\cdot|1)$ and γ_k.
- Renumbering the particles in S_k^κ yields a set of particles $\{\tilde{x}^{\kappa,i}, \tilde{\theta}^{\kappa,i}, \tilde{\omega}^{\kappa,i}\}_{i=1}^{N^\kappa}$ with N^κ the number of particles in S_k^κ.
- Weighted fraction $\tilde{\gamma}_k$ of the Q_k-arrived particles:

$$\gamma_k \approx \tilde{\gamma}_k = \sum_{\kappa\in\mathbb{K}, N^\kappa\neq 0} \sum_{i=1}^{N^\kappa} \tilde{\omega}^{\kappa,i}.$$

- If $N^\kappa = 0$ for all $\kappa \in \mathbb{K}$, then the algorithm stops with the estimate $P_{hit}(0, T) \approx 0$.

- If $k = m$, then stop HHIPS with the estimate $P_{hit}(0, T) \approx \prod_{k=1}^{m} \tilde{\gamma}_k$.
- For each $\kappa \in \mathbb{K}$ and $i = 1, \ldots, N^\kappa$,

$$\tilde{\omega}^{\kappa,i} := \tilde{\omega}^{\kappa,i} / \tilde{\gamma}_k.$$

- The estimated $p_{\xi_k | \chi_k}(\cdot|1)$ satisfies

$$p_{\xi_k|\chi_k}(dx, \theta|1) \approx \tilde{\pi}_k(dx, \theta) = \sum_{\kappa \in \mathbb{K}, N^\kappa \neq 0} \sum_{i=1}^{N^\kappa} \tilde{\omega}^{\kappa,i} \delta_{(\tilde{x}^{\kappa,i}, \tilde{\theta}^{\kappa,i})}(dx, \theta).$$

HHIPS Step 3. Copy the Q_k-arrived particles through κ_{τ_k+h}-conditional resampling.

- Evaluate aggregated mode probabilities at $\tau := \tau_k$ using (9.12):

$$p_{\kappa_{\tau+h}|\chi_k}(\kappa|1) \approx \varphi(\kappa) = \sum_{\kappa' \in \mathbb{K}, N^{\kappa'} \neq 0} \sum_{i=1}^{N^{\kappa'}} \sum_{\eta \in \mathbb{M}_{\kappa'}} p_{\theta_{\tau+h}|x_\tau, \theta_\tau}(\eta|\tilde{x}^{\kappa',i}, \tilde{\theta}^{\kappa',i})\tilde{\omega}^{\kappa',i}.$$

- For each $\kappa \in \mathbb{K}$, independently draw N_p random pairs $(x^{\kappa,i}, \theta^{\kappa,i})$, $i = 1, \ldots, N_p$, from the particle spanned empirical measure, using (9.11):

$$p_{x_\tau, \theta_\tau|\kappa_{\tau+h}, \chi_k}(dx, \theta|\kappa, 1) \approx \sum_{\kappa' \in \mathbb{K}, N^{\kappa'} \neq 0} \sum_{i=1}^{N^{\kappa'}} \sum_{\eta \in \mathbb{M}_{\kappa'}} p_{\theta_{\tau+h}|x_\tau, \theta_\tau}(\eta|\tilde{x}^{\kappa',i}, \tilde{\theta}^{\kappa',i}) \times \tilde{\omega}^{\kappa',i} \delta_{\{\tilde{x}^{\kappa',i}, \tilde{\theta}^{\kappa',i}\}}(dx, \theta)/\varphi(\kappa).$$

- This yields, for each $\kappa \in \mathbb{K}$, a set of particles $\{x^{\kappa,i}, \theta^{\kappa,i}, \omega^{\kappa,i}\}_{i=1}^{N_p}$ with $\omega^{\kappa,i} := \varphi(\kappa)/N_p$.
- If $k < m$, then repeat steps 1–3 for $k := k + 1$ and $N^\kappa := N_p$.

Remark 1. The key extensions of HHIPS over IPS for an SHS [8,9] are:

(i) embedding of an aggregation mode process;

(ii) particles are maintained per aggregation mode;

(iii) importance switching of aggregation mode is used for the conditional prediction of SHS particles;

(iv) hierarchical interaction is used for the resampling of particles that reached Q_k, $k = 1, \ldots, m - 1$.

9.7 Free flight air traffic example

We consider a specific free flight operational concept that has been developed within a recent European research project [37]. In the free flight air traffic example, the airspace is an en-route airspace without fixed routes and without support by air traffic control. All aircraft flying in this airspace are assumed to be properly equipped and enabled for free flight: the pilots can try to optimize their trajectory, due to the greater freedom to choose path and flight level. The pilots are only limited by their responsibility to maintain airborne separation, in which they are assisted by the so-called Airborne Separation Assistance System (ASAS). This system processes the information flows from the data-communication links between aircraft, the navigation systems and the aircraft guidance and control systems. ASAS detects conflicts, determines conflict resolution maneuvers and presents the relevant information to the aircrew. The number of agents involved in the free flight operation is huge and ranges from the Control Flow Management Unit to flight attendants. In the setting chosen for an initial risk assessment, the following agents are taken into account:

- one flying pilot in each aircraft;
- one non-flying pilot in each aircraft;
- various systems and entities per aircraft, such as the aircraft position evolution and the conflict management support systems;
- some global systems and entities, such as the communication frequencies and a satellite system.

The approach taken in developing the specific free flight concept of operation [37] is to avoid much information exchange between aircraft and to avoid dedicated decision-making by artificial intelligent machines. Although the conflict detection and resolution approach developed for this free flight concept has its roots in the modified potential field approach [30], it has some significant deviations from this. The main deviation is that conflict resolution is intentionally designed not to take the potential field of all aircraft into account. The resulting design can be summarized as follows:

- All aircraft are supposed to be equipped with Automatic Dependent Surveillance-Broadcast (ADS-B), which is a system that periodically broadcasts own aircraft state information, and continuously receives the state information messages broadcasted by aircraft that fly within broadcasting range ($\sim$ 100 Nm).

- To comply with pilot preferences, conflict resolution algorithms are designed to solve multiple conflicts one by one rather than in a fully concurrent way (see [30]).
- Conflict detection and resolution are state-based, that is, intent information, such as information at which point surrounding aircraft will change course or height, is assumed to be unknown.
- The vertical separation minimum is 1000 ft and the horizontal separation minimum is 5 Nm. A conflict is detected if these separation minima will be violated within 6 minutes.
- The conflict resolution process consists of two phases. During the first phase, one of the aircraft crews should make a resolution maneuver. If this does not work, then during the second phase, both crews should make a resolution maneuver.
- Prior to the first phase, the crew is warned when an ASAS alert is expected to occur if no preventive action would be implemented on time; this prediction is done by a system referred to as Predictive ASAS (P-ASAS).
- Conflict co-ordination does not take place explicitly, that is, there is no communication on when and how a resolution maneuver will be executed.
- All aircraft are supposed to use the same resolution algorithm, and all crew are assumed to use ASAS and to collaborate in line with the procedures.
- Two conflict resolution maneuver options are presented: one in the vertical and one in the horizontal direction. The pilot decides which option to execute.
- ASAS related information is presented to the crew through a cockpit display of traffic information.

In order to use the HHIPS algorithm for the estimation of collision risk in this free flight operation, we need to develop an MC simulator of this operation, such that the simulated trajectories constitute realizations of a hybrid state strong Markov process. Everdij and Blom [23,24,25,26] have developed a stochastically and dynamically colored Petri net (SDCPN) formalism that ensures the specification of a free flight MC simulation model which is of the appropriate class. In [9] it is explained how the SDCPN formalism has been used to develop an MC simulation model of a particular free flight design. The dimensionality of the resulting MC simulation model is very large, for example in simulating two aircraft there are about 10^{25} discrete mode combinations, and the Euclidean state may go up to $\mathbb{R}^{336}$ [9]. For this very large stochastic hybrid system we want to estimate the probability of collision between aircraft. This is practically infeasible using naive MC simulation.

9.8 Application of HHIPS to air traffic example

In [4,8,9] we developed a way to cast the air traffic SHS model within the setting of the IPS formulation, and used the IPS to evaluate demanding high risk bearing multi-aircraft scenarios. This IPS approach, however, does not work properly anymore for low risk bearing scenarios. The aim of this section is to demonstrate that the novel HHIPS works well for such a low risk bearing scenario, using the same SHS model.

In the low risk bearing scenario considered, two aircraft start at the same flight level, some 250 km away from each other, and fly on opposite direction flight plans head-on with a ground speed of 240 m/s. This means that collision may be reached after about 500 s simulation, hence we set $T = 600$ s. The collision reach probability is estimated by running the HHIPS algorithm ten times.[2] The aggregation modes chosen are all combinations of the following high-level mode values: global communication support is 'up' or 'down', and the decision-making loop of aircraft 1 is 'up' or 'down'. This leads to a total of four aggregation mode values.

The D_k's were identified through an iterative process of learning from conducting MC simulations. This quite easily led to the identification of a series of D_k's that appeared to work well. Although it is likely that further optimization of the D_k's may lead to a reduction in the variance and confidence interval of the estimates (see Chapter 3), we have not yet tried to do so.

Each identified D_k is defined by three parameters, the values of which are given in Table 9.1 for a sequence of eight nested subsets. Here d_k and h_k define a cylinder of diameter d_k and height h_k, respectively. Δ_k is the time period over which position and velocity differences between the two aircraft are compared. If within Δ_k the predicted position difference falls within the corresponding cylinder, then D_k is said to be reached. The three parameters of D_1, D_2, D_4, D_6 and D_8 are such that reaching them represents a type of conflict that is well known in air traffic, i.e. medium-term conflict, short-term conflict, conflict, near-collision and collision, respectively. The extra D_3, D_5 and D_7 appeared useful in avoiding too small fractions remaining from hitting D_4 after D_2, D_6 after D_4 and D_8 after D_6.

The number of particles used is 5000 per aggregation mode value; hence 20 000 particles are used per HHIPS run. The time step $h = 1$ s, and $\alpha = 0.001$.

Table 9.1 IPS conflict level parameter values

k	1	2	3	4	5	6	7	8
d_k (Nm)	4.5	4.5	4.5	4.5	2.5	1.25	0.50	0.054
h_k (ft)	900	900	900	900	900	500	250	131
Δ_k (min)	8	2.5	1.5	0	0	0	0	0

[2] In [7] a similar two-aircraft encounter scenario is simulated using an initial precursor of the current HHIPS.

Table 9.2 $\tilde{\gamma}_k$ values estimated by first five HHIPS runs. IPS based estimation typically yields values 0.0 for $k \geq 4$

k	Run 1	Run 2	Run 3	Run 4	Run 5
1	1.000	1.000	1.000	0.991	1.000
2	5.77E-04	5.64E-06	6.24E-06	5.04E-06	6.13E-06
3	6.40E-03	7.25E-01	7.20E-01	6.84E-01	7.66E-01
4	0.566	0.569	0.596	0.540	0.608
5	0.344	0.256	0.223	0.401	0.198
6	0.420	0.452	0.402	0.459	0.429
7	0.801	0.845	0.929	0.710	0.949
8	0.814	0.827	0.841	0.828	0.802
Π	1.97E-07	1.89E-07	1.89E-07	2.00E-07	1.85E-07

Table 9.2 presents the values for $\tilde{\gamma}_k$ which have been estimated during the first five HHIPS runs. The estimated mean probability of collision between the two aircraft equals 1.91×10^{-7}. The estimated standard deviation is 1.6×10^{-8}, which shows that the estimated value is quite accurate. It should be noticed that the variation in the fractions per level is significantly larger than the variation in the product of the fractions. Apparently, the dependency between the fractions $\tilde{\gamma}_k$ reduces the variation in the multiplication of these fractions.

Finally, we improved the availability/reliability of the ASAS related systems by a factor of 100, and then conducted the ten HHIPS runs again. This resulted in a 100-fold decrease of the collision reach probability. These results demonstrate that HHIPS works well for this large-scale SHS.

9.9 Concluding remarks

This chapter first presented the rare event estimation theory developed in [16,17,18,20,21] within the framework of probabilistic reachability analysis of SHS. Subsequently, the theory was extended with mode aggregation, importance switching and Rao–Blackwellization. This allows probabilistic reachability analysis theory to be applied to large-scale SHS, and in particular when the reachability probability considered receives significant contributions from combinatorially many rare modes. The power of the resulting novel sequential MC simulation approach was demonstrated through a successful application to collision risk estimation in a demanding future air traffic scenario. And, of course, there are several interesting directions for follow-up research, such as:

- extending the convergence proof for IPS to HIPS and HHIPS;
- incorporating parameter sensitivity assessment in IPS, HIPS and HHIPS;
- optimization of D_k's identification in the air traffic example.

Acknowledgement

The first author would like to thank Pierre Del Moral (INRIA, Bordeaux), Francois LeGland and Frederic Cérou (both of INRIA/IRISA, Rennes) and Pascal Lezaud (CENA/ENAC, Toulouse) for valuable discussions.

References

[1] A. Abate, S. Amin, M. Prandini, J. Lygeros, and S. Sastry. Probabilistic reachability for discrete time stochastic hybrid systems. In *Proceedings of the IEEE Conference on Decision and Control*, pp. 258–263, December 2006.

[2] S. Amin, A. Abate, M. Prandini, J. Lygeros, and S. Sastry. Reachability analysis for controlled discrete time stochastic hybrid systems. In J. Hespanha and A. Tiwari, eds, *Hybrid Systems: Computation and Control*, Lecture Notes in Computer Science 3927, pp. 49–63. Springer, Berlin, 2006.

[3] O. Atkin and J. K. Townsend. Efficient simulation of TCP/IP networks characterzed by non-rare events using DPR-based splitting. In *Proceedings of IEEE Globecom*, pp. 1734–1740, 2001.

[4] H. A. P. Blom, G. J. Bakker, B. Klein Obbink, and M. B. Klompstra. Free flight safety risk modelling and simulation. In *Proceedings of the International Conference on Research in Air Transportation (ICRAT)*, Belgrade, 26–28 June 2006.

[5] H. A. P. Blom, G. J. Bakker, and J. Krystul. Probabilistic reachability analysis for large scale stochastic hybrid systems. In *Proceedings of the IEEE Conference on Decision and Control*, pp. 3182–3189, New Orleans, December 2007.

[6] H. A. P. Blom and E. A. Bloem. Exact Bayesian and particle filtering of stochastic hybrid systems. In *IEEE Transactions on Aerospace and Electronic Systems*, **43**: 55–70, 2007.

[7] H. A. P. Blom, J. Krystul, and G. J. Bakker. Estimating rare event probabilities in large scale stochastic hybrid systems by sequential Monte Carlo simulation. In *Proceedings of the International Workshop on Rare Event SIMulation (RESIM)*, Bamberg, Germany, 9–10 October 2006.

[8] H. A. P. Blom, J. Krystul, and G. J. Bakker. A particle system for safety verification of free flight in air traffic. In *Proceedings of the IEEE Conference on Decision and Control*, pp. 1574–1579, San Diego, CA, 13–15 December 2006.

[9] H. A. P. Blom, J. Krystul, G. J. Bakker, M. B. Klompstra, and B. Klein Obbink. Free flight collision risk estimation by sequential MC simulation. In C. G. Cassandras and J. Lygeros, eds, *Stochastic Hybrid Systems*. CRC Press, Boca Raton, FL, 2007.

[10] T. E. Booth. Monte Carlo variance comparison for expected-value versus sampled splitting. *Nuclear Science and Engineering*, **89**: 305–309, 1985.

[11] J. A. Bucklew. *An Introduction to Rare Event Estimation*. Springer, New York, 2004.

[12] M. L. Bujorianu. Extended stochastic hybrid systems. In R. Alur and G. Pappas, eds, *Hybrid Systems: Computation and Control*, Lecture Notes in Computer Science 2993, pp. 234–249. Springer, Berlin, 2004.

[13] M. L. Bujorianu and J. Lygeros. General stochastic hybrid systems: modelling and optimal control. In *Proceedings of the 43rd IEEE Conference on Decision and Control*, pp. 1872–1877, Bahamas, 14–17 December, 2004.

[14] M. L. Bujorianu and J. Lygeros. Toward a general theory of stochastic hybrid systems. In H. A. P. Blom and J. Lygeros, eds, *Stochastic Hybrid Systems: Theory and Safety Critical Applications*, Lecture Notes in Control and Information Sciences 337, pp. 3–30. Springer, Berlin, 2006.

[15] G. Casella and C. P. Robert. Rao-Blackwellization of sampling schemes. *Biometrika*, **83**: 81–94, 1996.

[16] F. Cérou, P. Del Moral, F. Le Gland, and P. Lezaud. Genetic genealogical models in rare event analysis. Publications du Laboratoire de Statistiques et Probabilités, Toulouse III, 2002.

[17] F. Cérou, P. Del Moral, F. Le Gland, and P. Lezaud. Limit theorems for the multilevel splitting algorithm in the simulation of rare events. In M. E. Kuhl, N. M. Steiger, F. B. Armstrong, and J. A. Joines, eds, *Proceedings of the 2005 Winter Simulation Conference*, Orlando, December 2005.

[18] F. Cérou, P. Del Moral, F. Le Gland, and P. Lezaud. Genetic genealogical models in rare event analysis. *ALEA, Latin American Journal of Probability and Mathematical Statistics*, **1**: 181–203, 2006.

[19] M. H. A. Davis. *Markov Models and Optimization*. Chapman & Hall, London, 1993.

[20] P. Del Moral. *Feynman–Kac Formulae. Genealogical and Interacting Particle Systems with Applications*. Springer, New York, 2004.

[21] P. Del Moral and P. Lezaud. Branching and interacting particle interpretation of rare event probabilities. In H. A. P. Blom and J. Lygeros, eds, *Stochastic Hybrid Systems: Theory and Safety Critical Applications*, Lecture Notes in Control and Information Sciences 337, pp. 351–389. Springer, Berlin, 2006.

[22] A. Doucet, S. Godsill, and C. Andrieu. On sequential Monte Carlo sampling methods for Bayesian filtering. *Statistics and Computing*, **10**: 197–208, 2000.

[23] M. H. C. Everdij and H. A. P. Blom. Petri-nets and hybrid-state Markov processes in a power-hierarchy of dependability models. In *Proceedings of the IFAC Conference on Analysis and Design of Hybrid Systems (ADHS)*, pp. 355–360, Saint-Malo, France, June 2003.

[24] M. H. C. Everdij and H. A. P. Blom. Piecewise deterministic Markov processes represented by dynamically coloured Petri nets. *Stochastics*, **77**: 1–29, 2005.

[25] M. H. C. Everdij and H. A. P. Blom. Hybrid Petri nets with diffusion that have into-mappings with generalised stochastic hybrid processes. In H. A. P. Blom and J. Lygeros, eds, *Stochastic Hybrid Systems: Theory and Safety Critical Applications*, Lecture Notes in Control and Information Sciences 337, pp. 31–64. Springer, Berlin, 2006.

[26] M. H. C. Everdij and H. A. P. Blom. Enhancing hybrid state Petri nets with the analysis power of stochastic hybrid processes. In *Proceedings of the 9th International Workshop on Discrete Event Systems*, pp. 400–405, Gothenburg, May 2008.

[27] M. J. J. Garvels. The splitting method in rare event simulation. PhD thesis, Univ. Twente, 2000.

[28] M. J. J. Garvels, D. P. Kroese, and J. K. C. W. van Ommeren. On the importance function splitting simulation. *European Transactions on Telecommunications*, **13**: 363–371, 2002.

[29] P. Glasserman, P. Heidelberger, P. Shahabuddin, and T. Zajic. Multilevel splitting for estimating rare event probabilities. *Operations Research*, **47**(4): 585–600, 1999.

[30] J. Hoekstra. Designing for safety, the free flight air traffic management concept. PhD thesis, Delft University of Technology, November 2001.

[31] S. Juneja and P. Shahabuddin. Rare event simulation techniques: An introduction and recent advances. In S. Henderson and B. Nelson, eds, *Simulation, Handbooks in Operations Research and Management Science*, pp. 291–350. Elsevier, Amsterdam, 2006.

[32] J. Krystul. Modelling of stochastic hybrid systems with applications to accident risk assessment. PhD thesis, University of Twente, 2006.

[33] J. Krystul and H. A. P. Blom. Sequential Monte Carlo simulation of rare event probability in stochastic hybrid systems. In *Proceedings of the 16th IFAC World Congress*, Prague, June 2005.

[34] J. Krystul and H. A. P. Blom. Sequential Monte Carlo simulation for the estimation of small reachability probabilities for stochastic hybrid systems. In *Proceedings of IEEE-EURASIP International Symposium on Control, Communications and Signal Processing*, Marrakech, Morocco, March 2006.

[35] J. Krystul, H. A. P. Blom, and A. Bagchi. Stochastic hybrid processes as solutions to stochastic differential equations. In C. G. Cassandras and J. Lygeros, eds, *Stochastic Hybrid Systems*. CRC Press, Boca Raton, FL, 2007.

[36] P. L'Ecuyer, V. Demers, and B. Tuffin. Splitting for rare-event simulation. In *Proceedings of the 2006 Winter Simulation Conference*, pp. 137–148. IEEE, December 2006.

[37] F. Maracich. Flying free flight: pilot perspective and system integration requirement. *IEEE Aerospace and Electronic Systems Magazine*, pp. 3–7, July 2006.

[38] S. Prajna, A. Jadbabaie, and G. J. Pappas. A framework for worst-case and stochastic safety verification using barrier certificates. *IEEE Transactions on Automatic Control*, **52**(8): 1415–1428, 2007.

[39] M. Prandini and J. Hu. A stochastic approxmation method for reachability computations. In H. A. P. Blom and J. Lygeros, eds, *Stochastic Hybrid Systems: Theory and Safety Critical Applications*, Lecture Notes in Control and Information Sciences 337, pp. 107–139. Springer, Berlin, 2006.

10

Particle transport applications

Thomas Booth

10.1 Introduction

Historically, Monte Carlo nuclear particle transport simulations were the first large-scale use of Monte Carlo methods on digital computers. From the beginning, the field of transport has had some very important, and very difficult, problems to solve. The simulation of rare events arises almost immediately because nuclear particles (e.g., neutrons and photons) can be hazardous to human health at high doses. There are many rare event simulation applications besides shielding, but this chapter will focus on shielding because this application, and its importance, are widely understood. For instance, humans need to be shielded from the high-energy nuclear particles from a nuclear reactor, whether the reactor is a commercial power reactor, a naval reactor on a warship, or the sun. The penetration probabilities for nuclear shields are often 10^{-8} to 10^{-10} or smaller, so that it is an extremely rare particle that penetrates the shield.

This chapter focuses primarily on practical rare event simulations using the variance reduction techniques in the Los Alamos National Laboratory's Monte Carlo transport code MCNP [31]. The reasons for this focus are threefold. First, MCNP is by far the most widely used transport code in the world, both in terms of the number of users and the amount of computer time expended in MCNP calculations. Second, there are many Monte Carlo transport codes with many different variance reduction techniques and the author has neither the space nor the expertise to discuss every technique. Third, this chapter has been written as an introduction for non-transport Monte Carlo practitioners and not as a general

Rare Event Simulation using Monte Carlo Methods Edited by G. Rubino and B. Tuffin

survey of techniques in transport. (Because of this focus, the reader should note that the reference list is skewed toward publications associated with MCNP; see Lux and Koblinger [22] for a more representative reference list.)

The review process for this chapter indicated that transport Monte Carlo differs in significant ways from the Monte Carlo used in other fields, for example operations research. Not only are the terminology and methods quite different in many cases, there is sometimes even a conceptual difference in the viewpoints of what a Monte Carlo calculation is. This chapter first discusses these differences in terminology and viewpoint, to the extent that the author and reviewer have identified them.

In the major portion of this chapter, some of the methods used in rare event transport simulations are described and then demonstrated in the context of a sample transport problem. In addition to the methods illustrated on the sample problem, the 'comb' is discussed because few people are aware of the comb technique, despite some interesting theoretical and practical aspects. Inasmuch as the implications of using importance function information is not always well understood, some guidance is given. The chapter concludes with practical comments about when users should stop variance reduction efforts, followed by some comments on the future of Monte Carlo transport methods.

10.2 Scope of particle transport problems considered

The field of particle transport contains deterministic (i.e., non-stochastic) solution techniques as well as the Monte Carlo techniques discussed in this chapter. Almost all of the deterministic techniques solve the transport problem by solving the Boltzmann transport equation [1] for the particle flux (particle density times velocity) as a function of position, energy, angle and time. Monte Carlo is often viewed as an alternative approach to solving transport equations, so most of the texts implicitly assume in their discussions that the Monte Carlo codes are being used to estimate quantities such as particle densities, currents, and fluxes. This assumption is often invalid, as explained in the paragraph after next.

Those not familiar with transport can get some idea of the issue by considering the planar heat equation $\nabla^2 T = 0$ in a homogeneous medium with boundary conditions $T = 1$ on some part of the boundary and $T = 0$ on the rest of the boundary. The solution to the heat equation at a point P can be shown to be [2] the probability that a particle starting at $P = (x, y)$ with an isotropic random direction and randomly walking on circles reaches the $T = 1$ boundary rather than the $T = 0$ boundary. Thus, the random walk is 'solving' the heat equation. Note that if two particles are started at P with particle 2's direction opposite to particle 1's direction, the random walk of either one 'solves' the heat equation. If instead one changes the question to estimating the probability that *both* particles reach the $T = 1$ boundary, one needs to write a different equation involving the joint density of particles 1 and 2 because the heat equation gives no information about the joint density.

There are many transport problems for which solving the standard Boltzmann transport equation is irrelevant because it describes the behavior of individual particles. Any particle transport problems that depend on the collective behavior of several particles must be treated differently. From a Monte Carlo standpoint, the estimates made are dependent on collections of particles and therefore the *collection* of particles carries a statistical weight rather than the individual particles [4–6, 25]. One example is the coincident physical detection of a pair of gamma rays from an electron–positron annihilation event. The physical detector system responds only when both gamma rays enter the detector within a very short time interval ('in coincidence'). If two gamma rays enter the detector, but not in coincidence, the detector does not respond because the gamma rays could not be from the annihilation event. The Boltzmann equation provides no information about the probability of coincidence in the detector.

For simplicity, this chapter considers only those types of Monte Carlo calculations for which individual particles carry weights.

10.3 Transport terminology

A particle *history* is one independent sampling of the random walk process from the particle's source event (birth) to the termination of the particle (and progeny, if any). For instance, if a source particle subsequently produces fission progeny (e.g., as in a nuclear reactor), then the history includes the random walks of all fission particles related to the initial source particle. A *score* or *tally* is a contribution to a desired estimate such as the number of partices crossing a surface. For the sake of simplicity, the scores will always be assumed non-negative, as is almost always the case. The *history score*, s_n, or *history tally* is the sum of all scores for a given history. The history score is the basic statistical quantity in transport. If, for instance, one wishes to estimate the mean number of collisions per source particle, then one increments the history score, s_n, of particle n at each collision by $s_n + 1 \rightarrow s_n$. When the history is complete, the final s_n is then the history score.

A particle *track* is any particle associated with a particle history. For instance, in a nuclear chain reaction a particle history will consist of many particle tracks from fission. Tracks may also be created by variance reduction techniques (e.g., particle splitting, described later).

The sample mean and variance of the history score are estimated with N particle histories as

$$\hat{\mu} = \frac{1}{N}\sum_{n=1}^{N} s_n, \tag{10.1}$$

$$\hat{\sigma^2} = \frac{1}{N-1}\sum_{n=1}^{N}(s_n - \hat{\mu})^2. \tag{10.2}$$

The particle *weight* multiplies a score when a score is made. Note that a particle track can score many times and each time with a possibly different weight. For instance, if the desired estimate is the total energy crossing a surface, then the history score would be $w_1 E_1 + w_2 E_2 + w_3 E_3$, if the particle crossed the surface three times with energies E_i and weights w_i.

Note that weight in transport is simply a score multiplier that may change for many reasons. Some of the transport uses for weight seem to differ substantially from other Monte Carlo fields. For instance, consider estimating the number of collisions in disjoint regions 1 to 100. A naive estimate takes a particle of unchanging weight $w = 1$ and every time a collision occurs on sample n the current score is updated by $s_n + 1 \to s_n$. An equally valid estimate would be for the nth sample to consist of two particle tracks. Track 1 has weight $w = 1$ in odd regions and weight $w = 0$ in even regions. Conversely, track 2 has weight $w = 0$ in odd regions and weight $w = 1$ in even regions. Every time a collision occurs on either track 1 or track 2 the current score is updated by $s_n + w \to s_n$. The mean estimate is preserved because the weighted distribution of collisions is the same as in the analog case. (While this example is contrived solely for explanatory purposes, and is not effective as a variance reduction method, MCNP does, in fact, sometimes introduce additional weighted tracks in similar cases that are effective for variance reduction.) Note that a track's weight can go from 0 to non-zero in transport calculations; this is sometimes confusing to Monte Carlo practitioners who understand *weight* in the limited sense of a likelihood ratio correction. There are some transport situations (not discussed here) where the weight can be negative or even complex.

Probably most Monte Carlo transport practitioners understand and use the term 'analog Monte Carlo' in mostly the same way. That is, convenient probability densities are abstracted from the physical transport process. These convenient probability densities are then embedded in a transport code. For example, the neutron distance to collision is sampled conveniently from an exponential distribution without modeling the detailed interactions between the neutron and each nuclide along its path. For this chapter, the term 'analog' describes a direct sampling of these abstracted probability densities. For the most part, people have abstracted very similar probability densities from the physical transport process. Nonetheless, it is probably worthwhile to note that an analog sampling in the context of this chapter refers to the particular probability densities that MCNP has abstracted from the physical transport process. Roughly speaking, an analog Monte Carlo sampling of a transport problem is what one gets when no variance reduction techniques are used.

The physical transport process is a Markov process in the phase space $\mathbf{P} = (\mathbf{r}, \mathbf{\Omega}, E, t)$ because knowing the particle's position, direction, and energy at time t, the particle's behavior is independent of how it arrives at $\mathbf{P}$. Similarly, in an analog Monte Carlo simulation of nature, the future of a particle's random walk depends only on its current phase-space location $\mathbf{P}$, and not on how it arrived at $\mathbf{P}$.

Non-analog simulations of particle transport depart, in one way or another, from the analog process. Non-analog methods are also known as 'variance reduction methods' because the aim of using non-analog methods is to reduce the variance in the estimated mean for a given computer time. Note that non-analog simulations need not be Markov processes, despite the fact that the *physical* transport is a Markov process. Monte Carlo transport calculations typically use large numbers (say, 1000–10 000) of 'variance reduction parameters' as input to different variance reduction techniques. Note that the adjective 'variance reduction' is optimistic; it is quite possible to specify 'variance reduction' that actually increases the calculational variance for a fixed computing time.

10.4 Unbiased combinations of non-analog Monte Carlo techniques and fair games

A common Monte Carlo philosophy (not used for MCNP) is to consider non-analog techniques one, or perhaps two, at a time. The non-analog techniques are always explicitly known at the beginning of a Monte Carlo calculation. Typically, the set of possible random walk chains is thus known from the outset and one proves that a properly weighted estimator over the known set of chains will preserve the same mean estimate as an analog calculation. There are almost always some important restrictions, sometimes taken for granted and not even mentioned, associated with this approach. Consider, for example, the following discussion in [21]:

> In fact, $L(X_1, \ldots, X_j)$ can be viewed as a weight that the chain has accumulated so far. This weight will simply multiply the contribution of this chain to the estimator at the end (if $\tau_B < \tau_A$) and otherwise has no influence on the sample path of the chain after step j. If we decide to apply splitting or roulette to this chain at step j, then the weighting factors that these methods introduce can simply multiply the likelihood ratio.

Here, L is a likelihood ratio and τ_A and τ_B are the numbers of steps to reach (disjoint) regions A and B, respectively. Note the assumption 'otherwise has no influence on the sample path of the chain after step j'. The transport Monte Carlo in MCNP often samples from very different sets of chains that are explicitly dependent on the weight. From MCNP's perspective this assumption *unnecessarily* restricts the possible non-analog techniques available, and may lead to very suboptimal calculations compared to calculations that are not bound by this restriction.

Many Monte Carlo transport codes, in particular MCNP, may use many combinations of variance reduction techniques in a single calculation. Also, it is not uncommon for advanced users to introduce their own special Monte Carlo method designed solely for one specific difficult calculation. Users typically will combine their special method with some of the standard methods in MCNP. Having to write down an 'estimator' (see [28]) for each possible combination of variance reduction techniques, and prove unbiasedness, was not a practical thing to do for MCNP purposes. Spanier [28], for instance, has an estimator for sampling from biased transport kernels (p. 114) and a different estimator for splitting and Russian roulette (p. 128). Presumably if one uses both methods at the same time, one needs to define yet another estimator for the combination.

The philosophy for transport Monte Carlo is quite different. An example may help illustrate the difference. Suppose three people are working on a Monte Carlo calculation that is broken into three time intervals $0 \leq t < T_1$, $T_1 \leq t < T_2$, and $T_2 \leq t < T_3$. The ith person is responsible for the non-analog Monte Carlo methods used in the ith time interval. The first person does not need to know how the next two people will choose to do the sampling in their time intervals. The first person can use any non-analog techniques he likes, provided he guarantees the second person that if all samplings for $t \geq T_1$ are analog, then all mean estimates will be the same as if the entire calculation $0 \leq t < T_3$ were done analog. The second person need not know how the third person will choose to do the sampling. The second person can use any non-analog techniques he likes, provided he guarantees the third person that if all samplings for $t \geq T_2$ are analog, then all mean estimates will be the same as if the entire calculation $0 \leq t < T_3$ were done analog. The third person can use any non-analog techniques he likes, provided he too preserves the mean estimates that would occur if his part of the transport were purely analog.

The difficulties associated with proving unbiasedness for arbitrary combinations of variance reduction techniques were resolved by introducing the notion of variance reduction as a set of 'fair games'. A game (e.g., splitting) is said to be 'fair' if the mean values of all estimates are the same as a totally analog Monte Carlo procedure. That is, except for computation time, Monte Carlo transport practitioners would be happy with mean results from an analog calculation. The practitioners do not care what variance reduction games are played as long as the mean results are unaffected and the calculation is efficient. Transport results in MCNP do not rely on the 'estimator' concept. Instead, MCNP relies on the concept of a 'fair game' and a proof [12] that any combination of fair games is also a fair game.

One final comment about discarding the estimator approach is perhaps worthwhile. The author is not suggesting that it is *impossible* for someone to justify all possible combinations (current and future) of variance reduction techniques in MCNP with an estimator approach; the author is simply saying that it looks daunting and has never been done. Perhaps it is possible, perhaps not.

10.5 Weight-dependent vs weight-independent transport

Most theoretical Monte Carlo transport discussions assume that a particle's random walk is independent of the particle's weight. That is, two otherwise identical particles of weights w_1 and w_2 have the same distribution of possible random walks. For any particular random walk, particle 2's score is w_2/w_1 times particle 1's score. (Note that this assumption in transport is akin to the assumption in the previous section specified by the phrase 'otherwise has no influence on the sample path of the chain after step j'.) Under this assumption, a particle's score is directly proportional to its weight and the rth score moment for a particle of weight w is w^r times the rth score moment for a unit weight particle. [22, p. 163]. To give some idea of the ubiquitousness of this assumption, note that [22] first mentions this assumption in a footnote.

A cautionary note is perhaps worthwhile here. Because weight-independent simulations are more tractable mathematically, they account for almost all of the theoretical discussions in the Monte Carlo literature. (Two good exceptions can be found in [22, pp. 178 and 186].) One should not be misled into concluding that weight-independent simulations are more important, better, or more widely used than weight-dependent simulations. Many (probably most) of the large production Monte Carlo codes allow weight-dependent simulation. MCNP has always done weight-dependent simulation as a default. (To the author's knowledge, the predecessor codes to MCNP, as far back as the 1950s, always did weight-dependent simulation as a default as well.) There is often some distance between Monte Carlo theory and practice. Two examples are given below.

First, consider the weight window technique (described later in more detail) that enforces a range of acceptable weights by splitting if the weight is above the window and rouletting if the weight is below the window. After the enforcement, all weights are then within the window. The weight window is perhaps the most widely used variance reduction technique in transport Monte Carlo today, but it has received scant theoretical attention (Fox [19, pp. 213–233] gives an interesting discussion).

Second, consider Monte Carlo optimization techniques. There are numerous theoretical derivations about optimal parameters to minimize the variance; they almost always assume weight-independent transport. A favorite problem for theorists is optimizing the exponential transform [26, 27, 18, 24] (described later). (The reference list is not exhaustive; see [22, p. 487] for more.) Inasmuch as practical experience indicates that a weight window almost always improves the performance of the exponential transform, the usefulness of optimizing the exponential transform in the absence of a weight window is severely curtailed. An empirically optimized transform used with a weight window can give very good results. When the weight window is removed with the same transform parameter, the results are often disastrous. In one documented case [3, pp. 54–56], the efficiency decreased by a factor of 100.

The author's speculation as to the reason for the divergence between theory and practice is that:

(i) transport theoreticians focus on techniques that are more mathematically analyzable;

(ii) Monte Carlo transport practitioners focus on getting good results whether or not the techniques used are easily mathematically analyzable, provided the techniques provably preserve the mean estimates.

Note that if the state space is formally augmented to include the particle weight $A = ((\mathbf{r}, \mathbf{\Omega}, E, t), w)$, then most current Monte Carlo transport calculations are Markov processes in the augmented space. This augmentation does not remove the mathematical difficulties associated with analyzing weight-dependent random walks. The key mathematical advantage to analyzing weight-independent random walks is that the score from walks with one set of variance reduction parameters can often be written in terms of the score derived with another set of variance reduction parameters. With weight-dependent random walks, it is typically impossible, or at the very least impractical, to write the score from walks with one set of variance reduction parameters in terms of the score derived with another set of variance reduction parameters.

As a simple example of analyzing a weight-independent random walk process, let $p(s, w)ds$ be the probability that a particle of weight w at $\mathbf{P}$ contributes a score s in ds after it departs $\mathbf{P}$. (Note that $p(s, w)$ is generally the score density from a *non-analog* transport process after departing $\mathbf{P}$.) If the transport is weight-independent, then a particle of weight aw $(a > 0)$ will contribute a score as in the interval $d(as)$ with the same probability. That is,

$$p(as, aw)d(as) = p(s, w)ds. \tag{10.3}$$

The rth moment of the score distribution is

$$M_r = \int p(s, w)s^r\, ds. \tag{10.4}$$

Now consider splitting the particle into two particles, each of weight $w/2$. For the split case, the rth moment of the score distribution is

$$S_r = \int\int (s_1 + s_2)^r p(s_1, w/2)\ p(s_2, w/2)\ ds_1\ ds_2. \tag{10.5}$$

Without knowledge of what Monte Carlo techniques (e.g., a weight window) are used subsequently, it is impossible to know what $p(s, w/2)$ is. If the walks are known to be weight-independent, however, then

$$p(s, w/2)ds = p(2s, w)d(2s) \tag{10.6}$$

and, letting $x_i = 2s_i$,

$$S_r = \int\int \left(\frac{x_1 + x_2}{2}\right)^r p(x_1, w)\ p(x_2, w)\ dx_1\ dx_2. \tag{10.7}$$

From this (noting that $\int p(x, w)dx = 1$ because p is a probability density)

$$S_1 = \frac{1}{2}\int x_2 p(x_2, w)\, dx_2 + \frac{1}{2}\int x_1 p(x_1, w)\, dx_1 = \int xp(x, w)\, dx = M_1. \tag{10.8}$$

This shows that the mean is the same with or without the splitting. Similarly,

$$S_2 = \frac{1}{4}\int\int (x_1^2 + 2x_1x_2 + x_2^2)p(x_1, w)\, p(x_2, w)\, dx_1\, dx_2 = \frac{1}{2}\left(M_2 + M_1^2\right) \tag{10.9}$$

The knowledge of the statistical properties (M_r) of one Monte Carlo calculation thus allows one to predict the statistical properties (S_r) of a different Monte Carlo calculation. Thus, the statistical properties using one set of Monte Carlo variance reduction parameters can be empirically derived via simulation and then the statistical properties of calculations with other parameters can be inferred. Note that this analysis cannot be done without using the weight independence relationship of (10.3).

10.6 Analog Monte Carlo neutron transport steps

An analog Monte Carlo transport calculation consists of the following basic steps:

1. **Begin the nth source particle history.**

2. **Sample a source particle.**
   That is, sample a particle's phase-space coordinates $(\mathbf{r}, \mathbf{\Omega}, E, t) =$ (position, direction, energy, time) from $Q(\mathbf{r}, \mathbf{\Omega}, E, t)$.

3. **Sample the distance η to the particle's next event.**
   Let b be the distance to the boundary in direction $\mathbf{\Omega}$. The distance to the next event, η, is sampled using the exponential distribution $\sigma e^{-\sigma\eta}$. (Here, σ is the physical interaction probability per unit length derived from the nuclear data and the exponential distribution arises naturally from the physics.) If $\eta < b$, then the particle collides before reaching the boundary, otherwise the particle is put on the boundary. Move the particle a distance η in the particle direction $\mathbf{\Omega} = \mathbf{v}/v$, where $\mathbf{v}$ is the particle velocity and v is the particle's speed. That is, increment the position by $\mathbf{r} + \eta\mathbf{\Omega} \to \mathbf{r}$. Increment the time by $t + \eta/v \to t$. (Note that the energy, E, is constant between collisions.)

   If an interior boundary is reached (e.g., an interface between concrete and iron regions), then go to step 3.

   If an exterior boundary is reached (e.g., the particle escapes the solar system), go to 5.

   If no boundary is reached, the particle has collided. Go to step 4.

4. **Sample the collision.**
   Sample the collision nuclide i (e.g., carbon-12, carbon-13, nitrogen-14, iron-56).

   Sample for capture (termination of the particle when the collision nuclide removes the particle) vs survival (i.e., non-capture) on nuclide i. If capture occurs, go to step 5. If the particle survives the collision, then sample for the jth type of interaction on nuclide i (e.g., elastic scatter, inelastic scatter, fission).

   If necessary, sample the number of output particles K for interaction j on nuclide i. (For example, an elastic scatter always has exactly has one particle coming out of a collision, but the number of particles coming out of a fission event is random.) For interaction j (e.g., fission) on nuclide i (e.g., uranium-235), sample the output phase-space coordinates for the K output particles from a probability law

   $$C_{i,j}\left((\mathbf{r}, \mathbf{\Omega}, E, t) \rightarrow (\mathbf{r}_1, \mathbf{\Omega}_1, E_1, t), \ldots, (\mathbf{r}_\mathbf{K}, \mathbf{\Omega}_K, E_K, t)\right)$$

   (Note that the collision is instantaneous and the time does not change.)

   If $K > 1$, save $K - 1$ particles in a bank to process later.

   Go to step 3.

5. **Check for banked particles.**
   If there are particles waiting in the bank to be processed, take one from the bank and go to step 3.

   If there are no particles in the bank, go to step 6.

6. **End the nth source particle history.**
   Process the statistical results for the nth source particle. For example, if the nth source particle history contributes s_n, increment the m moment sums (for transport, usually $m = 1, 2, 3,$ and 4)

   $$S_m + s_n^m \rightarrow S_m$$

   so that error estimates can be made.

   Go to step 1 and sample a new source particle.

10.7 Intuitive ideas of variance reduction

In the analog simulation described in the previous section, every simulated particle represents one physical particle and the simulated particles are subject to the same probability laws as the physical particles. The number of particles penetrating a shield, for example, can be estimated by starting N simulated particles from the source and tallying the number penetrating the shield. Often, the number of particles in a physical source is very large, say more than 10^{15}, so N is typically much smaller than the physical number of particles. Because the process is a linear one, if N particles produce k penetrating particles, one can infer (at the end of the simulation) that a physical source of Q particles will produce $Q(k/N)$ penetrating particles.

It is often convenient to do this normalization at the beginning of the transport procedure rather than at the end. (The physical source strength, Q, might be the number of neutrons produced by a nuclear reactor over some time period, for example, or the number of X-rays from a medical imaging machine.) The user specifies a source strength Q and then, instead of assigning weight 1 to the computer particle, representing one physical particle, assigns a weight Q to the computer particle, representing Q physical particles.

Once the user starts viewing a computer particle as w physical particles, then it is a short intuitive leap to variance reduction. For instance, a 2:1 split can be interpreted as separating the w particles into two groups of particles each having $w/2$ particles. The average physics is preserved because there still are w physical particles after the split. In practice, this concept is generalized and as long as the *expected* weight (i.e., the expected number of physical particles) is preserved, then the average physics is preserved and the simulation means will be unbiased. This chapter will rely on this intuitive concept, though proving this intuitive notion is not always trivial [12, 8].

10.8 A sample rare event transport problem

Most Monte Carlo transport practitioners seem to learn variance reduction principles and methods most easily by applying them to sample transport problems. This chapter will consider some aspects of a simple rare event transport problem [3] that is commonly used for training purposes at Los Alamos.

The transport problem is shown in Figure 10.1. The geometry is cylindrically symmetric about the y-axis. There is a point isotropic neutron source at the bottom of a 200-cm diameter concrete cylinder of height 180 cm. The source energy spectrum is 95% at 2 MeV and 5% at 14 MeV. (Note that 2 MeV is typical of fission neutron energies and 14 MeV is typical of fusion neutron energies.) Any neutron exiting the cylindrical surface is immediately terminated by a perfect absorber in the region labeled 'zero importance'.

The primary tally (labeled F5) of interest is the time integrated neutron flux (particle density times speed per unit volume) at the detector point $D = (200, 0, 0)$. (Note that the probability that a neutron scatters directly at D is zero, but the 'point detector flux' estimate can be derived as the average flux in a small sphere that has shrunken to zero radius [31].) The detector only responds to particles above 0.01 MeV.

To contribute to the F5 tally, a neutron must:

(i) penetrate 180 cm of concrete;

(ii) leave the top of the concrete cylinder with a direction close enough to the cylinder axis that the neutron goes almost straight up the cylindrical void cell and crosses into the small low-density concrete (0.0203 g/cc) cell;

(iii) collide in the low density concrete cell (because point detector contributions are made only from collision/source points); and

(iv) have energy above 0.01 MeV.

These events are unlikely because:

(i) 180 cm of 2.03 g/cc concrete is difficult to penetrate;

(ii) there is only a small solid angle up the cylindrical void;

(iii) not many collisions will occur in 10 cm of the low-density (0.0203 g/cc concrete); and

(iv) particles lose energy penetrating the concrete.

10.9 The exponential transform

Consider a particle in a cell of constant material properties. The probability of going a distance η without collision is

$$e^{-\sigma\eta}, \tag{10.10}$$

where σ is the interaction probability per unit length for the material, called the 'cross-section' in nuclear parlance. That is, the probability of colliding in any interval $d\eta$ is $\sigma d\eta$.

If there is a boundary at a distance b along the flight path of a particle of weight w then the average weight reaching the boundary is

$$we^{-\sigma b}, \tag{10.11}$$

and the average weight colliding in interval $d\eta$ at a distance $\eta < b$ is

$$w\sigma e^{-\sigma\eta}d\eta. \tag{10.12}$$

The exponential transform biases the sampling of the distance the particle travels before collision. Instead of using the true cross-section σ, the sampling is done with a fictitious cross-section $\tilde{\sigma}$. Preserving the expected weight reaching the boundary requires a weight multiplication w_b such that

$$w_b we^{-\tilde{\sigma}b} = we^{-\sigma b}, \quad \text{i.e.,} \quad w_b = e^{-(\sigma-\tilde{\sigma})b}. \tag{10.13}$$

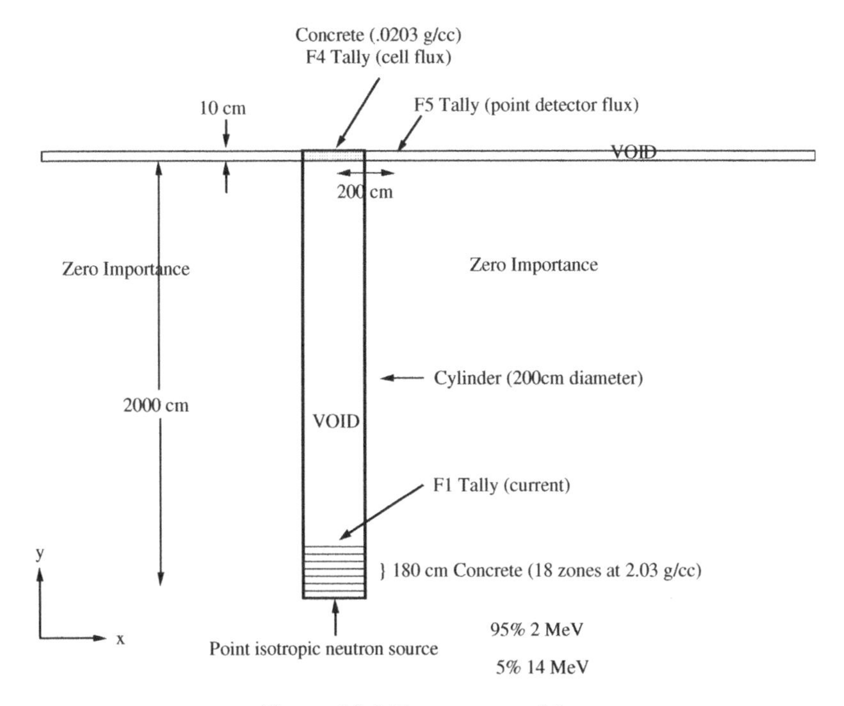

Figure 10.1 Transport problem.

Preserving the expected weight colliding in $d\eta$ about η requires a weight multiplication w_c such that

$$w_c w \tilde{\sigma} e^{-\tilde{\sigma}\eta} d\eta = w\sigma e^{-\sigma\eta} d\eta, \quad \text{i.e.,} \quad w_c = \frac{\sigma}{\tilde{\sigma}} e^{-(\sigma-\tilde{\sigma})\eta}. \tag{10.14}$$

The typical choice for a penetration problem like this is

$$\tilde{\sigma} = \sigma(1 - p\mu), \tag{10.15}$$

where p is the exponential transform parameter and μ is the cosine of the particle's direction relative to the penetration direction ($\hat{y}$ here). For $p > 0$, the cross-section is reduced for particles moving forward, resulting in longer jumps between collisions. Conversely, particles moving backward will have smaller jumps between collisions.

Table 10.1 shows results from MCNP (with 5 million samples) for the average weight penetrating the concrete cylinder (F1 tally in Figure 10.1). (To indicate when no estimate is available, MCNP displays a '0'.) Some things to note:

1. With no transform, or with $p = 0.1$, none of the particles penetrate, resulting in a zero mean.

2. The transform seems to be doing its job preferentially sampling particles deep into the shield.

Table 10.1 Results with different exponential transform biasing parameters

Transform parameter	Mean $\times 10^{-8}$	Fractional error	Variance of the variance	Tail slope	Figure of merit	Time (minutes)
0 (analog)	0.0000	0.0000	0.0000	0.0	0.0	32.17
0.1	0.0000	0.0000	0.0000	0.0	0.0	36.79
0.2	4.3760	0.6676	0.6682	0.0	0.06	35.82
0.3	3.9866	0.3713	0.2023	0.0	0.18	39.85
0.4	3.3257	0.2833	0.2219	0.0	0.32	38.93
0.5	3.2646	0.2515	0.3971	0.0	0.41	38.97
0.6	4.1875	0.4156	0.8391	0.0	0.14	41.86
0.7	2.9717	0.1502	0.1890	2.2	1.1	42.27
0.8	2.8228	0.2492	0.5888	1.9	0.37	43.71
0.9	2.3219	0.3310	0.8525	1.9	0.20	46.71
0.99 (poor)	1.7534	0.3639	0.4461	1.7	0.17	45.01
50 million						
0.7	4.5135	0.1670	0.3561	2.4	0.087	410.23

3. The time is increasing with p. This is understandable because fewer of the particles exit the $y = 0$ surface of the cylinder and thus the typical particle has more collisions in the biased game.

4. With one exception, the mean is decreasing with increasing p. This phenomenon occurred regularly enough that the exponential transform was sometimes called the 'dial an answer' technique at Los Alamos, because the mean seemed to depend on the user's choice of p. The sample mean will converge to the true mean after enough samples, but the user may not be able to afford that many samples; this is a rare event sampling problem that is occurring.

5. The fractional errors (standard deviation of the mean divided by the mean) are all quite high. The general rule of thumb is that means are not believable until the error is less than 10%.

6. The 'variance of the variance', $\mathrm{Var}(\mathrm{Var}(X)) / (\mathrm{Var}(X))^2$, values are all quite high. The general rule of thumb is that variance of the variance should be below 0.1 for a reliable calculation.

7. The tail slope [23, 31] is a measure of how many moments exist. A tail slope of k indicates that $k - 1$ moments exist. MCNP requires at least 500 non-zero history scores before attempting to estimate the slope; this accounts for the 0 slope estimate for the smaller p entries.

8. The figure of merit (fom) is $1/(error^2 \times time)$ and is a measure of the efficiency of the calculation.

9. The last line shows the result at 50 million histories for what looked like the best parameter at 5 million histories. Note that the fractional error estimate

has *increased* upon a tenfold increase in histories. This is indicative of poor sampling of rare events. Additionally, note that the variance of the variance has increased when it should be decreasing, and the figure of merit has decreased when it should be constant. Thus, this calculation would be deemed unreliable by almost all transport practitioners.

One of the features of a well-designed Monte Carlo transport code is that it provides large amounts of summary information about the sampling in addition to the tallies. As one example, MCNP gives the largest history score (for each tally) and its associated history number, so that the history can be rerun to produce an 'event log' that lists everything that happened during the history. The event log allows the user to investigate why a history scored so much. For the exponential transform, one finds an occasional particle that collided many times and was subject to numerous weight multiplications and now has a very large weight compared to other particles in the same phase-space region. This is not good.

10.10 The weight window

Consider two otherwise identical particles with weights $w_1 = w$ and $w_2 = 100w$ at the same phase-space point P. For weight-independent transport, particle 2 will contribute about 10 000 times as much to the second moment as particle 1. The time required to sample a collision is independent of the weight, so one is spending the same amount of time on particle 1 as particle 2, despite the fact that particle 2 contributes roughly 10 000 times as much to the variance. This makes no computational sense, so a 'weight window' is enforced so that all particles in a given region have roughly the same weight.

The weight window in a region consists of a lower weight w_l, a survival weight w_s, and an upper weight w_u such that $w_l < w_s < w_u$; typically, $w_s = 3w_l$ and $w_u = 5w_l$. If $w < w_l$ then the particle is rouletted and survives with probability w/w_s and weight w_s, or is killed. If $w_l \leq w \leq w_u$, no action is taken. If $w > w_u$ then the particle is split by the minimum integer m such that $w_l \leq w/m \leq w_u$. (Note that $w_u \geq 2w_l$ is also required.)

The weight window lower bound in some region R is typically chosen to be inversely proportional to the average score generated after the particle enters region R. The rationale behind this is that every particle in the problem then has roughly the same expected score independent of where it is. This is often a good choice, though later we shall discuss some limitations of this choice. For now, note that standard practice is either to attempt to estimate this expected score via bookkeeping on the Monte Carlo process or to deterministically solve equations [32, 30] for this expected score function, usually called the 'adjoint' or 'importance' function in the transport literature. The deterministic methods are beyond the scope of this chapter, but a quick explanation of the Monte Carlo bookkeeping is possible.

The importance of a particle at a point **P** in phase space equals the expected score a unit weight particle will generate. Imagine dividing the phase space into a number of phase space 'cells' or regions. The importance of a cell then can be defined as the expected score generated by a unit weight particle after entering the cell. Thus, with a little bookkeeping, the cell's importance can be estimated with the 'weight window generator' as

$$\textit{Expected Score (Importance)} = \frac{TS}{TW} \tag{10.16}$$

Here TS is the total score because of particles (and their progeny) entering the cell and TW is total weight entering the cell. References [11, 3] have a nice graphical illustration of a sample importance estimation process based on three source particles.

Ideally, weight window regions are chosen small enough that the expected score between adjacent regions does not vary more than a factor of 2–4. When the expected score between adjacent regions is larger than 4, it is suggested that the user further subdivide the problem into more zones.

10.11 Exponential transform with weight window

The weight window generator was run for 5 minutes with a transform parameter of $p = 0.7$ and produced the weight window lower bounds in the 18 zones (from the bottom of the concrete to the top of the concrete) of 5.000E-01, 1.007E-01, 4.147E-02, 1.492E-02, 5.008E-03, 1.645E-03, 5.485E-04, 1.913E-04, 7.117E-05, 2.844E-05, 1.199E-05, 5.253E-06, 2.356E-06, 1.078E-06, 4.891E-07, 2.217E-07, 1.007E-07, and 4.610E-08. The survival weight and upper window bounds were 3 and 5 times the lower bounds respectively.

Table 10.2 shows results for 500 000 history runs with different transform parameters. Comparing Tables 10.1 and 10.2, note that the particle histories per minute are roughly a factor of 10 less with the weight window. On the other

Table 10.2 Exponential transform with weight window

Transform parameter	Mean $\times 10^{-8}$	Fractional error	Variance of the variance	Tail slope	Figure of merit	Time (minutes)
0.0 (analog)	4.2811	0.0335	0.0034	10.0	28	32.00
0.1	4.0905	0.0323	0.0041	8.0	26	36.70
0.2	4.1190	0.0308	0.0033	10.0	27	38.26
0.3	4.2112	0.0281	0.0023	10.0	34	37.49
0.4	4.1524	0.0272	0.0031	10.0	35	38.80
0.5	4.1000	0.0277	0.0042	4.9	34	38.54
0.6	4.0929	0.0264	0.0028	10.0	36	39.93
0.7	4.2260	0.0290	0.0058	4.7	30	39.97
0.8	4.0316	0.0334	0.0076	6.6	23	39.53
0.9	4.0399	0.0387	0.0101	5.9	17	38.67
0.99 (poor)	4.0306	0.0670	0.0446	3.2	6.1	36.59

hand, the efficiency (fom) is dramatically improved and every calculation with $p \leq 0.9$ passes MCNP's ten statistical checks ([23], see also Section 10.19). In particular, note that the tail slope estimate ($k > 4$) is indicating at least three finite moments for all calculations, so that use of the central limit theorem is possible, unlike the calculations in Table 10.1. The last line shows that even with a very poor choice of a very high transform parameter, the tail slope indicates that the variance is finite.

Elaborating a bit more, using the central limit theorem requires at least two finite moments. Eventually, the high score tail of the density $f(x)$ must decrease faster than $1/x^3$ (slope 3), or else the second moment, $\int x^2 f(x)\, dx$, will not be finite. Note that it is even better if the high score tail of the density $f(x)$ decreases faster than $1/x^5$ (slope 5), so that four finite moments exist. When four moments exist, not only is the central limit theorem applicable, but also the sample variance is usually a good estimate of the true variance used in the central limit theorem. Valid confidence intervals are thus much more likely when four moments exist.

A complete print (event log) of the largest scoring history for $p = 0.6$ indicates that one of the (relatively rare) high-energy 14 MeV source neutrons is responsible for the largest score. At this point, one would attempt to bias the sampling of the source so that more 14 MeV neutrons were sampled, with correspondingly smaller weights. One could proceed by using the weight window generator to estimate an energy-dependent weight window, then adjust the source energy bias so that the source particles are within their space-energy weight windows. Space does not permit description here, but see [3] for an illustration of source energy bias and space-energy weight windows.

10.12 Collision biasing

This chapter has shown how to get particles through a bulk concrete shield, but almost none of the penetrating particles will have just the right angle to stream up the void in Figure 10.1. Let the analog collision kernel be $C(E, \mathbf{\Omega} \to E', \mathbf{\Omega}')$ for a particle entering collision at $E, \mathbf{\Omega}$ and exiting at $E', \mathbf{\Omega}'$. (The collision occurs at one spatial point in an instant of time, so space and time variables do not change upon collision.) Using the expected score function $I(E', \mathbf{\Omega}')$ one would like to sample from the biased collision density (given that exactly one particle is known to exit the collision)

$$\tilde{C}(E, \mathbf{\Omega} \to E', \mathbf{\Omega}') = \frac{C(E, \mathbf{\Omega} \to E', \mathbf{\Omega}')I(E', \mathbf{\Omega}')}{\int\int C(E, \mathbf{\Omega} \to E'', \mathbf{\Omega}'')I(E'', \mathbf{\Omega}'')dE''d\mathbf{\Omega}''}. \tag{10.17}$$

The trouble is that sampling $\tilde{C}$ efficiently is not straightforward. Recall from the analog sampling section that the collision process is a complicated procedure that is composed of a number of steps. Note that there are numerous ways to go from $(E, \mathbf{\Omega} \to E', \mathbf{\Omega}')$. One has to sum, over all possible reactions, on all possible collision nuclides, the probability densities for scattering to $(E', \mathbf{\Omega}')$. Although it is not too difficult to calculate C for any given $(E', \mathbf{\Omega}')$, calculating C for all $(E'', \mathbf{\Omega}'')$ required to evaluate the denominator in (10.17) is not practical. For this reason, it is problematical to sample $\tilde{C}$; in practice Monte Carlo transport codes use other approaches.

The TRIPOLI code [13] sometimes samples an approximation to $\tilde{C}$ by sampling C n times and evaluating $I_k = I(E''_k, \mathbf{\Omega}'_k)$. The jth sample is picked with probability and weight multiplier

$$p_j = \frac{I_j}{\sum_{k=1}^{n} I_k} = \frac{\frac{1}{n} I_j}{\frac{1}{n}\sum_{k=1}^{n} I_k}, \qquad w_j = \frac{1/n}{p_j}. \tag{10.18}$$

Note that the denominator in the rightmost equality for p_j is a Monte Carlo estimate of the integral in (10.17). This 'is computationally expensive and therefore rarely used in TRIPOLI-4 runs' [17].

The MCBEND sampling [15] ignores C altogether and samples the biased collision output from

$$\tilde{p}(E, \mathbf{\Omega} \to E', \mathbf{\Omega}') = \frac{I(E', \mathbf{\Omega}')}{\int\int I(E'', \mathbf{\Omega}'') dE'' d\mathbf{\Omega}''}, \tag{10.19}$$

and the scattering physics associated with C is incorporated with the weight multiplication

$$w = \frac{C(E, \mathbf{\Omega} \to E', \mathbf{\Omega}')}{\tilde{p}(E, \mathbf{\Omega} \to E', \mathbf{\Omega}')}. \tag{10.20}$$

(The I function is a piecewise constant function in MCBEND and the importance function used is actually energy-independent as well.)

MCNP's 'dxtran' method [31] accomplishes angle biasing in tandem with an expected value penetration technique. The basic dxtran method consists of defining a spherical region of interest. The collision sampling proceeds as in Section 10.6, except that (after the collision nuclide, interaction type, and energy have been sampled) the dxtran method splits a particle into a dxtran particle that crosses the dxtran sphere (before its next collision) and a non-dxtran particle that does not. The non-dxtran particle is sampled the same way, with the same weight, as it would have been without dxtran, except that the non-dxtran particle is killed if it reaches the sphere surface. Thus, for events not including crossing the sphere, the weight of particles executing any next event is identical. The dxtran particle's angle is sampled from an arbitrary density $\tilde{p}(\mathbf{\Omega})$ (usually a constant and always non-zero only in the cone of directions toward the sphere) with a weight multiplication such that the expected weight is preserved,

$$w_m \tilde{p}(\mathbf{\Omega}) = p(\mathbf{\Omega}), \tag{10.21}$$

where $p(\mathbf{\Omega})$ is the unbiased density. Of the particles scattered at $\mathbf{\Omega}$, the fraction that arrive at the surface of the sphere is

$$e^{-\int_0^{S(\mathbf{\Omega})} \sigma(\eta)d\eta}, \tag{10.22}$$

where $S(\mathbf{\Omega})$ is the distance to the sphere in the sampled direction $\mathbf{\Omega}$. Thus the dxtran particle's weight at the sphere is

$$w_{dxtran} = \frac{p(\mathbf{\Omega})}{\tilde{p}(\mathbf{\Omega})} e^{-\int_0^{S(\mathbf{\Omega})} \sigma(\eta)d\eta}. \tag{10.23}$$

Note that the dxtran particle has weight zero from the collision point until it reaches the sphere; it makes no tallies as these are already accounted for by the non-dxtran particle. (MCBEND has a similar method called 'forced flight' [15].)

10.13 Applying dxtran

For the sample problem here, a 100-cm radius dxtran sphere is placed at (0, 2000, 0); that is, at the top boundary of the void cylinder in Figure 10.1. All the runs in Table 10.3 were 60-minute runs with the same generated window and a transform parameter $p = 0.6$. Without dxtran, the fom is 12 and with dxtran the fom is 9.6. The history variance is decreasing with dxtran, but the fractional error in the mean reported in Table 10.3 has increased. The problem is that dxtran takes too much time, 193 113 compared to 811 398 histories per hour.

Transport codes contain a wealth of sampling information beyond just the estimates themselves, and a look at this information indicates that the zones near the source put little weight on the dxtran sphere and the zones at the top of the concrete put weights of about 10^{-10} on the sphere. It thus makes little sense to follow particles whose weights get lower than 10^{-10}. The exponential factor in (10.23) is accumulated zone by zone by moving the dxtran particle through each zone. When the dxtran particle has been exponentially attenuated through a distance $S'(\mathbf{\Omega}) < S(\mathbf{\Omega})$ its final weight at the dxtran sphere is known to be

Table 10.3 Applying dxtran to improve tally F4 (60-minute runs per window and $p = 0.6$)

Histories	Run type	Mean $\times 10^{-15}$	Fractional error	Variance of the variance	Tail slope	Figure of merit
811398	no dxtran	7.0365	0.0372	0.0102	3.3	12
193113	dxtran	7.4593	0.0417	0.0060	10.0	9.6
341834	dxtran/RR1	7.3325	0.0315	0.0036	8.5	17
562564	dxtran/RR 1,2	7.4355	0.0253	0.0043	5.3	26

less than

$$w' = \frac{p(\boldsymbol{\Omega})}{\tilde{p}(\boldsymbol{\Omega})} e^{-\int_0^{S'(\boldsymbol{\Omega})} \sigma(\eta) d\eta}. \tag{10.24}$$

To save time, a roulette game was played when $w' < 10^{-10}$ because this was consistent with the typical dxtran weights from collisions near the top of the concrete cylinder. That is, the particle either survived with probability $w'/10^{-10}$ with weight 10^{-10} or was killed. This saves time because far fewer zones have to be tracked through. When this roulette game (RR1) is played, the number of histories increases to 341 834 and the fom is now more than without dxtran.

In the zones near the source the particles will be subject to a large exponential attenuation, so it makes little sense to start a dxtran particle, track it through a number of zones until its weight becomes too low, and then roulette the particle. If one knows this is happening (this is known from summary information from the transport code), then one can play a roulette game even before producing a dxtran particle. For each cell the user supplies a roulette probability q. If the dxtran particle survives the roulette game, it's weight is multiplied by q^{-1}, otherwise it is killed.

Near the top of the concrete cylinder there will be little attenuation so there we take $q = 1$. At the bottom we expect the attenuation factor to be at least 0.001, so we take $q = .01$ there. In between, we geometrically space them resulting in the q probabilities in the concrete cylinder: 0.01, 0.013, 0.017, 0.023, 0.030, 0.039, 0.051, 0.067, 0.087, 0.11, 0.15, 0.20, 0.26, 0.34, 0.44, 0.58, 0.76, 1.0. Note from Table [10.3] that this second roulette game increases the particles per hour to 562 564 and the fom improves to 26.

10.14 Forced collisions

To estimate the flux at a point (the F5 tally), at each collision a tally of

$$\text{F5} = \frac{p(\boldsymbol{\Omega})}{R^2} e^{-\int_0^R \sigma(\eta) d\eta} \tag{10.25}$$

is made where R and $\boldsymbol{\Omega}$ are the distance and direction from the collision site to the detector point. To make any F5 estimate, a collision must occur in the tiny low density concrete cell at the top of Figure 10.1. To increase the number of collisions in this cell, the particle is split into an uncollided weight fraction $e^{-\sigma L}$ (that goes the full distance L through the cell) and a collided weight fraction $1 - e^{-\sigma L}$. The collision point is sampled from the conditional density (given that a collision is known to occur in $0 < \eta < L$)

$$p_c(s) ds = \frac{\sigma e^{-\sigma \eta} d\eta}{1 - e^{-\sigma L}}. \tag{10.26}$$

This ensures that every particle track that enters makes a contribution to the F5 tally. The result is given in Table 10.4. The reference F5 tally without forced

Table 10.4 Final results applying forced collisions to improve tally F5 (60-minute run)

Histories	Tally type	Mean	Fractional error	Variance of the variance	Tail slope	Figure of merit
562564	ref F5	4.3727×10^{-18}	0.0330	0.0087	4.3	15
526600	tally F1	4.3196×10^{-8}	0.0263	0.0028	10.0	24
	tally F4	7.5105×10^{-15}	0.0259	0.0027	10.0	25
	tally F5	4.4836×10^{-18}	0.0259	0.0027	10.0	25

collision (first line) is from the run reported on the last line of Table 10.3. The fom improved from 15 to 25. This concludes the illustration of rare event simulation on a sample problem.

10.15 The comb

MCNP does not use the comb technique, so the comb has not been demonstrated on the sample problem just concluded. Nonetheless, the comb is an old transport method and has some interesting aspects. The comb [16] inputs K tracks (of the same history) of varying weights $w_i > 0$ with sum $W = w_1 + \ldots + w_K$ and outputs M tracks of constant weight $L = W/M$.

Suppose that one has K particles and one decides that one would rather follow M particles. The K particles are 'combed' into M particles using an M-toothed comb. Figure 10.2 shows a comb with $K = 6$ and $M = 4$. The length of the comb is the sum of the particle weights

$$W = \sum_{i=1}^{K} w_i. \tag{10.27}$$

The comb teeth are equally spaced, with the position of the teeth randomly (ξ is a uniform random number on (0,1)) selected as

$$t_m = \xi \frac{W}{M} + (m-1)\frac{W}{M}, \qquad m = 1, \ldots, M. \tag{10.28}$$

Each time a tooth hits interval i, the ith particle is duplicated and assigned a *postcombed* weight

$$w_i' = \frac{W}{M}. \tag{10.29}$$

Defining the integer j by

$$j < \frac{w_i}{W/M} \leq j + 1, \tag{10.30}$$

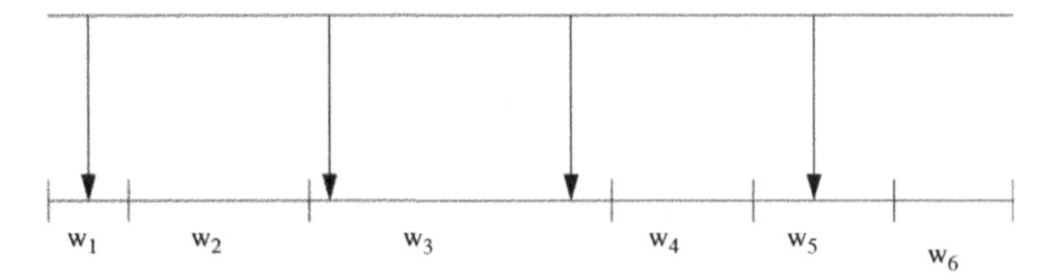

W = w1+w2+w3+w4+w5+w6

make 1 copy of particle 1 with weight W/M

make 0 copies of particle 2

make 2 copies of particle 3 with weight W/M

make 0 copies of particle 4

make 1 copy of particle 5 with weight W/M

make 0 copies of particle 6

Figure 10.2 Simple comb.

one sees that either j or $j+1$ teeth of a comb with a pitch of W/M will hit an interval of length w_i. In particular,

$$p_{i,j} = j + 1 - \frac{w_i}{W/M} \tag{10.31}$$

is the probability of j teeth in interval i, and

$$p_{i,j+1} = \frac{w_i}{W/M} - j \tag{10.32}$$

is the probability of $j+1$ teeth in interval i.

Let C_i be the total weight from the ith particle after combing. Then, using Eqs. 10.31 and 10.32, the expected weight after combing is

$$E[C_i] = p_{i,j} j \frac{W}{M} + p_{i,j+1}(j+1)\frac{W}{M} = \frac{w_i}{W/M}\frac{W}{M} = w_i. \tag{10.33}$$

Note that this combing *exactly* preserves the total weight because the output of the combing is always M particles of weight W/M.

One interesting aspect of combing is that the random walk of track i now depends on track j, whereas almost all Monte Carlo theory discussions assume that what happens to track i depends solely on track i's current phase-space position.

The comb is often employed to control populations of particles. For nuclear criticality problems, one often wants to set the number of particles in each fission generation to M. The comb provides a way to take the K particles at the end of a fission generation and produce M equally weighted particles whose total weight is the same as the precombed weight.

For rare event simulations, an importance weighted comb allows one to produce a set of particles where every particle in the set has equal expected tally. This is typically a much more computationally efficient distribution of weights for similar reasons as the weight window. Unlike weight windows, the comb has the additional advantage that the number of particles in the history can be constrained to any desired number.

10.16 Comments on using the importance function

Recall from Section 10.10 that the weight window at point $\mathbf{P}$ is typically chosen to be inversely proportional to the importance function at $\mathbf{P}$, which is the first moment of the score distribution for a particle at $\mathbf{P}$. This is interesting because one is attempting to minimize the second moment indirectly using information about the first moment.

Biasing using importance function information often works well if there is a relatively small spread in history scores. That is, each history contributes roughly the mean score. In this case, both the bulk of the mean and the bulk of the variance are produced by the same particles. That is, focusing on particles that contribute most to the second moment is similar to focusing on particles that contribute most to the mean. Indeed, in the limiting case of a zero variance solution all particles contribute exactly the mean score and there is no reason to consider the second moment at all.

The other side of the coin is that the available variance reduction techniques may not allow one to arrange the sampling so that there is a relatively small spread in history scores. In these cases, the set of particles contributing most of the variance may be very different from the set of particles contributing most of the mean. For instance, the set of typical particles that contribute 99% of the mean might only contribute 1% of the variance. In this case, focusing efforts on typical particles that score will not work very well because the typical particles are very different from the particles contributing most to the variance. One would do better to base the variance reduction on the second-moment equation rather than the importance (first-moment) equation.

It is sometimes useful to derive a second-moment equation (see [22, Ch. 5]) that is a function of some variance reduction games and their associated parameters. At this point, it is almost universal practice to consider only transport processes that are independent of particle weight. That is, suppose that $p(\mathbf{P}, w, s)ds$ is the probability that a particle of weight w at phase-space location **P** would contribute a score in ds about s in its subsequent random walk. One can write an equation for $p(\mathbf{P}, w, s)ds$ that depends on the variance reduction techniques used, but generically looks like

$$p(\mathbf{P}, w, s)ds = \int K(\mathbf{P} \to \mathbf{P}')p\left(\mathbf{P}', w', s - S(\mathbf{P} \to \mathbf{P}', w')\right) ds\, d\mathbf{P}'. \quad (10.34)$$

That is, the probability that a particle at **P** scores s in ds in its subsequent random walk is the probability $(K(\mathbf{P} \to \mathbf{P}')d\mathbf{P}')$ that the walk next goes to $\mathbf{P}'$, scoring $S(\mathbf{P} \to \mathbf{P}', w')$ times the probability, $p\left(\mathbf{P}', w', s - S(\mathbf{P} \to \mathbf{P}', w')\right) ds$, that the particle subsequently scores $s - S(\mathbf{P} \to \mathbf{P}', w')$ in ds so that the total score from **P** is $S(\mathbf{P} \to \mathbf{P}', w') + (s - S(\mathbf{P} \to \mathbf{P}', w') = s$. The rth-moment equation is then derived by integration of this probability function times s^r,

$$\begin{aligned} M_r(\mathbf{P}, w) &= \int s^r p(\mathbf{P}, w, s)ds \\ &= \int K(\mathbf{P} \to \mathbf{P}') \int s^r p\left(\mathbf{P}', w', s - S(\mathbf{P} \to \mathbf{P}', w')\right) ds\, d\mathbf{P}'. \end{aligned} \quad (10.35)$$

One can also derive an equation for the mean computing time as a function of the variance reduction games and their associated parameters. When this is done, one can maximize the fom directly by using a calculation with one set of variance reduction parameters to infer the behavior of the fom with different choices for those parameters. The paper by Burn [14], and many of the the references therein, maximize the fom by the 'direct statistical approach' (DSA method). The DSA method has been compared to the importance-based weight window method [9]. Both methods work well when there are no large changes in importance on a single transport step. When there is a large change in importance, importance-based splitting methods tend to oversplit because they do not factor in the time required to follow the extra split particles, as the DSA method does.

10.17 When to stop variance reduction efforts

There are a dozen or so commonly used variance reduction techniques in MCNP that can be used by themselves or in combination with other techniques. There are typically hundreds or thousands of parameters required to use some of these techniques (e.g., weight windows). Blindly changing parameters is almost never effective. Instead, code users are advised to get an 'event log' for the several largest scoring histories. The event log lists everything that happened in a history. If a history has made a particularly large score, then an event log can be used to identify the part of the random walk that is being poorly sampled. Once the user understands the poor sampling, the user can then make an informed guess at how to change the sampling via the variance reduction techniques. When none of the code's variance reduction techniques can impact the cause of the specific poor sampling, it is time to quit tinkering with variance reduction techniques and simply run the problem long enough.

10.18 The future

There are at least three areas, in various stages of development, that look promising for the future.

1. Deterministic transport methods can be used to determine variance reduction parameters for Monte Carlo transport codes [32, 30]. Currently people are using deterministic codes to solve for the importance function which is then used in the Monte Carlo calculation. In the future, deterministic codes presumably could also be used to solve second-moment and time equations, avoiding some of the pitfalls associated with optimizations based on the importance function alone (see Section [10.16]).
2. Adaptive Monte Carlo methods can accelerate the convergence rate instead of simply the reducing the coefficient C in the standard $C/\sqrt{N}$ convergence. These methods are being studied both inside and outside the field of transport (see [20, 10, 29] for an inexhaustive list). There has been limited success in the field of transport, but efforts continue.
3. The variance in Monte Carlo particle transport calculations is often dominated by a few particles whose importance increases manyfold on a single transport step. The 'ex post facto' method [7] is a novel variance reduction method that uses a large importance change as a trigger to resample the offending transport step. That is, the method is employed only after a random walk attempts a transport step that would otherwise introduce a large variance in the calculation. This method has been successfully tested in some difficult transport problems. The method is not limited to transport and appears generally applicable across many Monte Carlo fields.

10.19 Appendix: MCNP's ten statistical checks

MCNP records information on the mean, relative error, variance of the variance, figure of merit, and the tail slope behavior of the empirical score distribution to assist the user in evaluating the Monte Carlo results.

Statistical check for the mean:

- a non-monotonic behavior (no up or down trend) in the estimated mean as a function of the number histories N for the last half of the problem.

Statistical checks for the relative error (R) (i.e., standard deviation divided by the mean):

- an acceptable magnitude of the estimated R of the estimated mean (less than 0.05 for a point detector tally or less than 0.10 for a non-point detector tally);
- a monotonically decreasing R as a function of the number histories N for the last half of the problem;
- a $1/\sqrt{N}$ decrease in the R as a function of N for the last half of the problem.

Statistical checks for the variance of the variance (VOV):

- the magnitude of the estimated VOV should be less than 0.10 for all types of tallies;
- a monotonically decreasing VOV as a function of N for the last half of the problem;
- a $1/N$ decrease in the VOV as a function of N for the last half of the problem.

Statistical checks for the figure of merit (fom):

- a statistically constant value of the fom as a function of N for the last half of the problem;
- a non-monotonic behavior in the fom as a function of N for the last half of the problem.

Statistical check for the large score tail slope of the empirical score distribution ($f(x)$):

- the slope of the 25 to 201 largest positive history scores x should be greater than 3.0 so that the second moment $\int_{-\infty}^{\infty} x^2 f(x)dx$ will exist if the slope is extrapolated to infinity.

The seven N-dependent checks are for the last half of the problem. For example, one expects a roughly monotonic decrease in the relative error after 'enough' samples have been run. MCNP looks at the sample information from the last half of the samples ($N/2$ to N) and if the last half of the problem seems to have the correct dependence, then N is deemed large enough.

References

[1] G.I. Bell and S. Glasstone. *Nuclear Reactor Theory*. Van Nostrand, New York, 1970.

[2] T.E. Booth. Exact Monte Carlo solution of elliptic partial differential equations. *Journal of Computational Physics*, **39**: 396–404, 1981.

[3] T.E. Booth. A sample problem for variance reduction in MCNP. Technical Report LA-10363-MS, Los Alamos National Laboratory, 1985.

[4] T.E. Booth. Monte Carlo variance reduction approaches for non-Boltzmann tallies. Technical Report LA-12433, Los Alamos National Laboratory, 1992.

[5] T.E. Booth. A Monte Carlo variance reduction approach for non-Boltzmann tallies. *Nuclear Science and Engineering*, **116**: 113–124, 1994.

[6] T.E. Booth. Pulse height tally variance reduction in MCNP. Technical Report LA-13955, Los Alamos National Laboratory, 2002.

[7] T.E. Booth. Ex post facto Monte Carlo variance reduction. *Nuclear Science and Engineering*, **148**: 391–402, 2004.

[8] T.E. Booth. A transport process approach to understanding Monte Carlo transport methods. Technical Report LA-UR-04-1426, Los Alamos National Laboratory, 2004.

[9] T.E. Booth and K.W. Burn. Some sample problem comparisons between the DSA cell model and the quasi-deterministic method. *Annals of Nuclear Energy*, **29**(11): 733–765, 1993.

[10] T.E. Booth, J.A. Carlson, J.A. Favorite *et al.* Taking the next step with intelligent Monte Carlo. Technical Report LA-UR-00-4892, Los Alamos National Laboratory Report, October 1, 2001.

[11] T.E. Booth and J.S. Hendricks. Importance estimation in forward Monte Carlo calculations. *Nuclear Technology/Fusion*, **5**: 90–100, 1984.

[12] T.E. Booth and S.P. Pederson. Unbiased combinations of nonanalog Monte Carlo techniques and fair games. *Nuclear Science and Engineering*, **110**: 254–261, 1992.

[13] J.P. Both, A. Mazzolo, O. Petit, Y. Peneliau, and B. Roesslinger. User manual for version 4.3 of the TRIPOLI-4 Monte Carlo method particle transport computer code. Technical Report CEA-R-6044, CEA/Saclay, France, 2003.

[14] K.W. Burn. A new weight-dependent direct statistical approach model. *Nuclear Science and Engineering*, **125**: 128–170, 1997.

[15] S. Chucas and M. Grimstone. The acceleration techniques used in the Monte Carlo code MCBEND. *8th International Conference on Radiation Shielding, Arlington, TX, USA*, 25–28 April 1994.

[16] D.H. Davis. Critical-size calculations for neutron systems by the Monte Carlo method. Technical Report UCRL-6707, Lawrence Radiation Laboratory, December 12, 1961.

[17] C.M. Diop, O. Petit, E. Dumonteil, F.X. Hugot, Y.K. Lee, A. Mazzolo, and J.C. Trama. Tripoli-4: A 3D continuous-energy Monte Carlo transport code. *First International Conference on Physics and Technology of Reactors and Applications*, Marrakech, March 14–16, 2007, GMTR (2007).

[18] A. Dubi and D.J. Dudziak. Optimal choice of parameters for exponential biasing in Monte Carlo. *Nuclear Science and Engineering*, **70**: 1–13, 1979.

[19] B.L Fox. *Strategies for Quasi-Monte Carlo*. Kluwer Academic, Boston, 1999.

[20] S. Juneja and P. Shahabuddin. Rare event simulation techniques: An introduction and recent advances. In S.G. Henderson and B.L. Nelson, eds, *Simulation*, pp. 291–350. Elsevier, Amsterdam, 2006.

[21] P. L'Ecuyer and B. Tuffin. Splitting with weight windows to control the likelihood ratio in importance sampling. In *Proceedings of ValueTools 2006: International Conference on Performance Evaluation Methodologies and Tools*, Pisa, Italy, 2006. ACM Publications.

[22] I. Lux and L. Koblinger. *Monte Carlo Particle Transport Methods: Neutron and Photon Calculations*. CRC Press, Boca Raton, FL, 1991.

[23] S.P. Pederson, R.A. Forster, and T.E. Booth. Confidence intervals for Monte Carlo transport simulation. *Nuclear Science and Engineering*, **127**: 54–77, 1997.

[24] P.K. Sarkar and M.A. Prasad. Prediction of statistical error and optimization of biased Monte Carlo transport calculations. *Nuclear Science and Engineering*, **70**: 243–261, 1979.

[25] E Shuttleworth. The pulse height distribution tally in MCBEND. In *Proceedings of the International Conference on Radiation Shielding ICRS'9*, October 1999. Tsukuba, Japan.

[26] J. Spanier. An analytic approach to variance reduction. *SIAM Journal on Applied Mathematics*, **18**(1): 172–190, 1970.

[27] J. Spanier. A new multistage procedure for systematic variance reduction in Monte Carlo. *SIAM Journal on Numerical Analysis*, **8**(3): 548–554, 1971.

[28] J. Spanier and E.M. Gelbard. *Monte Carlo Principles and Neutron Transport Problems*. Addison-Wesley, Reading, MA, 1969.

[29] J. Spanier and R. Kong. A new adaptive method for geometric convergence. In H. Niederreiter, ed., *Proceedings of the Monte Carlo Quasi-Monte Carlo Methods 2002*, pp. 439–449., Springer, Berlin, 2004.

[30] J. Sweezy, F. Brown, T. Booth, J. Chiaramonte, and B. Preeg. Automated variance reduction for MCNP using deterministic methods. *Radiation Protection Dosimetry*, **116**: 508–512, 2005.

[31] X-5 Monte Carlo Team. MCNP: A general Monte Carlo N-particle transport code, Version 5. Technical Report LA-UR-03-1987, Los Alamos National Laboratory, April 24, 2003. Volume I: Overview and Theory.

[32] A. Wagner, J.C.; Haghighat. Automated variance reduction of Monte Carlo shielding calculations using the discrete ordinates adjoint function. *Nuclear Science and Engineering*, **128**(2): 186–208, 1998.

11

Rare event simulation methodologies in systems biology

Werner Sandmann

Compared to many other domains within which efficient Monte Carlo techniques for dealing with rare events are well established and have been successfully applied and advanced for decades, rare event simulation in systems biology is in its infancy. One major reason surely is that systems biology itself is still fairly young. Particularly relevant rare events in systems biology originate from the fact that different molecular reactions within the same biological or genetic network usually occur on multiple time scales, which means that their rates differ by orders of magnitude. Reactions with extremely small rates are rare events. This yields the problem of stiffness which poses serious difficulties in solving systems of differential equations as well as in performing efficient stochastic simulations. We give a prospective survey of approaches coping with the simulation of stiff Markovian models arising in systems biology. It becomes apparent that further investigation of rare event simulation in systems biology offers great potential for improvements, for instance by adapting techniques from other domains and possibly combining them with those that are already in use for biological systems.

Rare Event Simulation using Monte Carlo Methods Edited by G. Rubino and B. Tuffin

11.1 Introduction

Systems biology is a rapidly emerging field that investigates intra- and inter-cellular dynamics from a system-oriented point of view. As such, it is highly interdisciplinary and combines theoretical, experimental and computational techniques from mathematics, computer science, and engineering with those from physics, chemistry, molecular and cell biology. Biological systems, in the same way as artificial or technical ones, consist of mutually related components that interact with each other and the system environment. They are formidably complex, and developing computationally tractable models is desirable and necessary in order to gain insights and predict their behavior. The reader is referred to [3, 41, 42] for expositions of general concepts and methods in systems biology, and more specifically to [11, 74] for a focus on computational modeling and system-theoretical aspects.

The basic building blocks at the molecular level of biological systems are coupled biochemical reactions between different molecular species. The fundamental rules are given by stoichiometric equations defining which molecular species may react in order to result in a certain product and how many molecules are involved in the reaction. Quantitative timing aspects are specified by reaction rates assigned to each reaction. Sets of such molecular reactions constitute biological and genetic networks, also referred to as pathways, which are enormously complex as the set of reactions is huge.

Different (but related) modeling paradigms reflect different viewpoints or focuses. As usual, models are distinguished in terms of their states as well as by the rules driving the possible state changes and governing the system dynamics. They may be state-continuous or state-discrete, continuous or discrete in time, deterministic or stochastic. For a long time the most common, somehow canonical approach was deterministic, where the system state at any time is given by the concentrations (measured in moles per liter) of each molecular species. By the law of mass action, expressing the system dynamics in terms of the chemical rate equations yields a system of ordinary differential equations (ODEs) for the concentrations of molecular species. However, this properly reflects neither the discreteness of molecular quantities nor the stochastic nature of chemical reactions. The evidence of inherent randomness has been reported and pointed out in many studies, and stochastic models are today well established in system biology; see [6, 9, 32, 46, 47, 75, 79] to mention but a few particularly notable publications dealing specifically with stochastic models.

The most widespread stochastic modeling approach to biochemically reacting systems that we adopt in this chapter is by continuous-time Markov chains (CTMCs), whose dynamics are described in terms of the chemical master equation. As in other domains, the largeness of the models makes stochastic simulation a particular prominent analysis technique. Accelerating simulations is motivated by the fact that stochastic simulations in general are computationally costly, meaning that the advantage of being able to deal with large models has to be offset by the significant amount of computer time required. Rare events come into play since almost all biological systems possess stiffness. Hence, a major rare event simulation issue is accelerated simulation of stiff models.

11.2 Markovian modeling of biological systems

Markovian modeling of chemically reacting systems has a long tradition that can be traced back to the study of autocatalytic reactions in the 1940s [29]. In the 1950s, Singer [72] considered chain reactions and some types of coupled reactions, and Bartholomay [7] provided a large body of theory resulting in a series of papers on topics covering sequences of unimolecular and bimolecular reactions, reaction rate constants, and several applications. While using a different terminology, these early works already implicitly included the chemical master equation which is well known today and is equivalent to the Kolmogorov differential equations. Detailed reviews of the early literature with many more references can be found in [8, 49, 77]. It was also recognized quite early that in the thermodynamic limit, when the number of molecules and the volume approach infinity but the concentrations remain finite, the Markovian and the deterministic paradigm are equivalent [43, 51]. A physical justification of stochastic modeling of coupled chemical reactions by CTMCs was provided in the 1970s by Gillespie [34, 35] and later rigorously derived in [36] yielding that it is evidently in accordance with the theory of thermodynamics.

The basic assumption in stochastic modeling of biochemically reacting systems via CTMCs as well as in deterministic modeling by systems of ODEs is that the system is well stirred and thermally equilibrated, meaning that a well-stirred mixture of molecules inside some fixed volume will interact at constant temperature. To put it another way, the system is spatially homogeneous such that the numbers or concentrations of molecules do not depend on positions in space.

11.2.1 Model notation and terminology

We consider $d \in \mathbb{N}^+$ molecular species $S_1, \ldots, S_d$ and $M \in \mathbb{N}^+$ possible reactions $R_1, \ldots, R_M$, also referred to as reaction channels. Each reaction R_m is basically defined by a corresponding stoichiometric equation,

$$s_{m_1} S_{m_1} + \cdots + s_{m_r} S_{m_r} \longrightarrow s_{m_{r+1}} S_{m_{r+1}} + \cdots + s_{m_\ell} S_{m_\ell}, \quad r, \ell \in \mathbb{N},\ r \leq \ell, \tag{11.1}$$

where $S_{m_1}, \ldots, S_{m_r}$ are reactants, $S_{m_{r+1}}, \ldots, S_{m_\ell}$ are *products*, and the numbers $s_{m_1}, \ldots, s_{m_\ell} \in \mathbb{N}$ are *stoichiometric coefficients*. Such an equation simply expresses that the arrow's left-hand side can be transformed to its right-hand side if the required number of reactant molecules collide. The overall number of reactant molecules, $s_{m_1} + \cdots + s_{m_r}$, is called the order[1] of the reaction. Although equation (11.1) formally allows an arbitrary number of molecules, it is usually

[1] We consider discrete states described by the number of molecules of all species. If concentrations are considered, the order of a reaction is the sum of concentration exponents in the rate law for the reaction, and non-integer orders are possible.

not realistic to consider higher than third-order reactions as this would require a collision of more than three molecules. Even third-order reactions are often more realistically modeled as a pair of second-order reactions. Note that zero-order reactions do make sense. They are useful to model, for instance, synthesis without a stimulus, or any influence from 'outside' the system.

Modeling requires an appropriate specification of system states and system dynamics. In the CTMC approach, for each molecular species S_k at any time $t \geq 0$ a discrete random variable $X_k(t)$ describes the number of molecules of species S_k present at time t. The system state at time t is the discrete d-dimensional random vector $X(t) = (X_1(t), \ldots, X_d(t))$, and the set $\mathcal{X} \subseteq \mathbb{N}^d$ of all possible system states constitutes the system's state space. The conditional transient (time-dependent) probability that the system is in state $x \in \mathcal{X}$ at time t, given that the system starts in an initial state $x_0 \in \mathcal{X}$ at time t_0, is denoted by

$$p^{(t)}(x) := p^{(t)}(x \mid x_0, t_0) = P(X(t) = x \mid X(t_0) = x_0). \tag{11.2}$$

The system changes its state due to one of the possible reactions. For each reaction R_m the reaction rate is given by a state-dependent function α_m, called the *propensity function* of R_m where $\alpha_m(x)dt$ is the conditional probability that a reaction of type R_m occurs in the infinitesimal time interval $[t, t + dt)$, given that the system is in state x at time t. That is,

$$\alpha_m(x)dt = P\left(R_m \text{ occurs in } [t, t + dt) \mid X(t) = x\right). \tag{11.3}$$

Each reaction R_m has an assigned stochastic reaction rate constant[2] c_m such that the propensity function is simply given by c_m times the number of possible combinations of the required reactants and thus computes as

$$\alpha_m(x) = c_m \cdot \prod_{j=1}^{m_r} \binom{x_{m_j}}{s_{m_j}}, \tag{11.4}$$

where x_{m_j} denotes the number of molecules of species S_{m_j} present in state x, and s_{m_j} is the stoichiometric coefficient of S_{m_j} according to the relevant reaction's stoichiometric equation (11.1).

The probability that a reaction occurs within a specific time interval only depends on the length of this interval. Thus, the propensity functions are time-independent. Besides, given a current system state, the next state in the system's time evolution only depends on this current system state and neither on the specific time nor on the history of reactions that led to the current state. Hence, the system is in fact modeled as a (time-homogeneous, conservative) CTMC $(X(t))_{t \geq 0}$ with d-dimensional state space $\mathcal{X} \subseteq \mathbb{N}^d$.

The terminology and notation as introduced here and commonly used in biology as well as in chemistry and physics are quite different from those in

[2] Note that the CTMC modeling approach and the deterministic ODE approach are closely related, not only in the thermodynamic limit. The stochastic reaction rate constant can be easily converted to/from the rate constant provided by the law of mass actions. Only the system's volume and the Avogadro number must be appropriately taken into account; see [34, 80] for details.

mathematics, computer science, and engineering. In particular, they constitute a purely functional specification as opposed to an algebraic matrix specification. Consequently, expressions governing the system dynamics usually adhere to one of these specifications and at a first glance may appear to be rather different but of course they are equivalent. More specifically, the chemical master equation is one way to write the Kolmogorov differential equations.

11.2.2 Chemical master equation and Kolmogorov differential equations

For each reaction R_m denote by $v_m = (v_{m1}, \ldots, v_{md})$ the state change vector where v_{mk} is the change of molecules of species S_k due to a reaction of type R_m. Then, given that the system starts in an initial state $x_0 \in \mathcal{X}$ at time t_0, the system dynamics in terms of the state probabilities' time derivatives are described by the *chemical master equation*

$$\frac{\partial p^{(t)}(x)}{\partial t} = \sum_{m=1}^{M} \left(\alpha_m(x - v_m) p^{(t)}(x - v_m) - \alpha_m(x) p^{(t)}(x) \right). \tag{11.5}$$

In order to recognize the equivalence of the functional specification and a matrix specification note that the multidimensional discrete state space can be mapped to the set $\mathbb{N}$ of non-negative integers such that each state $x \in \mathcal{X}$ is uniquely assigned to an integer $i \in \{1, \ldots, |\mathcal{X}|\}$. The probability that a transition from state $i \in \mathbb{N}$ to state $j \in \mathbb{N}$ occurs within a time interval of length $h \geq 0$ is denoted by $p_{ij}(h)$, and correspondingly $\mathbf{P}(h) = (p_{ij}(h))_{i,j \in \mathbb{N}}$ is a stochastic matrix, where $\mathbf{P}(0)$ equals the unit matrix $\mathbf{I}$, since no state transitions occur within a time interval of length zero. It is well known (cf. [12, 77]) that a CTMC is uniquely defined by an initial probability distribution and a transition rate matrix, also referred to as infinitesimal generator matrix, $\mathbf{Q} = (q_{ij})_{i,j \in \mathbb{N}}$, consisting of transition rates q_{ij} where $\mathbf{Q}$ is the derivative at 0 of the matrix function $h \mapsto \mathbf{P}(h)$. The relation of each $\mathbf{P}(h)$ to $\mathbf{Q}$ is given by $\mathbf{P}(h) = \exp(h\mathbf{Q})$. In that way $\mathbf{Q}$ generates the transition probability matrices by a matrix exponential function which is basically defined as an infinite power series. Hence, all information on transition probabilities is covered by the single matrix $\mathbf{Q}$. In terms of $\mathbf{P}$ and $\mathbf{Q}$ the *Kolmogorov forward differential equations*, the *Kolmogorov backward differential equations*, and the *Kolmogorov global differential equations* can respectively be expressed as

$$\frac{\partial}{\partial t}\mathbf{P}(t) = \mathbf{P}(t)\mathbf{Q}, \quad \frac{\partial}{\partial t}\mathbf{P}(t) = \mathbf{Q}\mathbf{P}(t), \quad \frac{\partial}{\partial t} p^{(t)} = p^{(t)}\mathbf{Q}, \tag{11.6}$$

where $p^{(t)}$ denotes the vector of the transient state probabilities corresponding to (11.2). Explicitly writing the Kolmogorov global differential equations in terms of the coefficients and some algebra yields

$$\frac{\partial p_i^{(t)}}{\partial t} = \sum_{j:j \neq i} p_j^{(t)} q_{ji} - \sum_{j:j \neq i} p_i^{(t)} q_{ij} = \sum_{j:j \neq i} \left(p_j^{(t)} q_{ji} - p_i^{(t)} q_{ij} \right). \tag{11.7}$$

The equivalence of the chemical master equation and Kolmogorov differential equations can now be easily seen by interpreting $i \in \mathbb{N}$ as the number assigned to state $x \in \mathcal{X}$, that is, $p_i^{(t)} = p^{(t)}(x)$, $q_{ij} = \alpha_m(x)$ if j is the number assigned to state $x + v_m$, and $q_{ji} = \alpha_m(x - v_m)$ if j is the number assigned to state $x - v_m$.

11.2.3 Model structure and rare event issues

Although we are concerned with Markovian models, the specific structure of biological systems as well as the practically relevant topics largely disable direct one-to-one mappings from other domains to systems biology. Markovian models of biological and genetic networks are almost always potentially infinite–in either case the state space is multidimensional and extremely huge. They are typically not ergodic and contain absorbing states or absorbing classes since some molecular species might get exhausted, which often implies that no more reactions are possible at all or only reactions that do not lead out of a certain set of states. Absorption times can be important such as the time to apoptosis (cell death) or the time to viral infection. Stationary distributions are less relevant than in many other domains. They are relevant, for example, in the case of multiple absorbing classes, meta- or bistable systems such as bistable genetic switches, but transient phenomena are often of primary interest.

In general, transient analysis tends to be much more difficult than steady-state (limiting to infinite time) analysis. Moreover, in biological and genetic network models, except for zero-order reactions, we are always concerned with state-dependent rates given by the propensity functions. This is a significant difference from, for example, the vast majority of queuing models where constant arrival and service rates are rather usual. Inherent to biochemically reacting systems are multiple time scales. There are fast and slow reactions with rates differing by many orders of magnitude such that the system is stiff, which makes numerical analysis cumbersome both in the domain of ODEs and in the domain of CTMCs.

Stiffness also causes serious problems in stochastic simulation. Consequently, efficient simulation of stiff Markovian models is a major challenge in stochastic modeling and analysis of biological systems. Obviously, reactions with very small rates are rare events. Within the particular framework of huge and stiff Markovian models, some transition rates might be unknown. In practice, models are built according to real-world observations which in the context of biological systems often correspond to experiences and conclusions drawn from lab experiments. If the rates of specific reactions are unknown but empirical data is available, rare event simulation can be suitable for estimating small reaction rates. Therefore, estimating unknown rare event probabilities or rates is one topic of potentially high relevance. Even if the actual goal of the study is not the determination of the reaction rates, these rates are required for an appropriate model. Then rare event simulation becomes part of model building.

Almost all stochastic simulation techniques that have been applied in systems biology originate from chemical physics, which is quite natural against the background of molecular reactions driving biological and genetic networks. At present, stochastic simulation approaches to estimating unknown reaction rates or rates of reactive pathways between metastable regions almost exclusively appear in chemical physics. However, since the models are similar these methods are applicable in systems biology. Many of the algorithms in use are variations around importance sampling or closely related ideas; see [16, 28, 31]. For advanced, well-worked-out approaches we refer the interested reader in particular to transition path sampling [10, 30] and forward flux sampling [1, 2, 50, 76]. Presumably, sooner or later they will become more widespread and make their way into systems biology just like methods for simulating stiff models with known reaction rates.

Our major focus is on efficient simulation techniques for stiff models with all rates known. Rather than aiming to estimate unknown reaction rates or unknown probabilities of certain trajectories or sets of states, we consider accelerated simulation in the presence of rare events with known probabilities or rates, respectively. Some such techniques are established in systems biology and have been more or less extensively applied. After giving some background on stochastic simulation of biological systems in Section 11.3, we discuss in Section 11.4 the applicability of importance sampling. Section 11.5 outlines partitioning-based techniques as one important class of approximation approaches that are well suited for model reduction as well as for hybrid analysis. Section 11.6 provides a quite comprehensive description of adaptive tau-leaping, an approximate 'multistep method' that currently appears to be the most mature approach to accelerated simulation of stiff biochemically reacting systems.

11.3 Stochastic simulation: Background in systems biology

Stochastic simulation of biological systems is usually referred to as the Gillespie algorithm in the literature. It was Gillespie [34, 35] who, in addition to justifying Markovian models and the chemical master equation, also proposed the use of stochastic simulation for analyzing coupled chemical reactions. He formulated crude generation of CTMC trajectories in this framework. When the CTMC is in some state, it resides there for an exponentially distributed sojourn time the mean of which is given by the reciprocal of the sum of all outgoing transition rates. When a state transition occurs, the probability that the CTMC enters a particular state (the transition probability) is given by the transition rate to that state divided by the aforementioned sum of all outgoing transition rates. Hence, in the notation introduced in Section,

Init $t := t_0,\ x := x_0$ and t_{end}
while $t < t_{end}$
1. Compute all $\alpha_m(x)$ and $\alpha_0(x) := \alpha_1(x) + \cdots + \alpha_M(x)$;
2. Generate two random numbers u_1, u_2, uniformly distributed on $(0, 1)$;
3. Generate time τ to next reaction: $\tau = -\ln(u_1)/\alpha_0(x)$;

4. Determine reaction type R_m: $m = \min\{k : \alpha_1(x) + \cdots + \alpha_k(x) > u_2\alpha_0(x)\}$;
5. Set $t := t + \tau$; $x := x + v_m$;

Gillespie [34] referred to this method of trajectory generation as the 'direct method' and also discussed an equivalent method that he called the 'first reaction method' which is an equivalent interpretation of the CTMC dynamics often referred to as a race, for example in computer performance evaluation, particularly common in the context of stochastic Petri nets. Given any state, the tentative times until entering a particular state are all exponentially distributed with mean the reciprocal of the according transition rate. Hence, the next state will be entered according to the 'fastest' transition. Since the minimum of exponential distributions is again exponentially distributed with mean the reciprocal of the sum of the parameters (rates) of the exponential distributions involved, the equivalence to the former interpretation easily follows.

The pseudo-code above does not provide implementation details but expresses the properties to be computed mathematically. Basic skills and fundamental knowledge of data structures and algorithms immediately imply a binary search in Step 4. Astonishingly, for quite a long time much of the work and publications on stochastic simulation for biochemically reacting systems was concerned with 'improvements' by reducing the computation costs of determining the reaction type in the direct method, until finally the binary search version was formulated. Several mathematically equivalent implementations of the direct method all bear their own name; see [20, 45, 48]. For obvious reasons, we avoid the details here.

Applying the 'race version' of the CTMC dynamics, Gibson and Bruck [33] presented the 'next reaction method', now commonly referred to as the Gibson–Bruck algorithm. Their 'improvement' is simply a better implementation than a naive one of the direct method. Roughly speaking, only properties that change are recalculated (others are reused) after a reaction is simulated, and some standard advanced data structures that can be found in, for example, [24] are used. The algorithm was celebrated as more efficient than the direct method, but a simple count of the number of necessary operations leads to the conclusion that this cannot be true. Not surprisingly, [20] showed that a proper implementation of the direct method is indeed more efficient, though this seems hitherto to have gone largely unrecognized. The Gibson–Bruck algorithm is still often considered the best available method for statistically exact trajectory generation and it is therefore in widespread use in Monte Carlo studies of biological and genetic networks.

In contrast to the efforts expended on implementations of trajectory generation, hardly any investigation of statistical accuracy or robustness of estimators is available. A notable exception is [22] where distance measures for probability distributions are considered.

11.4 Can we apply importance sampling?

Given the methodological background provided in previous chapters of this book, the most obvious approach is to apply importance sampling. Indeed, as we are concerned with CTMCs, importance sampling can be appropriately formulated for biological networks [63, 64].

For CTMCs the relevant probability measures are path distributions, and absolute continuity corresponds to the condition that all paths that are possible under the original measure must remain possible under the importance sampling measure. This can be obviously achieved by the condition that for all positive rates in the original model the corresponding rates under importance sampling are positive. Since we are dealing with CTMCs given in terms of biochemical notation as described in Section 11.2.1, we need an appropriate framework for the application of importance sampling to that type of model specification. In particular, we need to express the distribution or density, respectively, of reaction paths.

Let $t_1 < t_2 < \ldots$ denote the successive time instants at which reactions occur and R_{m_i}, $m_i \in \{1, \ldots, M\}$ the reaction type that occurs at time t_i. Define $\tau_i := t_{i+1} - t_i$ to be the time between the ith and the $(i+1)$th reactions. Hence, state $x(t_i)$ is reached due to the ith reaction R_{m_i} at time t_i and remains unchanged for a sojourn time of τ_i, after which the $(i+1)$th reaction $R_{m_{i+1}}$ occurs at time t_{i+1} and changes the state to $x(t_{i+1})$. Thus, the time evolution of the system is completely described by the sequence of states and corresponding sojourn times, and in compact form $(x(t_0), \tau_0), (x(t_1), \tau_1), (x(t_2), \tau_2), \ldots$ describes a trajectory. For a trajectory up to the Rth reaction, considering the Markovian property which in turn implies exponentially distributed sojourn times, the reaction path density is given by

$$\begin{aligned} & dP((x(t_0), \tau_0), \ldots, (x(t_R), \tau_R)) \\ & \quad = p^{(t_0)}(x_0) \cdot \prod_{i=1}^{R} \alpha_{m_{i-1}}(x(t_{i-1})) \exp(\alpha_0(x(t_{i-1}))\tau_{i-1}), \end{aligned} \tag{11.8}$$

where $\alpha_0(x(t_{i-1})) := \alpha_1(x(t_{i-1})) + \cdots + \alpha_M(x(t_{i-1}))$. Note that for a given time horizon over which the system is observed (and should be simulated) the number R of reactions is not known in advance and not deterministic. Formally, it is a random stopping time, which is in accordance with the requirement of dP being a density of a probability measure P defined on the path space of the Markov process.

With importance simulation, the underlying probability measure determined by the propensity functions is changed. Since the only requirement is absolute continuity of the probability measures involved, there is great freedom of choice in changing the measure. It is only necessary that all reaction paths that are possible (have positive probability) under the original measure remain possible. Each probability measure on the path space that meets the aforementioned condition can be considered–even non-Markovian models are allowed as long as they assign positive probabilities to all possible reaction paths.

Nevertheless, we should avoid a large increase in trajectory generation efforts compared to the original measure. Thus, obviously the most natural (and valid) change of measure is to remain in the Markovian world. The easiest way is to change the original propensity functions to 'importance sampling propensity functions' α_m^* such that for all $m \in \{1, \ldots, M\}$ we have $\alpha_m^*(x) = 0 \Rightarrow \alpha_m(x) = 0,\ x \in \mathcal{X}$, or equivalently, starting with the original propensity functions, $\alpha_m(x) > 0 \Rightarrow \alpha_m^*(x) > 0,\ x \in \mathcal{X}$. One then generates trajectories according to the changed propensity functions and multiplies the results with the likelihood ratio to get unbiased estimates for the original system. The trajectory generation is thus carried out as before, for instance by applying the direct method, where now the changed propensity functions are used, yielding a sequence of states with associated sojourn times and reaction path density as in (11.8). Thus, denoting by $p^{*(t_0)}$ the initial distribution for the states, the likelihood ratio becomes

$$L(\omega) = \frac{p^{(t_0)}(x_0)}{p^{*(t_0)}(x_0)} \cdot \prod_{i=1}^{R} \frac{\alpha_{m_{i-1}}(x(t_{i-1})) \exp(\alpha_0(x(t_{i-1}))\tau_{i-1})}{\alpha_{m_{i-1}}^*(x(t_{i-1})) \exp(\alpha_0^*(x(t_{i-1}))\tau_{i-1})} \tag{11.9}$$

which can be efficiently computed during trajectory generation without much extra computational effort by successively updating its value after each simulated reaction according to the running product. In particular, the unbiased number of molecules can be obtained at any time.

Although arising naturally, the change of measure as described above may be too restrictive. In cases where more flexibility is needed, it is possible to use a different change of measure in each simulation step, or propensity functions that depend on the number of reactions that have already occurred (corresponding to a non-homogeneous model) or the history of the simulation steps just executed. Formally, define functions $\beta_m^{(r)}(x(t_0), \ldots, x(t_r))$, where, for all $m \in \{1, \ldots, M\}$, $\alpha_m(x(t_r)) > 0 \Rightarrow \beta_m^{(r)}(x(t_0), \ldots, x(t_r)) > 0$. Then the reaction path density under importance sampling is

$$\begin{aligned} &dP((x(t_0), \tau_0), \ldots, (x(t_R), \tau_R)) \\ &\quad = p^{(t_0)}(x_0) \cdot \prod_{i=1}^{R} \beta_{m_{i-1}}^{(i-1)}(x(t_0), \ldots, x(t_{i-1})) \exp(\beta_0(x(t_0), \ldots, x(t_{i-1}))\tau_{i-1}) \end{aligned} \tag{11.10}$$

and the corresponding likelihood ratio (leaving the initial distribution unchanged) becomes

$$L(\omega) = \prod_{i=1}^{R} \frac{\alpha_{m_{i-1}}(x(t_{i-1})) \exp(\alpha_0(x(t_{i-1}))\tau_{i-1})}{\beta_{m_{i-1}}^{(i-1)}(x(t_0), \ldots, x(t_{i-1})) \exp(\beta_0(x(t_0), \ldots, x(t_{i-1}))\tau_{i-1})}. \tag{11.11}$$

However, the latter form of the change of measure is more involved than the straightforward one.

In any case, importance sampling requires a target event such that the change of measure can be constructed with regard to the target event. This applies to situations where the interest is in the probability of a specific set of states, in the probability of reaching a rare set before returning to the (non-rare) origin, or in times until absorption in a rare set. One then often speaks of an attractor set and a rare set and the aim of importance sampling is to induce a drift towards the target event. Many examples can be found in previous chapters of this book. Obviously, if we are interested in transient or stationary probability distributions or in the moments of the numbers of molecules, there is no specific target event. The rare events are rare reactions, and there are many of them. As mentioned in Section 11.2.3, the rare event simulation problems in biological network models often occur because of stiffness rather than the rarity of certain states. Hence, it is not clear how to do importance sampling. At first glance, it might be reasonable to apply failure biasing techniques that are successful for highly reliable systems, but though the models for highly reliable systems are usually stiff, the property of interest then is the rate or probability of rare system failures, which constitute target events for importance sampling. It is currently an open question how far proper adaptations might be applied to accelerate the simulation of stiff models such that robust estimation becomes possible for whole distributions rather than specific target events. Finally, another well-known problem with importance sampling is that it is usually not suitable for transient analysis over large horizons. If the length of trajectories grows, then the likelihood ratio vanishes and importance sampling increases the variance of the estimators. While for steady-state simulations this problem can be often successfully circumvented by regenerative simulation, there is no satisfactory solution for transient analysis over a large time horizon.

On the positive side, there are other useful applications of likelihood ratios. Recently, they have been used to compare system behaviors by simultaneous simulation of multiple parameter settings [64]. Given that rate constants are often experimentally obtained and thus naturally perturbed, such comparisons are useful to demonstrate the sensitivity of the system dynamics to (small) parameter perturbations. Similarly, it is reasonable to estimate sensitivities by likelihood ratio gradient estimation techniques. Further speed-up can be provided by combination with uniformization [65, 66].

11.5 Partitioning-based techniques

One important class of techniques for large and stiff models is based on partitioning. In order to reduce its size and computational complexity, the overall system model is partitioned into computationally tractable, non-stiff submodels which are analyzed separately. The key is to define a reasonable partitioning such that appropriately combining the results for submodels yields good approximations for the overall model. Applied to biological systems driven by coupled molecular reactions, usually either molecular species or reactions are classified as 'fast' or 'slow' and the model is partitioned accordingly.

The underlying rationale in the context of chemical kinetics is based on assuming a partial equilibrium or quasi-steady state, which are well-known notions in deterministic reaction kinetics; see [55, 56, 61, 68–70]. Partial equilibrium means the systems exhibits fast reactions that are always in equilibrium, which gives rise to partition into fast and slow reactions. Quasi-steady state applies similarly to molecular species. For many stiff systems both notions are equivalent. In stochastic models, equilibrium and steady state are different in that they refer to transition probabilities or rates and to time-invariance properties of probability distributions. Stochastic interpretations of deterministic partial equilibrium and quasi-steady state are similar to the prominent concepts of quasi-stationary distributions, lumpability, and nearly completely decomposable systems, which have been investigated at great length in applied probability and exploited to develop algorithms based on decomposition, aggregation, and disaggregation. Comprehensive treatments of the theoretical foundations as well as particularly relevant applications to the numerical solution of CTMCs can be found in [5, 13, 25–27, 44, 53, 71, 73] and the references therein. However, these concepts and methods are not (yet) prevalent in systems biology or chemical kinetics. Instead, only quite recently, some pioneering work within the context of chemically reacting systems has been done in that modified chemical master equations and corresponding analysis algorithms were derived. More specifically, the notion of deterministic partial equilibrium has been adapted in [40] and the notion of deterministic quasi-steady state in [57].

Assume that the molecular species can be classified into fast and slow, and renumber the d species such that $S_1, \ldots, S_s$ are slow and $S_{s+1}, \ldots, S_d$ are fast. Then the system state at any time t is $X(t) = (Y(t), Z(t))$ where $Y(t)$ and $Z(t)$ are the state vectors for the slow and the fast species, respectively. Denote the state change vector by $v_m = (v_m^y, v_m^z)$ accordingly. Now, the chemical master equation (11.5) reads

$$\frac{\partial p^{(t)}(y,z)}{\partial t} = \sum_{m=0}^{M} (\alpha_m(y - v_m^y, z - v_m^z) p^{(t)}(y - v_m^y, z - v_m^z) - \alpha_m(y,z) p^{(t)}(y,z)). \tag{11.12}$$

Assuming that Z conditional on Y is Markovian and adapting the quasi-steady steady assumption to the Markovian model yields

$$\frac{\partial p^{(t)}(z \mid y)}{\partial t} = \sum_{m=0}^{M} (\alpha_m(y - v_m^y, z - v_m^z) p^{(t)}(z - v_m^z \mid y) - \alpha_m(y,z) p^{(t)}(z \mid y)) = 0, \tag{11.13}$$

which implies that the distribution of $(Z \mid Y)(t)$ is time-invariant, and one gets

$$\frac{\partial p^{(t)}(y)}{\partial t} = \sum_{m=0}^{M} (\beta_m(y - v_m^y) p^{(t)}(y - v_m^y) - \beta_m(y) p^{(t)}(y)), \tag{11.14}$$

where

$$\beta_m(y) = \sum_z \alpha_m(y, z) p(z \mid y). \tag{11.15}$$

The above equations have been provided by [57] as a stochastic version of the deterministic quasi-steady state assumption. Similar equations have been obtained in [40] for partitioning according to fast and slow reactions, hence essentially based on the partial equilibrium assumption. Closely related to both approaches is the *slow-scale stochastic simulation algorithm* [17] where the partitioning is according to fast and slow species and an alternative approximate equation for the slow species is derived. An extension to three time scales is proposed in [14].

Given any of these sets of assumptions and equations, once a partitioning is defined the submodels can be tackled by any suitable solver. In fact, subsequently to the work in [40, 57], various algorithmic versions appeared that mainly differ in the techniques used to solve and combine the submodels, either purely simulative or hybrid. Quite often, the slow submodel is solved by simple direct stochastic simulation and only the fast submodel is subject to different solvers. Advanced stochastic simulation techniques such as tau-leaping (see Section 11.6) are applied, or the submodel is treated by ODE solvers for the deterministic reaction rate equations. Hence, the submodels are sometimes even handled with different modeling paradigms. It is beyond our scope to go into the details and we refer the reader to the original papers, [14, 54, 62, 78] to mention but a few.

All these algorithms are quite young. Case studies and applications are limited to a few relatively simple models. Analyses of their accuracy or statistical robustness for general model classes are still lacking, and the important issues of checking the validity of the underlying assumptions and automated partitioning require further investigation. Also, almost all algorithms proposed so far apply stochastic simulation at least to parts of the model, though the reduced size and complexity render non-simulative solutions possible. In the conclusion of [57], Rao and Arkin emphasize that they believe the true strength of the quasi-steady-state assumption is as a tool for model reduction. We fully agree and note that the same holds for all partitioning techniques. Concerning solution techniques, purely numerical, non-simulative methods, if efficiently applicable, should be clearly preferred as they eliminate the statistical uncertainty from the results so that the question of the robustness of estimators becomes obsolete. One step in this direction is [15] where a stiff enzymatic reaction set is numerically solved via aggregation, which has been demonstrated to be much faster than the slow-scale stochastic simulation algorithm. Further investigations are forthcoming [67]. Undoubtedly, integrating the knowledge from the numerical solution of Markovian models is very promising. More generally, combining methods from different domains – which is essentially the goal of systems biology – is highly desirable and offers a great potential for significant advances in stochastic chemical kinetics.

11.6 Adaptive tau-leaping

Tau-leaping was originally proposed by Gillespie [37] as an approach to accelerate the inherently slow trajectory generation at the cost of statistical exactness. The basic idea is, instead of simulating every reaction, to determine at any time t a time step size τ by which the simulation is advanced. Given the system state (number of molecules of each species) $X(t)$ at time t and the selected step size τ, the system state $X(t')$ at time $t' := t + \tau$ is approximated. Then at time t' a new step size τ' is selected and so on until the simulation time has reached the time horizon of interest. Hence, if the step sizes are chosen such that many reactions are likely to occur within a time interval $[t, t + \tau)$, then the trajectory generation becomes significantly faster. On the other hand, much care must be taken with a proper choice of the time steps in order to avoid too rough an approximation that would result from excessively large time steps. Furthermore, the original tau-leaping approach, which is nowadays referred to as explicit tau-leaping, is not suitable for stiff systems, and implicit tau-leaping was proposed in [59] as a modification in order to deal particularly with stiffness. Recently, in an attempt to combine the advantages of both explicit and implicit tau-leaping, Cao *et al.* [19] invented an adaptive explicit–implicit tau-leaping algorithm which we describe in what follows. We take a top-down approach starting with the fundamental underlying rationale, after which we successively detail the specific steps.

Denote by K_m the random variable describing the number of times that a reaction of type R_m occurs in the time interval interval $[t, t + \tau)$. Then

$$X(t + \tau) = X(t) + \sum_{m=1}^{M} v_m K_m. \tag{11.16}$$

Accordingly, a basic scheme for any algorithm that advances the simulation by predefined time steps instead of simulating every single reaction is as follows:

Init $t := t_0,\ x := x_0$ and t_{end};
while $t < t_{end}$
1. Compute all $\alpha_m(x)$ and $\alpha_0(x) := \alpha_1(x) + \cdots + \alpha_M(x)$;
2. Choose a step size τ according to some appropriate rule;
3. Compute suitable estimates $\hat{k}_1, \ldots, \hat{k}_M$ for $K_1, \ldots, K_M$;
4. Set $t := t + \tau$ and update the system state x according to (11.16);

Tau-leaping assumes that all propensity functions are at least approximately constant in $[t, t + \tau)$, which must be formally specified and is referred to as the *leap condition*. Then, handling all propensity functions as if they were indeed constant gives an appropriate rule for Step 2 of the above algorithm. An essential difference between explicit and implicit tau-leaping lies in the updating of the system state, that is, in obtaining the estimates $\hat{k}_1, \ldots, \hat{k}_M$. In addition, the choice of the step size τ differs.

If all propensity functions are constant in $[t, t+\tau)$, the random variable K_m is Poisson distributed with mean $\tau\alpha_m(X(t))$. Consequently, explicit tau-leaping proceeds by simply computing the estimates $\hat{k}_1, \ldots, \hat{k}_M$ as realizations of the corresponding Poisson distributed random variables. Obviously, (11.16) then becomes an explicit deterministic expression for $X(t+\tau)$ as a function of x and obeys similarities to the explicit (forward) Euler method for solving systems of deterministic ODEs. More specifically, if the number of molecules of each species is large and the Poisson random variates are approximated by their means, (11.16) becomes the explicit Euler formula for the deterministic reaction rate equations. However, explicit ODEs solvers become unstable for stiff systems, and the same holds for explicit tau-leaping in the case of stiff Markovian systems.

Implicit tau-leaping is inspired by the implicit (backward) Euler method which is known to be well suited for stiff ODE systems. A completely implicit version of tau-leaping would require random variates to be generated according to the Poisson distribution with parameters $\tau\alpha_m(X(t+\tau))$, $m = 1, \ldots, M$, which depend on the unknown random state $X(t+\tau)$. Instead, a partially implicit version is considered. Rewriting K_m as $K_m - \tau\alpha_m(X(t)) + \tau\alpha_m(X(t))$ and evaluating all propensity functions in the last term at $X(t+\tau)$ instead of $X(t)$ yields

$$X(t+\tau) = X(t) + \sum_{m=1}^{M} v_m\Big(K_m - \tau\alpha_m(X(t)) + \tau\alpha_m(X(t+\tau))\Big). \qquad (11.17)$$

Then, in a first step, all K_m are again approximated by Poisson random variables as with explicit tau-leaping. Once the realizations, denoted by $k_1, \ldots, k_M$, have been generated and given $X(t) = x$, (11.17) becomes an implicit deterministic equation that is solved by Newton iteration. Typically, the resulting estimate $\hat{x}(t+\tau)$ for $X(t+\tau)$ is not integer-valued. Therefore, in practice, the estimates to be used for the updating in Step 4 of the above algorithm are obtained by rounding the corresponding term in (11.17) to the nearest integer,

$$\hat{k}_m = \mathrm{round}(k_m - \tau\alpha_m(x) + \tau\alpha_m(\hat{x}(t+\tau))). \qquad (11.18)$$

It has been empirically demonstrated that implicit tau-leaping significantly speeds up the simulation of *some* stiff systems. As an alternative to (11.17), motivated by the properties of the trapezoidal rule for solving systems of deterministic ODEs, [21] proposed substituting (11.17) by the *trapezoidal tau-leaping formula*

$$X(t+\tau) = X(t) + \sum_{m=1}^{M} v_m\Big(K_m - \frac{\tau}{2}\alpha_m(X(t)) + \frac{\tau}{2}\alpha_m(X(t+\tau))\Big), \qquad (11.19)$$

which sometimes yields higher accuracy. However, it depends on the specific problem at hand whether (11.17) or (11.19) should be preferred.

The crucial point in both explicit and implicit tau-leaping is an appropriate choice of the step size τ by an automated procedure. For accuracy both methods require that the propensity functions must be approximately constant in the time interval $[t + \tau)$. The step size must be efficiently computed and the time steps must be significantly larger than in a single-reaction simulation such that the computational overhead is negligible compared to the simulation speed-up. Furthermore, it is obviously possible that an updating step results in a negative number of molecules of some species if certain *critical reactions* occur too often and exhaust one or more of its reactants, which must be avoided. Since the invention of tau-leaping, the step size selection procedure has been modified several times until finally in [18] it reached its current state for the explicit version. However, with implicit tau-leaping for stiff systems the step size can often be chosen much larger than suggested by [18], which motivated the adaptive version in [19].

In state x, the expected time to the next reaction is $1/\alpha_0(x)$. Consequently, if a candidate step size is less than $n_a/\alpha_0(x)$ it is considered inefficient and n_b single reactions are simulated according to the standard direct method, where both n_a and n_b are parameters to be specified. A reaction is taken as critical if the maximum number of times it can occur before exhausting one of its reactants is less than some threshold n_c, another parameter to be specified. It may not be known in advance whether or not the problem at hand is stiff. In particular, the system dynamics may be such that in some time periods the system possesses stiffness and in others it does not. With adaptive tau-leaping at each updating step during the simulation either explicit or implicit tau-leaping is chosen dynamically. Hence, a decision rule is necessary. Adaptive tau-leaping applies the simple rule that the system is considered to be stiff if the tentative step size for explicit tau-leaping is more than n_d times smaller than the tentative step size for implicit tau-leaping, which introduces another parameter to be specified.

We are now ready to formulate an algorithm that, given state x at time t and step size selection procedures for explicit and for implicit tau-leaping, dynamically chooses one of the two methods with an appropriate step size. Our formulation here streamlines the formulation in [19, Section 4] and is much more concise but equivalent to it.

1. Define set $\mathcal{C}$ of indices of critical reactions:

$$\mathcal{C} := \left\{ m \in \{1, \ldots, M\} : \alpha_m(x) \rangle 0 \wedge \min_{i:v_{im}<0} \left\lfloor \frac{x_i}{|v_{im}|} \right\rfloor < n_c \right\};$$

2. Compute candidate step sizes $\tau^{(ex)}, \tau^{(im)}$ for explicit and implicit tau-leaping;

3. If $\tau^{(ex)} < n_a/\alpha_0(x) \wedge \tau^{(im)} < n_a/\alpha_0(x)$
then simulate n_b single reactions, update t and x, and goto 1;

4. Compute candidate step size $\tilde{\tau}$ as expected time to next critical reaction:

$$\text{Generate } \tilde{\tau} \sim \text{Exponential}\left(\sum_{m \in C} \alpha_m(x)\right);$$

5. If $\tau^{(ex)} \rangle \min(\tau^{(im)}/n_d; \tilde{\tau})$
   then use explicit tau-leaping with $\tau := \min(\tau^{(ex)}; \tilde{\tau})$;
   else use implicit tau-leaping with $\tau := \min(\tau^{(im)}; \tilde{\tau})$;

6. If $x + \sum_m \hat{k}_m v_m$ has negative components
   then reduce $\tau^{(ex)}$ and $\tau^{(im)}$, and goto 3;

Note that the last step is required because there is still a positive probability of generating negative population sizes, though this probability should be small for appropriately chosen parameters. Hence, altogether the step size selection procedure can be interpreted as an acceptance–rejection method. The inventors more specifically reduce $\tau^{(ex)}$ and $\tau^{(im)}$ by half, but this seems rather arbitrary and may be subject to changes.

It remains to precisely specify Step 2, which has been subject to various improvements that we shall briefly outline. In order to formalize the leap condition of approximately constant propensity functions, an error control parameter $\varepsilon > 0$ is required. In early versions the goal was to bound for every reaction the expected change in its propensity function during a time step of size τ by $\varepsilon\alpha_0(x)$, hence by ε times the sum of all propensity functions evaluated at state x.

The original tau-selection procedure in [37] does not always yield an appropriate step size that satisfies this condition, but later in [38] it was shown that the largest value of τ that indeed satisfies it can be obtained by bounding the mean and the standard deviation of the expected change in the propensity function of each reaction by $\varepsilon\alpha_0(x)$. It was also recognized that instead of bounding the change in the propensity function for all reactions by $\varepsilon\alpha_0(x)$ it is more appropriate to bound the change in the propensity function individually for every reaction R_m by $\varepsilon\alpha_m(x)$, which corresponds to bounding the relative changes in each propensity function by ε. Strictly applied, this implies that τ becomes zero and the simulation does not advance at all if any of the propensity functions evaluated at state x is very small. But, as noted in [18], if α_m changes at all, then according to equation (11.4) it changes by at least c_m such that a change of less than c_m does not make sense, and consequently the change in α_m can be bounded by the maximum of $\varepsilon\alpha_m(x)$ and c_m. Furthermore, [18] presented a procedure that approximately enforces this bound, which is much faster than estimating the mean and the standard deviation according to [38]. The essential underlying rationale is, instead of directly considering propensity functions, to bound the relative changes in populations of certain molecular species such that the relative changes in the propensity functions will be all approximately bounded by ε. It differs for explicit and implicit tau-leaping only in the species that are taken into account and is the current 'state of the art' in step size selection for tau-leaping. Here, we present it in a compact form. The details of the derivation can be found in [18].

In either case, it suffices to consider reactant species. Denote by $\mathcal{R}$ the set of indices of all reactant species and define, for all $i \in \mathcal{R}$,

$$\hat{\mu}_{i,\mathcal{M}}(x) := \sum_{m \in \mathcal{M}} v_{mi}\alpha_m(x), \quad \hat{\sigma}^2_{i,\mathcal{M}}(x) := \sum_{m \in \mathcal{M}} v^2_{mi}\alpha_m(x), \tag{11.20}$$

where $\mathcal{M}$ denotes a set of indices of reactions. Then the candidate step size dependent on $\mathcal{M}$ is

$$\tau_{\mathcal{M}} = \min_{i \in \mathcal{R}} \left(\frac{\max(\varepsilon x_i/g_i(x), 1)}{|\hat{\mu}_{i,\mathcal{M}}(x)|}, \frac{\max(\varepsilon x_i/g_i(x), 1)^2}{|\hat{\sigma}^2_{i,\mathcal{M}}(x)|} \right). \tag{11.21}$$

where g_i is a function defined in order to guarantee that bounding the relative change of states is sufficient for bounding the relative change of propensity functions. Denote by $h(i)$ the highest order of reactions in which species S_i appears as reactant and by $n(i)$ the maximum number of S_i molecules required by any of the highest-order reactions. Then

$$g_i(x) = h(i) + \frac{h(i)}{n(i)} \sum_{j=1}^{n(i)-1} \frac{j}{x_i - j}. \tag{11.22}$$

The step size depends on $\mathcal{M}$ only through $\hat{\mu}_{i,\mathcal{M}}(x)$ and $\hat{\sigma}^2_{i,\mathcal{M}}(x)$, and the only difference in the step size selection for explicit and implicit tau-leaping is in the choice of $\mathcal{M}$ which defines exactly those reactions that are considered in the step size selection. For explicit tau-leaping, these are simply the non-critical reactions. Hence, $\tau^{(ex)} = \tau_{\{1,\ldots,M\}\setminus\mathcal{C}}$.

For implicit tau-leaping the partial equilibrium assumption (see Section 11.5) is exploited. As it is difficult to identify all reactions that are in partial equilibrium, only reversible reactions are checked for partial equilibrium, which means that their propensity functions evaluated at state x must be approximately equal. More specifically, it is assumed in [19] that two reactions R_{m_1}, R_{m_2}, where R_{m_1} reverses R_{m_2} and vice versa, are in partial equilibrium at state x if the difference in their propensity functions is less than δ times the minimum of the propensity functions. That is,

$$|\alpha_{m_1}(x) - \alpha_{m_2}(x)| \leq \delta \min(\alpha_{m_1}(x), \alpha_{m_2}(x)), \quad \delta > 0. \tag{11.23}$$

Then in the step size selection for implicit tau-leaping only reactions that are neither critical nor in partial equilibrium (checking only reversible reactions) are considered. Hence, denoting by $\mathcal{E}$ the set of indices of (reversible) partially equilibrated reactions, we have $\tau^{(im)} = \tau_{\{1,\ldots,M\}\setminus\mathcal{C}\setminus\mathcal{E}}$.

It has been empirically demonstrated that tau-leaping can significantly accelerate the simulation of chemically reacting systems. The consistency and stability are formally addressed in [60]. However, further investigation of various issues is desirable. First of all, the choice of the parameters $\varepsilon, n_a, n_b, n_c, n_d$ is currently rather informal and they are chosen heuristically. The inventors state that they 'normally' or 'usually' take ε in the range 0.03–0.05, $n_a = 10$, $n_b = 10$ if the previous step uses implicit tau-leaping and $n_b = 100$ otherwise, $n_c = 10$, $n_d = 100$, and δ 'around' 0.05. This seems to rely only on empirical comparisons of tau-leaping with the direct single-reaction method. Obviously, in the huge models constructed in practice, such a comparison will not be feasible as the direct method does not provide accurate results in reasonable time. Similarly, the statistical robustness of tau-leaping requires further investigation. Most often, the statistical accuracy of the estimates was only shown by comparison to the direct method. The first formal approach given in [22] considers the Kolmogorov distance, the histogram distance and the newly introduced self distance for probability measures. Finally, another issue of interest is how the probability of rejecting a candidate step size depends on the parameters mentioned above.

Tau-leaping is an active field of research and diverse variants exist. Indeed, in recent years the literature on tau-leaping has grown exponentially. We refer the interested reader to [4, 23, 39, 52, 58, 81] for some recent developments.

References

[1] R. J. Allen, D. Frenkel, and P. R. ten Wolde. Forward flux sampling-type schemes for simulating rare events: Efficiency analysis. *Journal of Chemical Physics*, **124**(19): 194111/1–194111/17, 2006.

[2] R. J. Allen, D. Frenkel, and P. R. ten Wolde. Simulating rare events in equilibrium or nonequilibrium stochastic simulations. *Journal of Chemical Physics*, **124**(2): 024102/1–024102/16, 2006.

[3] U. Alon. *An Introduction to Systems Biology: Design Principles of Biological Circuits*. Chapman & Hall, Boca Raton, FL, 2007.

[4] D. F. Anderson. Incorporating postleap checks in tau-leaping. *Journal of Chemical Physics*, **128**(5): 054103, 2008.

[5] W. J. Anderson. *Continuous-Time Markov Chains: An Application-Oriented Approach*. Springer, New York, 1991.

[6] A. Arkin, J. Ross, and H. H. McAdams. Stochastic kinetic analysis of developmental pathway bifurcation in phage λ-infected Escherichia coli cells. *Genetics*, **149**: 1633–1648, 1998.

[7] A. F. Bartholomay. A stochastic approach to chemical reaction kinetics. PhD thesis, Harvard University, 1957.

[8] A. T. Bharucha-Reid. *Elements of the Theory of Markov Processes and Their Applications*. McGraw-Hill, New York, 1960.

[9] W. J. Blake, M. Kaern, C. R. Cantor, and J. J. Collins. Noise in eukaryotic gene expression. *Nature*, **422**: 633–637, 2003.

[10] P. G. Bolhuis, D. G. Chandler, C. Dellago, and P. L. Geissler. Transition path sampling: Throwing ropes over mountain passes, in the dark. *Annual Reviews of Physical Chemistry*, **59**: 291–318, 2002.

[11] J. M. Bower and H. Bolouri, eds. *Computational Modeling of Genetic and Biochemical Networks*. MIT Press, Cambridge, MA, 2001.

[12] P. Brémaud. *Markov Chains: Gibbs Fields, Monte Carlo Simulation, and Queues*. Springer, New York, 1999.

[13] L. A. Breyer and A. G. Hart. Approximations of quasi-stationary distributions for Markov chains. *Mathematical and Computer Modelling*, **31**: 69–79, 2000.

[14] K. Burrage, T. Tian, and P. Burrage. A multi-scaled approach for simulating chemical reaction systems. *Progress in Biophysics and Molecular Biology*, **85**: 217–234, 2004.

[15] H. Busch, W. Sandmann, and V. Wolf. A numerical aggregation algorithm for the enzyme-catalyzed substrate conversion. In *Proceedings of the 2006 International Conference on Computational Methods in Systems Biology*, Lecture Notes in Computer Science 4210, pp. 298–311. Springer, Berlin/Heidelberg/New York, 2006.

[16] W. Cai, M. H. Kalos, M. de Koning, and V. V. Bulatov. Importance sampling of rare transition events in Markov processes. *Physical Review E*, **66**: 046703, 2002.

[17] Y. Cao, D. T. Gillespie, and L. R. Petzold. The slow-scale stochastic simulation algorithm. *Journal of Chemical Physics*, **122**(1): 014116/1–014116/18, 2005.

[18] Y. Cao, D. T. Gillespie, and L. R. Petzold. Efficient stepsize selection for the tau-leaping simulation. *Journal of Chemical Physics*, **124**: 044109/1–144109/11, 2006.

[19] Y. Cao, D. T. Gillespie, and L. R. Petzold. The adaptive explicit-implicit tau-leaping method with automatic tau selection. *Journal of Chemical Physics*, **126**(22): 224101/1–224101/9, 2007.

[20] Y. Cao, H. Li, and L. R. Petzold. Efficient formulation of the stochastic simulation algorithm for chemically reacting systems. *Journal of Chemical Physics*, **121**(9): 4059–4067, 2004.

[21] Y. Cao and L. R. Petzold. Trapezoidal tau-leaping formula for the stochastic simulation of biochemical systems. In *Proceedings of the 1st Conference on Foundations of Systems Biology in Engineering*, pp. 149–152, 2005.

[22] Y. Cao and L. R. Petzold. Accuracy limitations and the measurement of errors in the stochastic simulation of chemically reacting systems. *Journal of Computational Physics*, **212**(1): 6–24, 2006.

[23] Y. Cao and L. R. Petzold. Slow-scale tau-leaping method. *Computer Methods in Applied Mechanics and Engineering*, **197**(43–44): 3472–3479, 2008.

[24] T. H. Cormen, C. E. Leiserson, and R. L. Rivest. *Introduction to Algorithms*. MIT Press, Cambridge, MA, 1990.

[25] P.-J. Courtois. *Decomposability, Queueing and Computer Applications*. Academic Press, New York, 1977.

[26] P.-J. Courtois. On time and space decomposition of complex structures. *Communications of the ACM*, **28**(6): 590–603, 1985.

[27] J. N. Darroch and E. Seneta. On quasistationary distributions in absorbing continuous-time finite Markov chains. *Journal of Applied Probability*, **4**: 192–196, 1967.

[28] M. de Koning, W. Cai, B. Sadigh, M. H. Kalos, and V. V. Bulatov. Adaptive importance sampling Monte Carlo simulation of rare transition events. *Journal of Chemical Physics*, **122**(7): 074103, 2005.

[29] M. Delbrück. Statistical fluctuations in autocatalytic reactions. *Journal of Chemical Physics*, **8**: 120–124, 1940.

[30] C. Dellago, P. G. Bolhuis, and P. L. Geissler. Transition path sampling. *Advances in Chemical Physics*, **123**: 1–78, 2002.

[31] R. Elber. Long-timescale simulation methods. *Current Opinion in Structural Biology*, **15**: 151–156, 2005.

[32] N. Fedoroff and W. Fontana. Small numbers of big molecules. *Science*, **297**: 1129–1131, 2002.

[33] M. A. Gibson and J. Bruck. Efficient exact stochastic simulation of chemical systems with many species and many channels. *Journal of Physical Chemistry A*, **104**: 1876–1889, 2000.

[34] D. T. Gillespie. A general method for numerically simulating the time evolution of coupled chemical reactions. *Journal of Computational Physics*, **22**: 403–434, 1976.

[35] D. T. Gillespie. Exact stochastic simulation of coupled chemical reactions. *Journal of Physical Chemistry*, **71**(25): 2340–2361, 1977.

[36] D. T. Gillespie. A rigorous derivation of the chemical master equation. *Physica A*, **188**: 404–425, 1992.

[37] D. T. Gillespie. Approximate accelerated stochastic simulation of chemically reacting systems. *Journal of Chemical Physics*, **115**: 1716–1732, 2001.

[38] D. T. Gillespie and L. R. Petzold. Improved leap-size selection for accelerated stochastic simulation. *Journal of Chemical Physics*, **119**: 8229–8234, 2003.

[39] L. A. Harris and P. Clancy. A partitioned leaping approach for multiscale modeling of chemical reaction dynamics. *Journal of Chemical Physics*, **125**(14): 144107, 2006.

[40] E. L. Haseltine and J. B. Rawlings. Approximate simulation of coupled fast and slow reactions for stochastic chemical kinetics. *Journal of Chemical Physics*, **117**(15): 6959–6969, 2002.

[41] H. Kitano. *Foundations of Systems Biology*. MIT Press, Cambridge, MA, 2001.

[42] E. Klipp, R. Herwig, A. Kowald, C. Wierling, and H. Lehrach. *Systems Biology in Practice*. Wiley-VCH, Weinheim, 2005.

[43] T. G. Kurtz. The relationship between stochastic and deterministic models for chemical reactions. *Journal of Chemical Physics*, **57**(7): 2976–2978, 1972.

[44] J. Ledoux, G. Rubino, and B. Sericola. Exact aggregation of absorbing Markov processes using the quasi-stationary distribution. *Journal of Applied Probability*, **31**: 626–634, 1994.

[45] H. Li and L. R. Petzold. Logarithmic direct method for discrete stochastic simulation of chemically reacting systems. Technical Report, 2006. http://www.engineering.ucsb.edu/~cse/Files/ldm0513.pdf.

[46] H. H. McAdams and A. Arkin. Stochastic mechanisms in gene expression. *Proceedings of the National Academy of Science USA*, **94**: 814–819, 1997.

[47] H. H. McAdams and A. Arkin. It's a noisy business! *Trends in Genetics*, **15**(2): 65–69, 1999.

[48] J. M. McCollum, G. D. Peterson, C. D. Cox, M. L. Simpson, and N. F. Samatova. The sorting direct method for stochastic simulation of biochemical systems with varying reaction execution behavior. *Computational Biology and Chemistry*, **30**(1): 39–49, 2006.

[49] D. A. McQuarrie. Stochastic approach to chemical kinetics. *Journal of Applied Probability*, **4**: 413–478, 1967.

[50] M. J. Morelli, R. J. Allen, S. Tanase-Nicola, and P. R. ten Wolde. Eliminating fast reactions in stochastic simulations of biochemical networks: A bistable genetic switch. *Journal of Chemical Physics*, **128**(4): 045105/1–045105/13, 2008.

[51] I. Oppenheim, K. E. Shuler, and G. H. Weiss. Stochastic and deterministic formulation of chemical rate equations. *Journal of Chemical Physics*, **50**(1): 460–466, 1969.

[52] M. F. Pettigrew and H. Resat. Multinomial tau-leaping for stochastic kinetic simulations. *Journal of Chemical Physics*, **126**(8): 084101, 2007.

[53] P. K. Pollett. The determination of quasistationary distributions from the transition rates of an absorbing Markov chain. *Mathematical and Computer Modelling*, **22**: 279–287, 1995.

[54] J. Puchalka and A. M. Kierzek. Bridging the gap between stochastic and deterministic regimes in the kinetic simulations of the biochemical reaction networks. *Biophysical Journal*, **86**: 1357–1372, 2004.

[55] J. D. Ramshaw. Partial chemical equilibrium in fluid dynamics. *Physics of Fluids*, **23**(4): 675–680, 1980.

[56] J. D. Ramshaw and L. D. Cloutman. Numerical method for partial equilibrium flow. *Journal of Computational Physics*, **39**(2): 405–417, 1981.

[57] C. V. Rao and A. Arkin. Stochastic chemical kinetics and the quasi-steady-state-assumption: Application to the Gillespie algorithm. *Journal of Chemical Physics*, **118**: 4999–5010, 2003.

[58] M. Rathinam and H. El Samad. Reversible-equivalent-monomolecular tau: A leaping method for small number and stiff stochastic chemical systems. *Journal of Computational Physics*, **224**: 897–923, 2007.

[59] M. Rathinam, L. R. Petzold, Y. Cao, and D. T. Gillespie. Stiffness in stochastic chemically reacting systems: The implicit tau-leaping method. *Journal of Chemical Physics*, **119**: 12784–12794, 2003.

[60] M. Rathinam, L. R. Petzold, Y. Cao, and D. T. Gillespie. Consistency and stability of tau-leaping schemes for chemical reaction systems. *Multiscale Modeling and Simulation*, **4**(3): 867–895, 2005.

[61] M. Rein. The partial-equilibrium approximation in reacting flows. *Physics of Fluids A*, **4**(5): 873–886, 1992.

[62] H. Salis and Y. Kaznessis. Accurate hybrid stochastic simulation of a system of coupled chemical or biochemical reactions. *Journal of Chemical Physics*, **122**: 054103, 2005.

[63] W. Sandmann. Applicability of importance sampling to coupled molecular reactions. In *Proceedings of the 12th International Conference on Applied Stochastic Models and Data Analysis*, 2007.

[64] W. Sandmann. Simultaneous stochastic simulation of multiple perturbations in biological network models. In *Proceedings of the 2007 International Conference on Computational Methods in Systems Biology*, Lecture Notes in Computer Science 4695, pp. 15–31. Springer, Berlin/Heidelberg/New York, 2007.

[65] W. Sandmann. Stochastic simulation of biochemical systems via discrete-time conversion. In *Proceedings of the 2nd Conference on Foundations of Systems Biology in Engineering*, pp. 267–272. Fraunhofer IRB Verlag, Stuttgart, 2007.

[66] W. Sandmann. Discrete-time stochastic modeling and simulation of biochemical networks. *Computational Biology and Chemistry*, **32**(4): 292–297, 2008.

[67] W. Sandmann and V. Wolf. Computational probability for systems biology. In *Proceedings of the Workshop on Formal Methods in Systems Biology*, Lecture Notes in Computer Science 5054, pp. 33–47. Springer, Stuttgart, 2008.

[68] M. Schauer and R. Heinrich. Analysis of the quasi-steady state approximation for an enzymatic one-substrate reaction. *Journal of Theoretical Biology*, **79**: 425–442, 1979.

[69] L. A. Segel. On the validity of the steady-state assumption of enzyme kinetics. *Bulletin of Mathematical Biology*, **50**: 579–593, 1988.

[70] L. A. Segel and M. Slemrod. The quasi-steady state assumption: A case study in perturbation. *SIAM Review*, **31**(3): 446–477, 1989.

[71] E. Seneta. *Non-Negative Matrices and Markov Chains*. Springer, New York, 1981.

[72] K. Singer. Application of the theory of stochastic processes to the study of irreproducible chemical reactions and nucleation processes. *Journal of the Royal Statistical Society, Series B*, **15**(1): 92–106, 1953.

[73] W. J. Stewart. *Introduction to the Numerical Solution of Markov Chains*. Princeton University Press, Princeton, NJ, 1994.

[74] Z. Szallasi, J. Stelling, and V. Periwal, eds. *System Modeling in Cellular Biology*. MIT Press, Cambridge, MA, 2006.

[75] T. E. Turner, S. Schnell, and K. Burrage. Stochastic approaches for modelling in vivo reactions. *Computational Biology and Chemistry*, **28**: 165–178, 2004.

[76] C. Valeriani, R. J. Allen, M. J. Morelli, D. Frenkel, and P. R. ten Wolde. Computing stationary distributions in equilibrium and nonequilibrium systems with forward flux sampling. *Journal of Chemical Physics*, **127**(11): 114109/1–114109/11, 2007.

[77] N. van Kampen. *Stochastic Processes in Physics and Chemistry*. Elsevier, Amsterdam, 1992.

[78] K. Vasudeva and U. S. Bhalla. Adaptive stochastic-deterministic chemical kinetics simulation. *Bioinformatics*, **20**(1): 78–84, 2004.

[79] D. J. Wilkinson. *Stochastic Modelling for Systems Biology*. Taylor & Francis, Boca Raton, FL, 2006.

[80] O. Wolkenhauer, M. Ullah, W. Kolch, and K.-H. Cho. Modelling and simulation of intracellular dynamics: Choosing an appropriate framework. *IEEE Transactions on NanoBioscience*, **3**(3): 200–207, 2004.

[81] Z. Xu and W. Cai. Unbiased tau-leaping methods for stochastic simulation of chemically reacting systems. *Journal of Chemical Physics*, **128**(15): 154112, 2008.

Index

Rare Event Simulation using Monte Carlo Methods Edited by G. Rubino and B. Tuffin

Printed and bound by CPI Group (UK) Ltd, Croydon, CR0 4YY

16/04/2025

14658544-0002